Geodetic and Geophysical Observations in Antarctica

Alessandro Capra · Reinhard Dietrich (Eds.)

Geodetic and Geophysical Observations in Antarctica

An Overview in the IPY Perspective

Springer

Prof. Dr. Alessandro Capra
Università Modena e Reggio Emilia
Dipto. Ingegneria Meccanica e Civile
Via Vignolese, 905
41100 Modena
Italy
alessandro.capra@unimore.it

Dr. Reinhard Dietrich
TU Dresden
Inst. Planetare Geodäsie
Mommsenstr. 13
01062 Dresden
Germany
reinhard.dietrich@tu-dresden.de

ISBN: 978-3-540-74881-6   e-ISBN: 978-3-540-74882-3

Library of Congress Control Number: 2008932856

© 2008 Springer-Verlag Berlin Heidelberg

This work is subject to copyright. All rights are reserved, whether the whole or part of the material is concerned, specifically the rights of translation, reprinting, reuse of illustrations, recitation, broadcasting, reproduction on microfilm or in any other way, and storage in data banks. Duplication of this publication or parts thereof is permitted only under the provisions of the German Copyright Law of September 9, 1965, in its current version, and permission for use must always be obtained from Springer. Violations are liable to prosecution under the German Copyright Law.

The use of general descriptive names, registered names, trademarks, etc. in this publication does not imply, even in the absence of a specific statement, that such names are exempt from the relevant protective laws and regulations and therefore free for general use.

Cover design: deblik, Berlin

Printed on acid-free paper

9 8 7 6 5 4 3 2 1

springer.com

To Oto and Otino
To my mother and my family
Alessandro

Preface

The polar regions with their unique geophysical and geodynamic environment are characterized by close interactions between solid earth, cryosphere, hydrosphere and athmosphere. They are directly linked to the global climate system. Geodetic and geophysical observations play an essential role in the scientific investigation of the polar regions, especially with regard to permanent and temporary observatories.

For several years geodetic observatories have been realized in Antarctica by continuous or periodic data acquisitions.

The observation techniques include: Global Navigation Satellite Systems (GNSS, in particular GPS and GLONASS, in futre also GALILEO); further geodetic space techniques (DORIS, VLBI); tide gauges observations; absolute, tidal and relative gravimetry; seismology; geomagnetometry; meteorology.

Within the Scientific Commettee on Antarctic Research (SCAR) the SCAR group of experts on Geodetic Infrastructure of Antarctica (GIANT) spent great effort to coordinate the international activities and to enhance the development of geodetic and geophysical observatories.

In order to gain a better understanding of the polar environment long time series as recorded by permanent observatories should be integrated together with data acquired by field surveys like geological and glaciological sampling, geophysical investigations, and with satellite data (remote sensing, satellite gravimetry).

Focussed on the geodetic-geodynamic aspects of the IPY project "Polar Earth Observing Network" (POLENET, IPY full proposal no. 185) special attention should be given to regional and global model generation and validation (regional geodynamics, plate tectonics, postglacial rebound, global climate system, global change).

Special sessions at international symposia have been dedicated to Antarctic geodetic and geophysical observations and data analysis; among others we would like to mention Session G10 "Short and long-term observations in polar region" - EGU06, Vienna, April 2006; and the International Workshop "GPS in the IPY: The POLENET Project" - Dresden, Germany, October 4-6, 2006. relevant contributions have been presented at the ISAES X (X. International Symposium on Antarctic Earth Science), Santa Barbara, California, USA, August 2007.

The book is intended to give an overview on all aspects of the scientific utilization of geodetic-geophysical observations in Antarctica, of data analysis and geodynamic interpretation as well as of the various technological aspects in setting up autonomous observatories in Antarctica.

Contents

A Precise Reference Frame for Antarctica from SCAR GPS Campaign Data and Some Geophysical Implications 1
Reinhard Dietrich and Axel Rülke

Technologies to Operate Year-Round Remote Global Navigation Satellite System (GNSS) Stations in Extreme Environments 11
Michael J. Willis

VLNDEF Project for Geodetic Infrastructure Definition of Northern Victoria Land, Antarctica 37
A. Capra, M. Dubbini, A. Galeandro, L. Gusella, A. Zanutta, G. Casula, M. Negusini, L. Vittuari, P. Sarti, F. Mancini, S. Gandolfi, M. Montaguti and G. Bitelli

Communications Systems for Remote Polar GNSS Station Operation 73
Bjorn Johns

Current Status and Future Prospects for the Australian Antarctic Geodetic Network 85
Gary Johnston, Nicholas Brown and Michael Moore

Geodetic Research on Deception Island and its Environment (South Shetland Islands, Bransfield Sea and Antarctic Peninsula) During Spanish Antarctic Campaigns (1987–2007) 97
M. Berrocoso, A. Fernández-Ros, M.E. Ramírez, J.M. Salamanca, C. Torrecillas, A. Pérez-Peña, R. Páez, A. García-García, Y. Jiménez-Teja, F. García-García, R. Soto, J. Gárate, J. Martín-Davila, A. Sánchez-Alzola, A. de Gil, J.A. Fernández-Prada and B. Jigena

Validation of the Atmospheric Water Vapour Content from NCEP Using GPS Observations Over Antarctica 125
Sibylle Vey and Reinhard Dietrich

Geodynamics of the Tectonic Detachment in the Penola Strait (Antarctic Peninsula, Archipelago of Argentina Islands) 137
K.R. Tretyak, Y.I. Golubinka, A.J. Kulchytskyy and L.V. Babiy

GPS and Radiosonde Derived Precipitable Water Vapour Content and its Relationship with 5 Years of Long-Wave Radiation Measurements at "Mario Zucchelli" Station, Terra Nova Bay, Antarctica 145
Pierguido Sarti, Monia Negusini, Christian Lanconelli, Angelo Lupi, Claudio Tomasi and Alessandra Cacciari

Results of the Investigations of the GNSS Antennae in the Framework of SCAR GIANT Project "In Situ GNSS Antenna Tests and Validation of Phase Centre Calibration Data" 179
Jan Cisak and Yevgen M. Zanimonskiy

Atmospheric Impact on GNSS Observations, Sea Level Change Investigations and GPS-Photogrammetry Ice Cap Survey at Vernadsky Station in Antarctic Peninsula 191
J. Cisak, G. Milinevsky, V. Danylevsky, V. Glotov, V. Chizhevsky, S. Kovalenok, A. Olijnyk and Y. Zanimonskiy

A Validation of Ocean Tide Models Around Antarctica Using GPS Measurements 211
Ian D. Thomas, Matt A. King and Peter J. Clarke

Continuous Gravity Observation with the Superconducting Gravimeter CT #043 at Syowa Station, Antarctica 237
K. Doi, K. Shibuya, H. Ikeda and Y. Fukuda

Tide Gauges in the Antarctic and sub-Antarctic: Instrumentation and Calibration Issues from an Australian Perspective 249
Christopher Watson, Roger Handsworth and Henk Brolsma

Tidal Gravimetry in Polar Regions: An Observation Tool Complementary to Continuous GPS for the Validation of Ocean Tide Models 267
Mirko Scheinert, Andrés F. Zakrajsek, Sergio A. Marenssi, Reinhard Dietrich and Lutz Eberlein

Joint Geophysical Observations of Ice Stream Dynamics 281
S. Danesi, M. Dubbini, A. Morelli, L. Vittuari and S. Bannister

Geomagnetic Observatories in Antarctica; State of the Art and a Perspective View in the Global and Regional Frameworks 299
L. Cafarella, D. Di, S. Lepidi and A. Meloni

Structure of the Wilkes Basin Lithosphere along the ITASE01 Geotraverse 319
F. Coren, N. Creati and P. Sterzai

Report on Photogrammetric Research Conducted at the Antarctic Station "Academician Vernadskyy" 333
V. Hlotov

The Contribution of Russian Geodesists and Topographers to Antarctic Mapping 347
Alexander V. Yuskevich

Contributors

L.V. Babiy
National University "Lvivska Politechnika", Stepana Bandery street, 12, 79013 Lviv, Ukraine

S. Bannister
GNS Science, Fairway Drive, PO 30368 Lower Hutt 5040, New Zealand,
e-mail: s.bannister@gns.cri.nz

M. Berrocoso
Laboratorio de Astronomía, Geodesia y Cartografía. Departamento de Matemáticas. Facultad de Ciencias. Campus de Puerto Real. Universidad de Cádiz. 11510 Puerto Real (Cádiz-Andalucía). España,
e-mail: manuel.berrocoso@uca.es

G. Bitelli
ISTART, Universita' di Bologna, Bologna, Italy

H. Brolsma
Australian Antarctic Division, Channel Highway, Kingston, Tasmania, Australia 7050,
e-mail: Henk.Brolsma@aad.gov.au

N. Brown
Geoscience Australia, GP0 Box 378, Canberra, Australia 2601

A. Cacciari
Carlo Gavazzi Space SpA c/o Istituto di Scienze dell'Atmosfera e del Clima – CNR, Via P. Gobetti 101, 40129 Bologna, Italy

L. Cafarella
Istituto Nazionale di Geofisica e Vulcanologia, Rome, Italy;
e-mail: cafarella@ingv.it

A. Capra
DIMeC, Universita' di Modena e Reggio Emilia

G. Casula
INGV, Roma

V. Chizhevsky
National University Lvivska Politechnica, Lviv, Ukraine

J. Cisak
Institute of Geodesy and Cartography, Warsaw, Poland,
e-mail: jcisak@igik.edu.pl

P.J. Clarke
School of Civil Engineering and Geosciences, Newcastle University, Newcastle upon Tyne, NE2 1JU, UK

F. Coren
Istituto Nazionale di Oceanografia e Geofisica Sperimentale (OGS), Borgo Grotta Gigante, 34010 Sgonico (Trieste), Italy

N. Creati
Istituto Nazionale di Oceanografia e Geofisica Sperimentale (OGS), Borgo Grotta Gigante, 34010 Sgonico (Trieste), Italy;
e-mail: ncreati@inogs.it

S. Danesi
Istituto Nazionale di Geofisica e Vulcanologia, Sezione Bologna, Via Donato Creti, 12 – 40128 Bologna Italy,
e-mail: danesi@bo.ingv.it

V. Danylevsky
National Taras Shevchenko University of Kyiv, Kyiv, Ukraine

A. de Gil
Laboratorio de Astronomía, Geodesia y Cartografía. Departamento de Matemáticas. Facultad de Ciencias. Campus de Puerto Real. Universidad de Cádiz. 11510 Puerto Real (Cádiz-Andalucía). España

D. Di
Istituto Nazionale di Geofisica e Vulcanologia, Rome, Italy,
e-mail: dimauro@ingv.it

R. Dietrich
Technische Universität Dresden, Institut für Planetare Geodäsie, D-01062 Dresden, Germany,
e-mail: dietrich@IPG.geo.tu-dresden.de

K. Doi
National Institute of Polar Research, Kaga 1-9-10, Itabashi-ku, Tokyo 173-8515, Japan

M. Dubbini
Università di Modena e Reggio Emilia, DiMec, Via Vignolese 905, 41100 Modena Italy,
e-mail: marco.dubbini@unimo.it

Contributors

L. Eberlein
Technische Universität Dresden, Institut für Planetare Geodäsie, 01062 Dresden, Germany,
e-mail: scheinert@ipg.geo.tu-dresden.de

J.A. Fernández-Ros
Laboratorio de Astronomía, Geodesia y Cartografía. Departamento de Matemáticas. Facultad de Ciencias. Campus de Puerto Real. Universidad de Cádiz. 11510 Puerto Real (Cádiz-Andalucía). España

Y. Fukuda
Graduate School of Science, Kyoto University, Kyoto 606-8502, Japan

A. Galeandro
DIASS, Politecnico di Bari, Taranto, Italy

S. Gandolfi
DISTART, Universita' di Bologna, Bologna, Italy

J. Gárate
Servicio de Satélites. Sección de Geofísica. Real Instituto y Observatorio de la Armada. San Fernando (Cádiz, España)

A. García-García
Departamento de Volcanología. Museo Nacional de Ciencias Naturales. Consejo Superior de Investigaciones Científicas. C/ José Gutiérrez Abascal, 2. Madrid

F. García-García
Escuela Superior de Ingeniería Cartográfica y Geodésica. Universidad Politécnica de Valencia

V. Glotov
National University Lvivska Politechnica, Lviv, Ukraine

Y.I. Golubinka
National University "Lvivska Politechnika", Stepana Bandery street, 12, 79013 Lviv, Ukraine

L. Gusella
DISTART, Universita' di Bologna, Bologna, Italy

R. Handsworth
Australian Antarctic Division, Channel Highway, Kingston, Tasmania, Australia 7050,
e-mail: Roger.Handsworth@aad.gov.au

V. Hlotov
National University "Lvivska Politechnika", Ukraine

H. Ikeda
Research Facility Center for Science and Technology Cryogenics Division, University of Tsukuba, Tennodai 1-1-1, Tsukuba 305-8577, Japan

B. Jigena
Laboratorio de Astronomía, Geodesia y Cartografía. Departamento de Matemáticas.
Facultad de Ciencias. Campus de Puerto Real. Universidad de Cádiz. 11510 Puerto
Real (Cádiz-Andalucía). España

Y. Jiménez-Teja
Laboratorio de Astronomía, Geodesia y Cartografía. Departamento de Matemáticas.
Facultad de Ciencias. Campus de Puerto Real. Universidad de Cádiz. 11510 Puerto
Real (Cádiz-Andalucía). España

B. Johns
UNAVCO, 6350 Nautilus Dr., Boulder, Colorado, 80301, USA,
e-mail: johns@unavco.org

G. Johnston
Geoscience Australia, GP0 Box 378, Canberra, ACT 2601, Australia

A. King Matt
School of Civil Engineering and Geosciences, Newcastle University, Newcastle
upon Tyne, NE2 1JU, UK

S. Kovalenok
Ministry Education and Science of Ukraine, Kyiv, Ukraine

A.J. Kulchytskyy
National University "Lvivska Politechnika", Stepana Bandery street, 12, 79013
Lviv, Ukraine

C. Lanconelli
Istituto di Scienze dell'Atmosfera e del Clima – CNR, Via P. Gobetti 101, 40129
Bologna, Italy

S. Lepidi
Istituto Nazionale di Geofisica e Vulcanologia, Rome, Italy,
e-mail: lepidi@ingv.it

A. Lupi
Istituto di Scienze dell'Atmosfera e del Clima – CNR, Via P. Gobetti 101, 40129
Bologna, Italy

F. Mancini
DAU, Politecnico di Bari, Bari, Italy

A. Marenssi Sergio
Dirección Nacional del Antártico, Instituto Antártico Argentino, Cerrito 1248
(C1010AAZ), Buenos Aires, Argentina

J. Martín-Davila
Servicio de Satélites. Sección de Geofísica. Real Instituto y Observatorio de la
Armada. San Fernando (Cádiz, España)

A. Meloni
Istituto Nazionale di Geofisica e Vulcanologia, Rome, Italy,
e-mail: meloni@ingv.it

G. Milinevsky
National Taras Shevchenko University of Kyiv, Kyiv, Ukraine,
e-mail: genmilinevsky@gmail.com

M. Montaguti
DISTART, Universita' di Bologna; IRA – INAF, Bologna, Bologna, Italy

M. Moore
Geoscience Australia, GP0 Box 378, Canberra, ACT 2601, Australia

A. Morelli
Istituto Nazionale di Geofisica e Vulcanologia, Sezione Bologna, Via Donato Creti, 12 – 40128 Bologna Italy,
e-mail: morelli@bo.ingv.it

M. Negusini
Istituto di Radioastronomia – INAF, Via P. Gobetti 101, 40129 Bologna, Italy

A. Olijnyk
National Scientific Center "Institute of metrology", Kharkiv, Ukraine

R. Páez
Laboratorio de Astronomía, Geodesia y Cartografía. Departamento de Matemáticas. Facultad de Ciencias. Campus de Puerto Real. Universidad de Cádiz. 11510 Puerto Real (Cádiz-Andalucía). España

A. Pérez-Peña
Laboratorio de Astronomía, Geodesia y Cartografía. Departamento de Matemáticas. Facultad de Ciencias. Campus de Puerto Real. Universidad de Cádiz. 11510 Puerto Real (Cádiz-Andalucía). España

M.E. Ramírez
Laboratorio de Astronomía, Geodesia y Cartografía. Departamento de Matemáticas. Facultad de Ciencias. Campus de Puerto Real. Universidad de Cádiz. 11510 Puerto Real (Cádiz-Andalucía). España

A. Rülke
TU Dresden, Institut für Planetare Geodäsie, D-01062 Dresden, Germany

J.M. Salamanca
Laboratorio de Astronomía, Geodesia y Cartografía. Departamento de Matemáticas. Facultad de Ciencias. Campus de Puerto Real. Universidad de Cádiz. 11510 Puerto Real (Cádiz-Andalucía). España

A. Sánchez-Alzola
Laboratorio de Astronomía, Geodesia y Cartografía. Departamento de Matemáticas. Facultad de Ciencias. Campus de Puerto Real. Universidad de Cádiz. 11510 Puerto Real (Cádiz-Andalucía). España

P. Sarti
IRA – INAF, Bologna, Bologna, Italy

M. Scheinert
Technische Universität Dresden, Institut für Planetare Geodäsie, 01062 Dresden, Germany

K. Shibuya
National Institute of Polar Research, Kaga 1-9-10, Itabashi-ku, Tokyo 173-8515, Japan

R. Soto
Servicio de Satélites. Sección de Geofísica. Real Instituto y Observatorio de la Armada. San Fernando (Cádiz, España)

P. Sterzai
Istituto Nazionale di Oceanografia e Geofisica Sperimentale (OGS), Borgo Grotta Gigante, 34010 Sgonico (Trieste), Italy

B. Thomas Ian
School of Civil Engineering and Geosciences, Newcastle University, Newcastle upon Tyne, NE2 1JU, UK,
e-mail: Ian.Thomas@newcastle.ac.uk

C. Tomasi
Istituto di Scienze dell'Atmosfera e del Clima – CNR, Via P. Gobetti 101, 40129 Bologna, Italy

C. Torrecillas
Laboratorio de Astronomía, Geodesia y Cartografía. Departamento de Matemáticas. Facultad de Ciencias. Campus de Puerto Real. Universidad de Cádiz. 11510 Puerto Real (Cádiz-Andalucía). España

K.R. Tretyak
National University "Lvivska Politechnika", Stepana Bandery street, 12, 79013 Lviv, Ukraine

S. Vey
Institut für Planetare Geodäsie, Technische Universität Dresden, Germany

L. Vittuari
Università di Bologna, DISTART, Viale del Risorgimento, 2 – 40136 Bologna Italy,
e-mail: luca.vittuari@mail.ing.unibo.it

C. Watson
Surveying and Spatial Science Group, School of Geography and Environmental Studies, University of Tasmania, Private Bag 76, Hobart, Australia 7001,
e-mail: cwatson@utas.edu.au

M.J. Willis
Byrd Polar Research Center and School of Earth Sciences, Ohio State University.
275 Mendenhall, 125 South Oval Mall, Columbus, Ohio, USA. 43210,
e-mail: willis.146@osu.edu

A.V. Yuskevich
Department of foreign relations, Neftegazgeodeziya, 195112 St. Petersburg,
Prospekt Utkina 15, Russia,
e-mail: yuskevich@ngg.ru

F. Zakrajsek Andrés
Dirección Nacional del Antártico, Instituto Antártico Argentino, Cerrito 1248
(C1010AAZ), Buenos Aires, Argentina

Y. Zanimonskiy
Institute of Radio Astronomy National Academy of Sciences of Ukraine, Kharkiv,
Ukraine

M. Zanimonskiy Yevgen
Institute of Radio Astronomy National Academy of Sciences of Ukraine, Kharkiv,
Ukraine,
e-mail: yzan@poczta.onet.pl

A. Zanutta
DISTART, Universita' di Bologna, Bologna, Italy

A Precise Reference Frame for Antarctica from SCAR GPS Campaign Data and Some Geophysical Implications

Reinhard Dietrich and Axel Rülke

Abstract The SCAR GPS Campaigns have been carried out since 1995. Based on these data a densification of the Terrestrial Reference Frame (ITRF) in Antarctica has been performed. The results can be used for precise positioning, navigation and validation purposes. Furthermore, the station velocities were obtained. Based on these vectors the kinematics of the Antarctic plate could be determined and an active rifting of 7 mm/year across the Bransfield Strait was measured. Vertical rates are important for the validation of models on glacial isostatic adjustment. The activities are being continued and extended during the International Polar Year within the POLENET project.

1 Introduction

The realization of a consistent global Terrestrial Reference System (TRS) is a fundamental need for global geodetic and geodynamic research. The International Terrestrial Reference System (ITRS) meets the conventions of the International Association of Geodesy (IAG) and the International Union of Geodesy and Geophysics (IUGG) in order to provide the basic principle of coordinates on Earth (McCarthy and Petit, 2004). Observations of geodetic space techniques such as the Global Positioning System (GPS), Satellite Laser Ranging (SLR), Very Long Baseline Interferometry (VLBI) and Doppler Orbitography and Radiopositioning Integrated by Satellite (DORIS) are used to realize the ITRS which means to assign coordinates and velocities to physical markers. Such a realization is represented by the International Terrestrial Reference Frame (ITRF) with its recent release ITRF2005 (Altamimi et al., 2007). In the former release ITRF2000, regional densifications of the global frame became an official part of the ITRF for the first time (Altamimi et al., 2002). Within the IAG, corresponding working

---

Reinhard Dietrich
TU Dresden, Institut für Planetare Geodäsie, D-01062 Dresden, Germany,
e-mail: dietrich@IPG.geo.tu-dresden.de

Axel Rülke
TU Dresden, Institut für Planetare Geodäsie, D-01062 Dresden, Germany

groups were formed in order to provide the input for the regional densifications. For the Antarctic continent, the Group of Experts on Geodetic Infrastructure (GoE on GIANT) within the Scientific Committee on Antarctic Research (SCAR) fulfilled this task.

2 Measurement Campaigns

Since 1995, SCAR GPS Campaigns have been organized by the GIANT group on a broad international basis. The participating countries with the responsible contact persons are summarized in Table 1. Figure 1 shows the geographical distribution of the observation sites. Many sites are located on the Antarctic Peninsula and the South Shetland Islands (Fig. 2). The permanently observing stations of the International GNSS Service (IGS, Dow et al., 2005) form the backbone of the campaigns (large white squares in Fig. 1). The amount of stations involved and the archived observation files are shown in Fig. 3. The campaigns were carried out on an annual basis, lasting 20 days in every Antarctic summer season. With an increasing permanent network in Antarctica the formerly strict observation schedule became more flexible. All campaign observations are compiled and archived in the SCAR GPS data base at TU Dresden/Germany. The data policy allows all participating partners to use all data of the database for their own investigations.

Table 1 The SCAR GPS Crustal Movement Campaigns: Participating countries and contact persons

Argentina	A. Zakrajsek
Australia	J. Manning, G.Johnston
Brazil	E. Simões de Fonseca Jr.
Bulgaria	
Chile	C. Iturrieta, R. Barriga Vargas
China	E. Dongchen
France	C. Vigny
Germany	R. Dietrich
India	E.C. Malaimani
Italy	A. Capra
Japan	K. Shibuya
Norway	T. Eiken
Poland	J. Cisak
Russia	A. Yuskevich
South Africa	R. Wonnacott
Spain	M. Berrocoso
Sweden	E. Asenjo
U.K.	A. Fox
USA	L. Hothem
Ukraine	K. Tretyak, G. Milinevski
Uruguay	H. Rovera

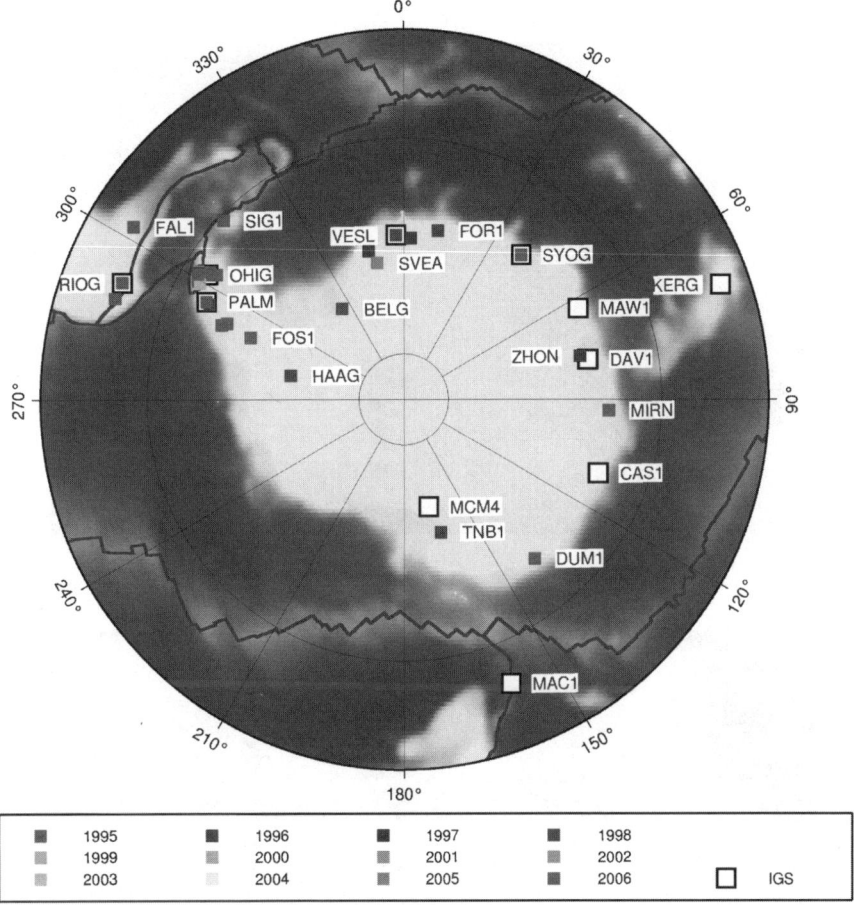

Fig. 1 Distribution of GPS stations in Antarctica. The year of the first occupation is color-coded. Large squares indicate permanent sites of the IGS. The brown lines show boundaries of tectonic plates (Bird, 2003)

3 Data Analysis

As the responsible center for the regional densification the TU Dresden group performed regular data analyses using the Bernese GPS Software (Dach et al., 2007). For the ITRF2000, the corresponding input (SINEX files) was provided to the ITRF combination center for the combination solution (Altamimi et al., 2002). In parallel, also own solutions for the SCAR GPS network were computed using the ITRF as a reference frame (Dietrich et al., 2001, 2004). The analysis strategy has been continuously improved and new modeling components were implemented: Now, the 2nd and 3rd order ionospheric corrections (Fritsche et al., 2005) are considered and the isobaric hydrostatic mapping function based on numerical weather data (Niell, 2000) is used for the tropospheric corrections (Vey et al., 2006). The

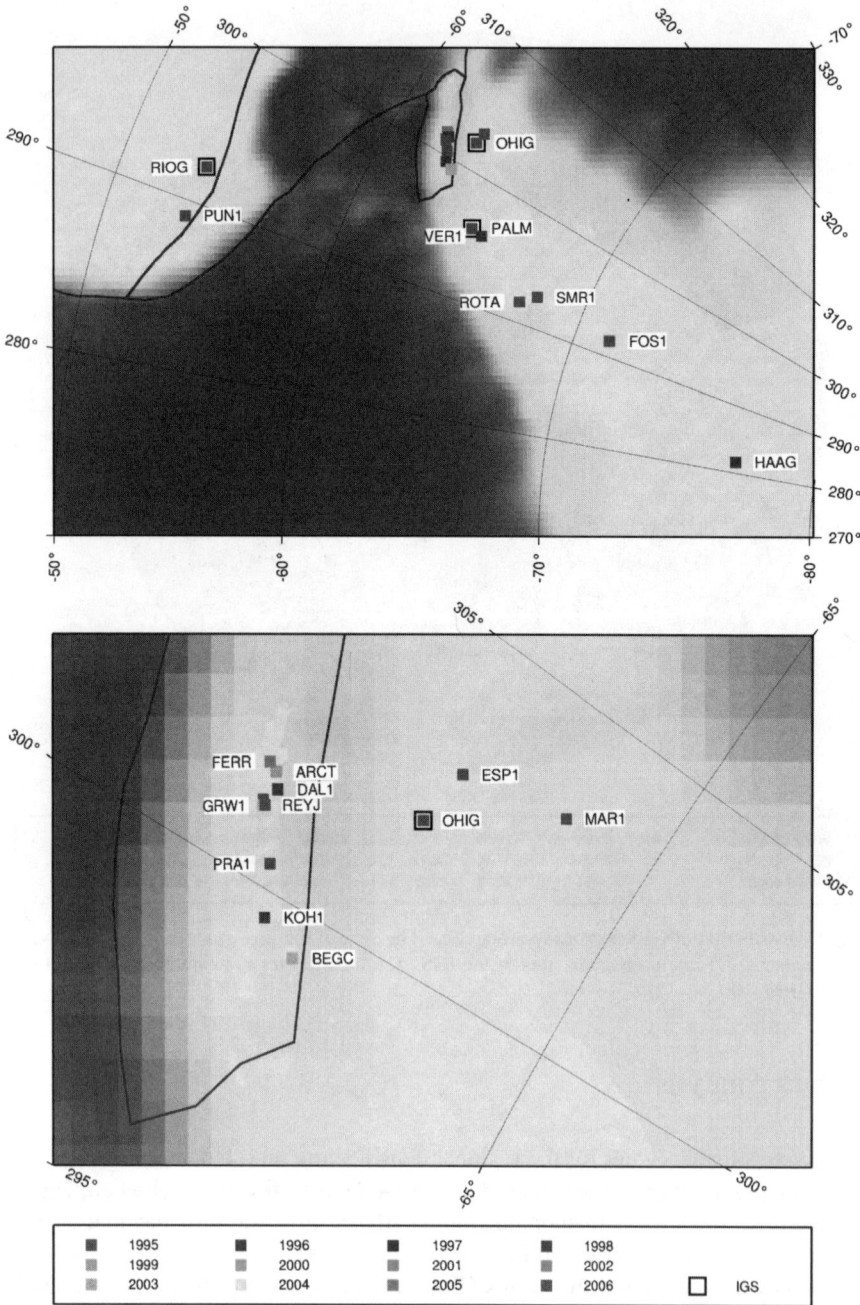

Fig. 2 Distribution of GPS stations in Antarctica: Detail of the Antarctic Peninsula (*top*) and Bransfield Strait/South Shetland Islands (*bottom*). The year of the first occupation is color-coded. Large squares indicate permanent sites of the IGS

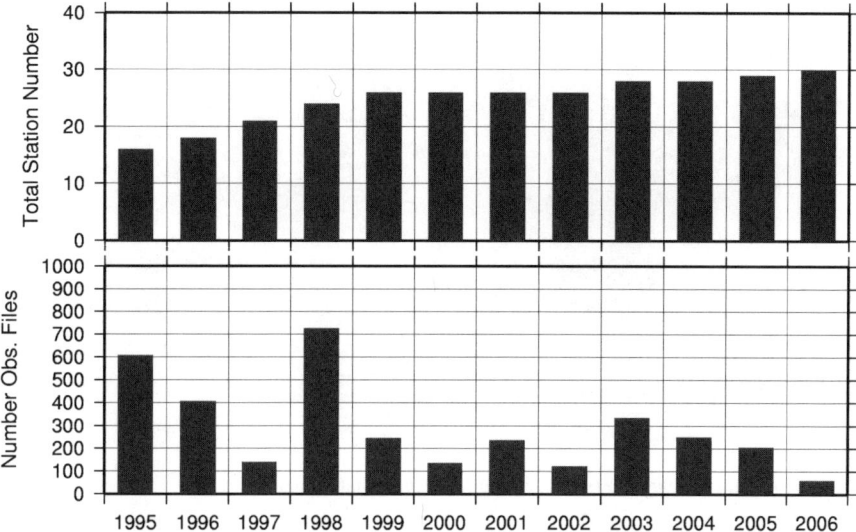

Fig. 3 SCAR GPS Campign database: Cumulative number of stations hold in the database (*top panel*) and total number of observation files provided (*bottom panel*)

antenna phase center variations are corrected using an absolute phase center model for the satellite and the receiver antennas (Schmid et al., 2007). The introduction of satellite orbits and Earth rotation parameters from a homogeneously reprocessed global GPS network further improved the homogeneity of the results (Steigenberger et al., 2006).

4 Results

For about 30 stations, precise coordinates and site velocities were obtained. The daily repeatability as a measure for precision of the station coordinates is in the order of 2...4 mm for the horizontal components and 5...8 mm for the station height. This means, that the station coordinates in Antarctica form a sound basis for different applications like navigation and reference for airborne surveys or ground truthing for satellite altimeter missions. Additionally, an accurate vertical reference could be determined for the Antarctic tide gauges.

5 Geophysical Implications

The repeated determination of station coordinates allowed the estimation of site velocities (Fig. 4). The horizontal velocities were used to determine the kinematics of the Antarctic plate. The residual velocities (Fig. 5) clearly show that the horizontal

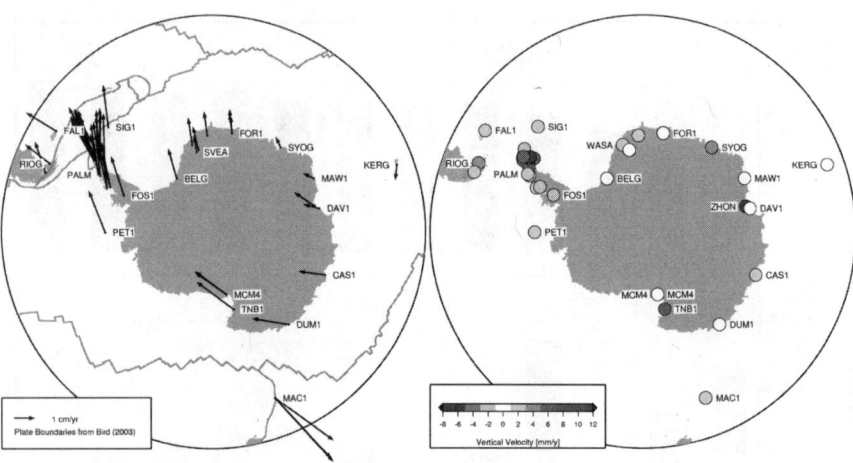

Fig. 4 Horizontal (*right panel*) and vertical (*left panel*) station velocities

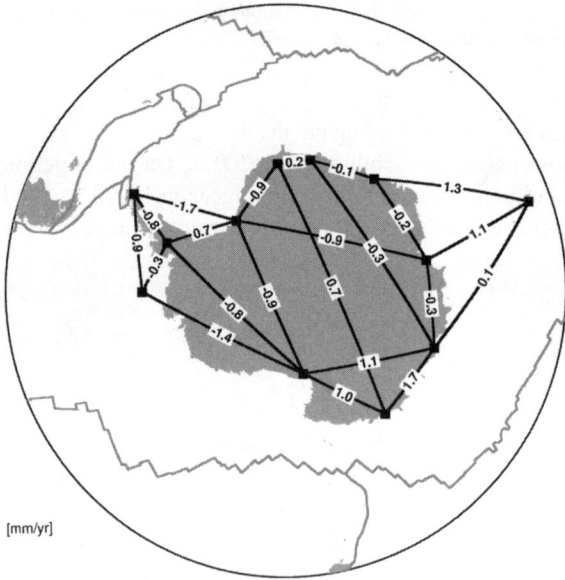

Fig. 5 Changes of spherical distances within the Antarctic Plate in mm/year. The plate boundaries are taken from Bird (2003)

movements in Antarctica including the Antarctic Peninsula can be well described by a rigid plate rotation. The remaining deformations (Fig. 6) are small and do not exceed the noise level which is in the order of 1 ... 2 mm/year. For the Bransfield strait area, an active rifting in the order of 7 mm/year could be determined (Fig. 7).

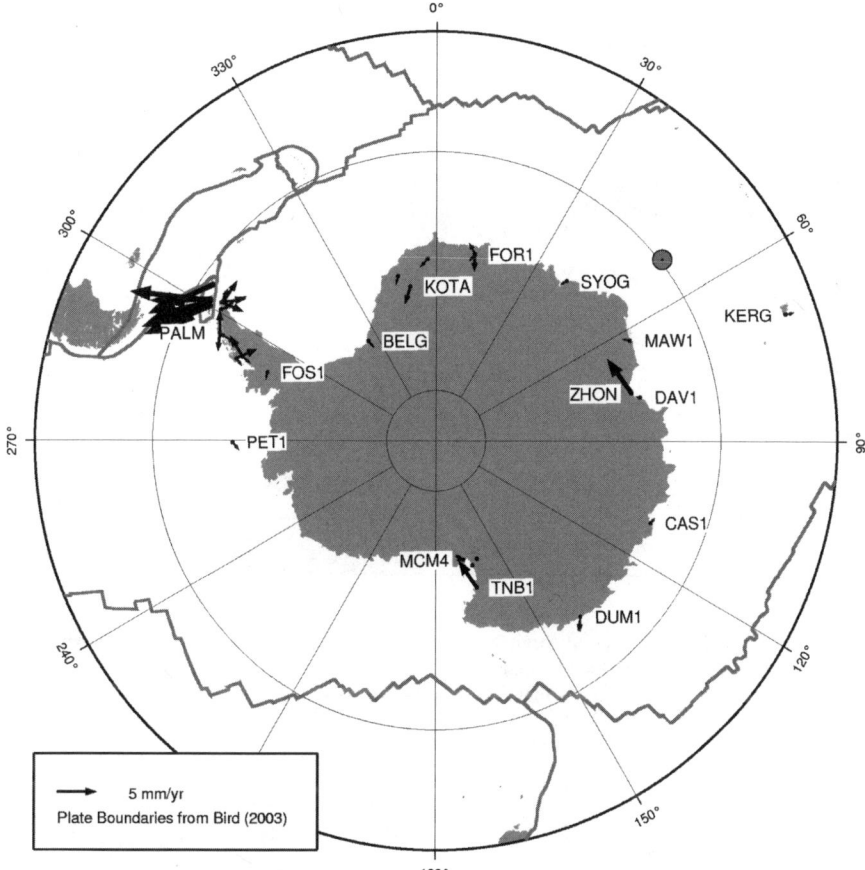

Fig. 6 Residual vectors of the station velocities after removing the rigid plate model for the Antarctic plate. The displayed rotation pole is a left-hand rotation pole. The plate rotation estimate is 0.221 + −0.012 deg./My

The vertical rates can be used to validate models on glacial isostatic adjustment (GIA, Fig. 8). A general agreement to the model of Ivins and James (2005) becomes visible. However, there remain large gaps in the GPS station distribution which impedes a more detailed validation of GIA models.

6 Conclusions and Outlook

The SCAR GPS Campaigns represent a valuable densification of the permanent GPS network in Antarctica in terms of reference system realization. Hence they provide a high quality reference frame for all kinds of GPS based investigations in Antarctica

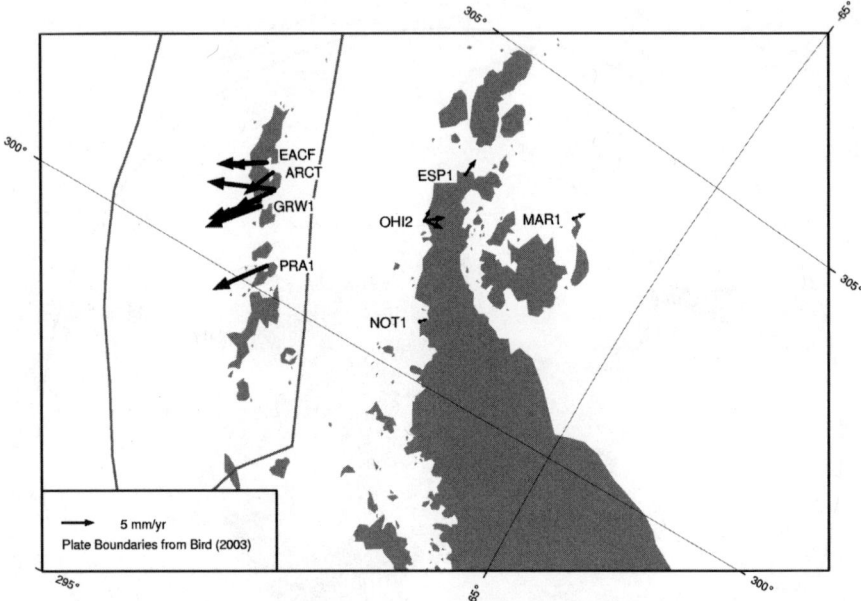

Fig. 7 Opening of the Bransfield Strait between the South Shetland Islands and the Antarctic Peninsula. The vectors show the residuals of the horizontal station velocities after removing the rigid plate model of the Antarctic Plate (cf. Fig. 5). The opening rate of the Bransfield Strait is about 7 mm/yr

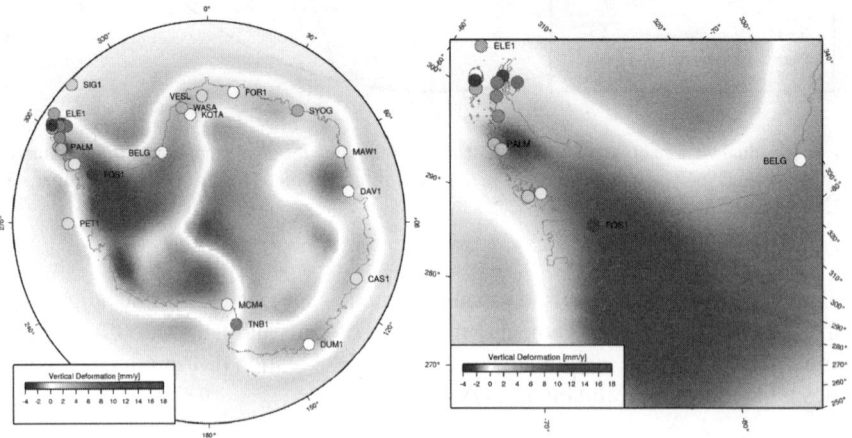

Fig. 8 Vertical station velocities in Antarctica and comparison with model predictions. The laminar color-coding represents the model IJ05 [Ivins and James, 2005]

(e.g. Scheinert et al., 2006, Capra et al., 2007). The determined horizontal station motions can be used for geophysical interpretations, such as plate kinematics of the Antarctic Plate. The long time span of more than 12 years of observations allows to determine reliable vertical station motions which provide useful information to validate GIA models.

Within the International Polar Year 2007/2008 (IPY) also the activities in Antarctic GPS geodesy will be intensified. The IPY project POLENET (Polar Earth Observing Network) with its strong focus on GPS will promote these goals. Here, autonomous GPS stations at remote locations will help to further close the gaps still existing in the Antarctic GPS network. The GIANT consortium is actively participating in the POLENET field work.

Acknowledgments We gratefully thank all individuals and agencies who supported the field projects, collected and provided the data. An anonymous reviewer is gratefully acknowledged.

References

Altamimi, Z., P. Sillard, and C. Boucher (2002), ITRF2000: A new release of the International Terrestrial Reference Frame for earth science applications, *J. Geophys. Res., 107*(B10), doi:10.1029/2001JB000561.

Altamimi, Z., X. Collilieux, J. Legrand, B. Garayt, and C. Boucher (2007), ITRF2005: A new release of the International Terrestrial Reference Frame based on time series of station positions and Earth orientation parameters, *J. Geophys. Res.*, 112(B09401), doi:10.1029/2007JB004949.

Bird, P. (2003), An updated digital model of plate boundaries, *GGG*, 4(3), 1027, doi:10.1029/2001GC000252.

Capra, A., F. Mancini, and M. Negusini (2007), GPS as a geodetic tool for geodynamics in northern Victoria Land, Antarctica. *Antarctic Sci.*, 19(1), 107–114, doi:10.1017/S0954102007000156.

Dach, R., U. Hugentobler, P. Fridez, and M. Meindl (Eds.) (2007), Bernese GPS Software Version 5.0, Astronomical Institute, University of Berne, Switzerland.

Dietrich, R., R. Dach, G. Engelhardt, J. Ihde, W. Korth, H.-J. Kutterer, K. Lindner, M. Mayer, F. Menge, H. Miller, C. Müller, W. Niemeier, J. Perlt, M. Pohl, H. Salbach, H.-W. Schenke, T. Schöne, G. Seeber, A. Veit, and C. Völksen (2001), ITRF coordinates and plate velocities from repeated GPS campaigns in Antarctica – an analysis based on different individual solutions. *J. Geod.*, 74(11/12), 756–766, doi:10.1007/s001900000147.

Dietrich, R., A. Rülke, J. Ihde, K. Lindner, H. Miller, W. Niemeier, H.-W. Schenke, G. Seeber, (2004 July), Plate Kinematics and Deformation Status of the Antarctic Peninsula based on GPS. *Global Planet. Change*, Ice sheets and neotectonics, 42(1–4), 313–321, doi:10.1016/j.gloplacha.2003.12.003.

Dow, J. M., R. E. Neilan, and G. Gendt (2005), The International GPS Service (IGS): Celebrating the 10th Anniversary and Looking to the Next Decade, *Adv. Space Res.*, 36(3), 320–326, doi: 10.1016/j.asr.2005.05.125.

Fritsche, M., R. Dietrich, C. Knöfel, A. Rülke, S. Vey, M. Rothacher, and P. Steigenberger (2005), Impact of higher-order ionospheric terms on GPS estimates, *Geophys. Res. Lett.*, 32, L23311, doi:10.1029/2005GL024342.

Ivins, E. R., and T. S. James (2005), Antarctic glacial isostatic adjustment: a new assessment, *Antarctic Sci.*, 14(4), 541–553, doi:10.1017/S0954102005002,968.

McCarthy, D. D., and G. Petit (Eds.) (2004), IERS Conventions (2003), IERS Technical Note No. 32, IERS Conventions Center, Verlag des Bundesamtes für Kartographie und Geodäsie, Frankfurt am Main.

Niell, A. E. (2000), Improved atmospheric mapping functions for VLBI and GPS, *Earth Planets Space*, 52, 699–702.

Scheinert, M., E. Ivins, R. Dietrich, and A. Rülke (2006), Vertical Crustal Deformations in Dronning Maud Land, Antarctica: Observations versus Model Predictions. In: Fütterer, D., D. Damaske, G. Kleinschmidt, H. Miller, F. Tessensohn (eds.), *Antarctica – Contributions to Global Earth Sciences* (Proc. of the IX ISAES, Potsdam, September 8–12, 2003), pp. 357–360. Springer Berlin – Heidelberg – New York.

Schmid, R., P. Steigenberger, G. Gendt, M. Ge, and M. Rothacher (2007), Generation of a consistent absolute phase center correction model for GPS receiver and satellite antennas, *J. Geod.*, doi: 10.1007/s00190-007-0148-y.

Steigenberger, P., M. Rothacher, R. Dietrich, M. Fritsche, A. Rülke, and S. Vey (2006), Reprocessing of a global GPS network, *J. Geophys. Res.*, 111, B05402, doi:10.1029/2005JB003747.

Vey, S., R. Dietrich, M. Fritsche, A. Rülke, M. Rothacher, and P. Steigenberger (2006), Influence of mapping function parameters on global GPS network analyses: Comparisons between NMF and IMF, *Geophys. Res. Lett.*, 33, L01814, doi:10.1029/2005GL024361.

Technologies to Operate Year-Round Remote Global Navigation Satellite System (GNSS) Stations in Extreme Environments

Michael J. Willis

Abstract The POLar Earth observing NETwork (POLENET) is an ambitious international project to deploy geophysical instruments at very remote high-latitude sites during the International Polar Year (IPY). One of the goals of the project is to run instruments year round with as little maintenance as possible. POLENET will be installed using robust lightweight systems, minimizing the need for heavy battery banks as much as possible. This weight reduction is needed in order to meet logistical constraints on deployment at very remote sites. New and established technologies for Global Navigation Satellite System (GNSS) stations are critically examined here in order to determine the best balance between reliable power generation and storage and the logistical cost to deploy such a system. Best practices are summarized from successful projects that have run reliably in extreme polar environments.

1 Introduction

Large areas of the polar ice sheets are changing at surprisingly rapid rates. Satellite observations (Rignot et al. 1995; Joughin et al. 2003; Chen et al. 2006a,b; Luthcke et al. 2006; Llubes et al. 2007), airborne measurements (Abdalati et al. 2001; Thomas et al. 2006) and glacial isostatic adjustment model predictions (James and Ivins 1998; Fleming and Lambeck 2004; Ivins and James 2005) indicate that most of the changes are occurring at locations far from existing infrastructure and that the changes have been poorly observed, or even missed, in the past. Satellites measurements are hampered as they are unable to separate the signal caused by ice sheet mass change from the signal caused by the solid Earth response to both long and short-term ice mass changes. In order to measure the changes in the mass of the ice sheet, the effect of bedrock motion must be removed.

Michael J. Willis
Byrd Polar Research Center and School of Earth Sciences, Ohio State University.
275 Mendenhall, 125 South Oval Mall, Columbus, Ohio, USA. 43210,
e-mail: willis.146@osu.edu

Data from remote continuously running GNSS sites are uniquely able to measure the bedrock response on all temporal scales, through daily and seasonal elastic motions to multi-year secular trends. GNSS measurements are however spatially sparse at present and far removed from the areas where changes are occurring. More remote stations at a greater density are required to allow the mass changes of the ice sheets to be weighed and to calculate the transfer of mass from the ice sheets into the global ocean.

Several remote GNSS sites are operating successfully in Antarctica, but the logistical cost to install such stations at distances of more than a few hundred kilometers from a manned base remains high due to the weights, time required onsite and distances involved. Current stations are either heavy or functionally limited, such as reduced to working several months out of the full year. This paper will review technologies in current use and their operational performance and will highlight some promising new technologies that will help increase functionality and reduce the logistical cost of remotely operating GNSS stations for the IPY and beyond.

Issues surrounding remote GNSS stations were first discussed formally by the US community at the "Autonomous Systems in Extreme Environments" workshop at the Jet Propulsion Laboratory in 1999. The challenges raised at this meeting have since been further examined (Antarctic Remote Observatories Workshop, 2004; Antarctic Seismic and GPS Technologies, Challenges, and Opportunities Workshop, 2005) and have led those in the US involved in Antarctic geodesy and seismology to support a successful Major Research Instrumentation (MRI) proposal to the US National Science Foundation (NSF) titled "Development of a power and communication system for remote autonomous GPS and seismic stations in Antarctica." Award details are found at www.nsf.gov/awardsearch/showAward.do?AwardNumber=0619908.

The goal of the award is "to design, integrate, and test a scalable power and communication system optimized for ease of deployment and reliable multiyear operation in severe polar environments" (Prescott et al. 2006). The project will result in the design and fabrication of a system made from largely off-the-shelf components that will operate continuously year-round and provide data on to the internet. This capability will allow many more sites to be installed in remote polar regions. At present the largest constraint on deploying remote sites is the need to repeatedly visit them, at great logistical cost, to pick up data and to perform maintenance.

In this paper remote stations are defined as being unreachable from any year-round station with permanent personnel during the winter period. As an operational goal, remote stations should only be visited once per year, and preferably less often. "Extreme" refers to polar regions which experience extreme low temperatures, very high wind speeds, have a limited amount or no daylight in the winter and which present severe challenges in deploying equipment at locations far away from established logistical hubs.

This paper draws extensively on experiences of the TAMDEF (TransAntarctic Mountain DEFormation) project which, since 2000, established six continuously running remote sites in Antarctica (Willis et al. 2006). TAMDEF is an NSF funded project to measure glacial isostatic adjustment and neotectonics in the interior of Antarctica, NSF grant# 0230285.

2 Science Payloads

The science payload of a remote station will include the GNSS receiver, with its associated power draw and operating temperature range, and may also include ancillary devices such as state of health monitors and meteorology packages. Adding communications to the science payload is desirable to meet some science objectives and to reduce the need for site visits. Each addition to the payload has a specific power cost and operating temperature limits.

2.1 GNSS Receiver

The GNSS receiver forms the most critical part of the remote system and needs to be powered as reliably as possible. It is essential that the receiver is robust and has a proven reliability record under conditions likely to be experienced at high latitudes, including extreme cold and high ionospheric activity (Skone and de Jong 2000; Skone 2001) It is critical that the GNSS receiver uses as little power as possible. A receiver that draws more than ~5 watts becomes hard to power year round at remote high latitude locations due to the complexity and weight of the necessary power systems.

Specifications of modern GNSS receivers should include those outlined in the "additionally desired characteristics" IGS guidelines (Dow et al. 2005; Moore 2007): dual-frequency, all-in-view capability, multiple constellation capability, a capability to record at 1 Hz or better, and the ability to use an external geodetic antenna. Furthermore, the receiver should have IP connectivity, a capability to record more than 512 onboard files and produce small, preferably compressed, daily data files.

A recent survey of 73 GNSS receiver manufactures in the January 2007 edition of GPS World found only 9 manufacturers provide capabilities even near those required. Most other receivers are unsuitable due to their high power requirements. Several low-power receivers are unable to record more than 512 files internally and/or create large 24-h data files. Whether these secondary characteristics are drawbacks depends on communications requirements, which vary depending on the location of the station. Low-power micro-controllers provide an avenue that allows a broader range of receivers to be considered, however their deployment, at present, is limited. Of further consideration is the power draw of the GNSS antenna, which varies depending on the specifications of the amplifier within the antenna. The addition of a geodetic grade antenna can change the power draw of the receiver by between 0.1 and 1 watt.

2.1.1 TAMDEF Experience

The TAMDEF project has used three different receivers for its continuous remote sites (summarized in Table 1). The power draw has varied depending on the communications set up being used. Three sites have had no communications and used

Table 1 Power configurations at TAMDEF remote CGNS sites

Style	Receiver Type	Receiver draw (W)	Communications Type	Comms. Draw* (W)	PV size (W)	Batteries (Amp Hr)
1	JNS E-GGD	2.5	None	0	160	1200
2	TRIMBLE 5700	2.7	Radio Link	2	400	2400
3	TRIMBLE NetRS	3.5	Satellite modem	1	400	2748

*daily average value

GPS receivers with a power draw of between 2.0 and 2.5 W. These receivers ran successfully year round (Fig. 1) with 160 W solar panels and 1200 Amp Hours of Sealed Lead Acid (SLA) batteries.

Two sites have used radio modems to communicate through radio repeater infrastructure back to the US Antarctic Program (USAP) base at McMurdo (77.8°S, 166.7°E). Radio repeaters are typically used around settlements or scientific bases to communicate out to about 200 km. In Antarctica, the cold, dry air reduces attenuation of the radio signal and long range communications are possible using low-power systems. Two TAMDEF sites have run successfully year round, using large 400 W solar arrays and large (2000+ Amp Hours) battery banks. These systems have a total daily average load of 4.5 watts. The extra power requirements above those of the GNSS receivers alone are due to the radio modems used.

A third TAMDEF system configuration draws 5.25 W and has run successfully year round (Fig. 1). This system uses a Trimble NetRS® GPS receiver connected to a NAL Research Iridium L-Band Data Transceiver (satellite modem). The NetRS receiver has the capability of internally storing more than 512 files and uses onboard data compression to produce small hourly data files, about 30 Kb in size when recording at a 30 s rate. Small files are essential when transmitting data through the slow, 2400 baud rate Iridium data service as calls are sometimes dropped (akin to using a cell phone in an area of poor coverage). Any missing small files can be quickly retrieved, a time and cost saving preferable to downloading an entire day-long file a

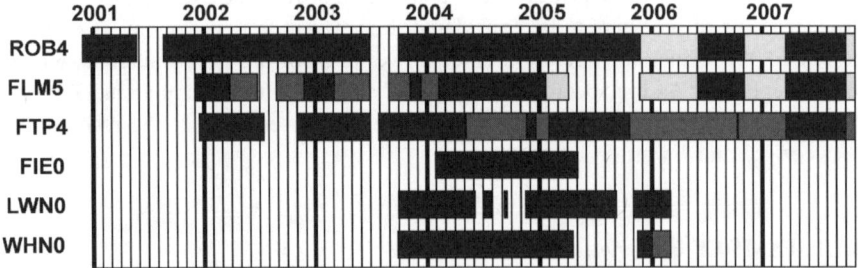

Fig. 1 Performance of TAMDEF remote GPS sites. Blue indicates fully functioning, red indicates powered but non-usable, yellow indicates fully functional and data sent by radio modem, green indicates fully functional and data sent by satellite modem

second time. The hourly files are automatically concatenated at the downstream end by a Linux computer system. The ability to record more than 512 files is essential. If an Iridium modem at a remote site were to fail completely and the receiver used was set to record 24 one-hour duration files per day but could only store 512 files, then only just over 21 days of data could be stored internally before the receiver either stops logging or over-writes older data. The NetRS is limited only by its ~900 Mb internal memory size, that will allow several years of data to be stored onboard if necessary.

2.2 State of Health and Temperature Monitoring

Logging temperatures at various positions in and around equipment enclosures at TAMDEF remote sites and at a test site built by UNAVCO in McMurdo, has resulted in an extremely valuable data set. The temperature logging was performed at TAMDEF sites using 9V Lithium battery powered Avatel datascribe® stand alone temperature loggers. At the UNAVCO test site a very low power Campbell Scientific data logger was used to record voltages, power draw, power production and temperatures. From the multiple sites we see that receivers operate reliably and batteries provide sufficient charge to run low-power receivers to an ambient outside temperature of $-50°C$. At the TAMDEF sites the battery banks supplying the receiver are split between several enclosures. Six batteries are kept at the same temperature as the receiver, the rest are in plywood boxes and closely follow the outside ambient temperature. The GPS receivers themselves typically produce enough heat so that if insulated well, they stay about $10°C$ warmer than ambient temperatures (see Sect. 4.1).

At present a low resolution temperature sensor is integrated into some GNSS receivers, but this sensor must be interrogated through an active communications link. It is also possible to check voltage on the receiver power ports when connected. These capabilities provide some rudimentary diagnostic information at sporadic intervals. A preferred rigorous state of health and temperature-logging system would be integrated with the receiver and be able to store and send its data alongside the GPS message back to the user. Information logged would include the meteorological conditions at the site, receiver temperature, battery temperatures, power generation system voltages, receiver voltage, file system status, modem status and, if possible, battery capacitance. These data would allow more robust diagnostics should the system fail in the field while providing valuable data for the meteorological community.

3 Power Generation

Powering a remote GNSS system year round at polar latitudes is hard to achieve. The polar nights, ranging from very short days at around $60°$ of latitude to complete darkness for 6 months of the year at $90°$ latitude, preclude the year-round use of photovoltaic cells. Solar radiation (insolation) varies through time at different latitudes. The high winds, common in polar regions, are often unpredictable in duration and in strength, which has led to severe problems using wind turbines as reliable power sources. Furthermore, the cold temperatures of the polar night reduce the capacity of most batteries. Therefore, very large battery banks are required to store enough energy to power a GNSS system continuously through the dark period. This section details the power systems available and in use at remote sites in polar regions and some that are being developed for use in the near future. The focus will first be on power generation, then power control, then batteries.

3.1 Photovoltaic Cells

Passive photovoltaic cells (PV) are an ideal power source during the long polar summer days. The systems are reliable, have no moving parts and work more efficiently the colder it gets (Fig. 2) (Akhmatov et al. 1998; Nishioka et al. 2003). PV cells have no lower operating temperature limit. As a rule of thumb the solar panels should be tilted at an angle off vertical equal to the colatitude of the site (see Fig. 3). This will maximize the power produced by the panels at low sun periods. The PV should point towards true north in the southern hemisphere and true south in the northern hemisphere. This again allows the panels to recharge the power storage system as late and as early in the year as possible, when the sun skirts the horizon. Solar panels with high tilt angles have two further benefits, they can be energized by the diffuse sunlight reflected off surrounding snow (Yoshioka et al. 2002) and the high angle stops snow from accumulating on the solar panels. Figure 4 shows the performance of different size solar panel arrays as solar insolation changes with latitude for the southern hemisphere. More powerful panels provide useful electricity later into the year than small panels, as shown by the left-hand intercept between a horizontal line (indicating panel size) and the insolation curve specific to the latitude. Intercepts on the right hand side similarly show the first day in springtime in which useable energy

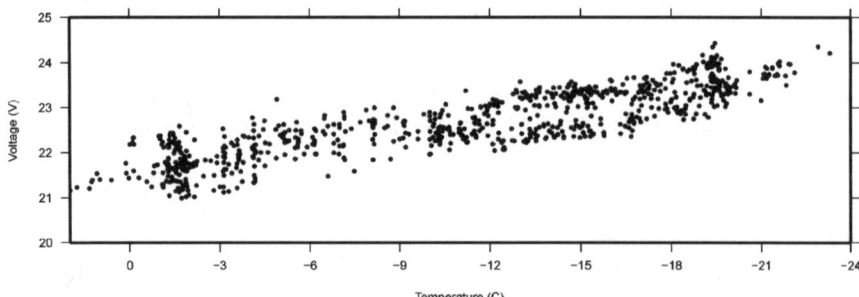

Fig. 2 Increasing solar panel voltage plotted against decreasing air temperature at McMurdo. Data collected between February and May 2006. Provided by Observation Hill test site. Data courtesy of UNAVCO

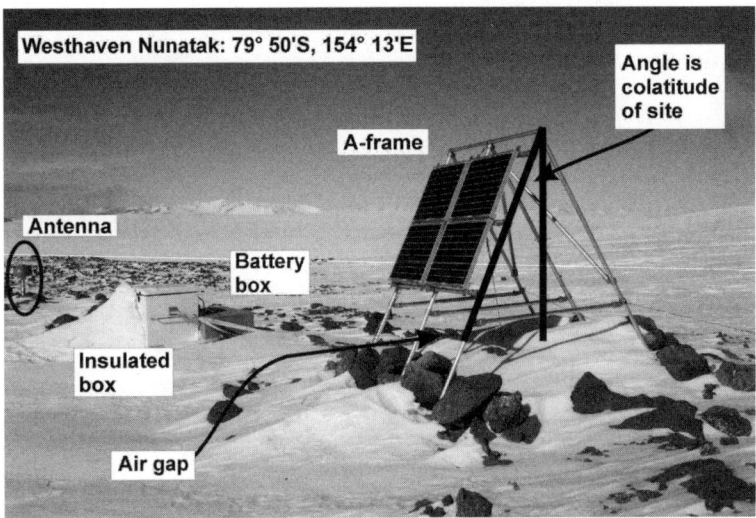

Fig. 3 Westhaven Nunatak GPS site showing configuration of 160 Watts of solar panels on A-frame. Vacuum-panel insulated box and GPS antenna visible in background. The air gap beneath the solar panels stopped drifting snow from building up. The frame is weighed down by 200 kg or more of rocks and lightly anchored using aircraft wire and 3 expansion bolts set into bedrock

is produced. For the assumed 5 W draw system, a larger solar panel array provides significant advantages over a small array. As the size of the PV system is increased there is a corresponding decrease in the number of days when the solar panels produce no usable power. Therefore fewer batteries are required for systems with large PV arrays. For example, at 75°S, a 215 watt PV system will require enough batteries to run ∼144 days when the solar power is insufficient to recharge the system. A 95 watt PV system will require enough batteries to run ∼230 days at the same latitude, a difference of 114 days.

Physical damage to the solar panel array is the main problem experienced. Occasionally during the TAMDEF project we have seen solar panels destroyed during extreme wind events. Bracing and reinforcement of solar panels will be discussed in the environmental hardening section.

New solar cell technologies are under investigation by several companies. An increase in the efficiency of PV arrays is possible through new chemistries and the use of concentrators (Deb 1998; Rannels 2000; Delahoy et al. 2004; Ryu et al. 2006); currently most PVs operate at about 10–20% efficiency in converting incoming solar radiation to useable electricity. It should be noted that amorphous silicone/thin film panels are not suitable for long-term use in polar regions. They work poorly in cold conditions compared to mono- or poly-crystalline panels, show a strong tendency towards lower efficiency after only a short period of operation and are sensitive to a narrower frequency of radiation, which reduces their ability to make electricity during low sun periods. Large rigid crystalline PV arrays do have their drawbacks, however, as they are more prone to wind loading and damage.

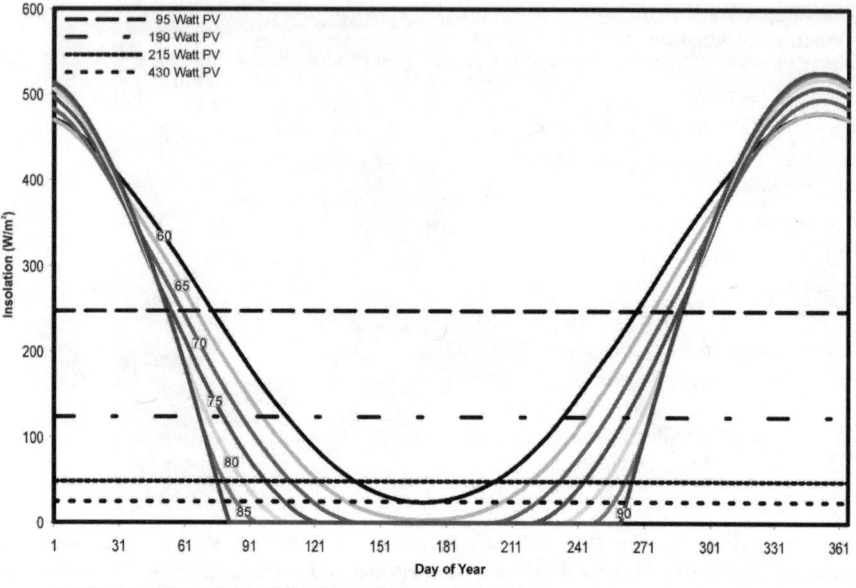

Fig. 4 Insolation at selected latitudes for southern hemisphere. Horizontal lines show insolation required by different photovoltaic panel sizes to run a GNSS system with a 5 watt draw. Higher power panels provide significantly more days of power throughout the year. The intercept between a horizontal line (panel power) and a latitudinally dependent insolation curve on the left hand side of the diagram (between day 55 and day 160 depending on panel size and latitude) indicates the last day that a panel of size X can produce more usable power than the GNSS system draws at that latitude. The intercept between the horizontal line and the insolation curve on the right similarly indicates the first day of spring in which the panels can produce more power than the GNSS receiver draws. The time period between these two intercepts is the number of days the GNSS system must run on batteries or alternative power

For further information concerning the installation and use of photovoltaic arrays in polar regions the reader is referred to http://www.polarpower.org/technologies/solar_power.

3.2 Wind Turbines

Even though the polar regions are consistently the windiest regions on the planet, the wind still provides a highly variable energy source. Small wind generators have been used with varying degrees of success on the polar ice sheets where low wind speeds are often more of a hindrance than extreme events; see Anandakrishnan et al. (2000) for successful turbine application, and see Sterling (2005) and Rachelson (2006) for an application with mixed results. In the mountainous areas of Antarctica, the use of small horizontal axis turbines has been extremely problematic. Although some turbines last two or three years, there are inevitable failures during periodic extreme wind events (Zwartz and Helsen 2002; Donnellan and Luyendyk 2004; Kyle 2007). Figure 5 highlights the temperature and wind regime at six sites in the polar regions. Winds during storm events reach speeds of up to 180 km/h.

Development and deployment of wind turbines is still desirable, however, because wind energy provides a useful complement to solar panels. A wind turbine potentially provides the bulk of its power during the dark months when conditions are typically windier, as shown by the average wind speeds in Fig. 5. Solar panels can only provide power during daylight. A system powered solely by PV panels requires a large battery bank to run year round at high latitudes. If the weight of a wind turbine, with its associated tower and bracing, is lighter than the weight of the PV-only battery bank and if the wind regime at the site is known, then a wind generator should be considered as it will reduce the overall logistical cost of deployment of the system.

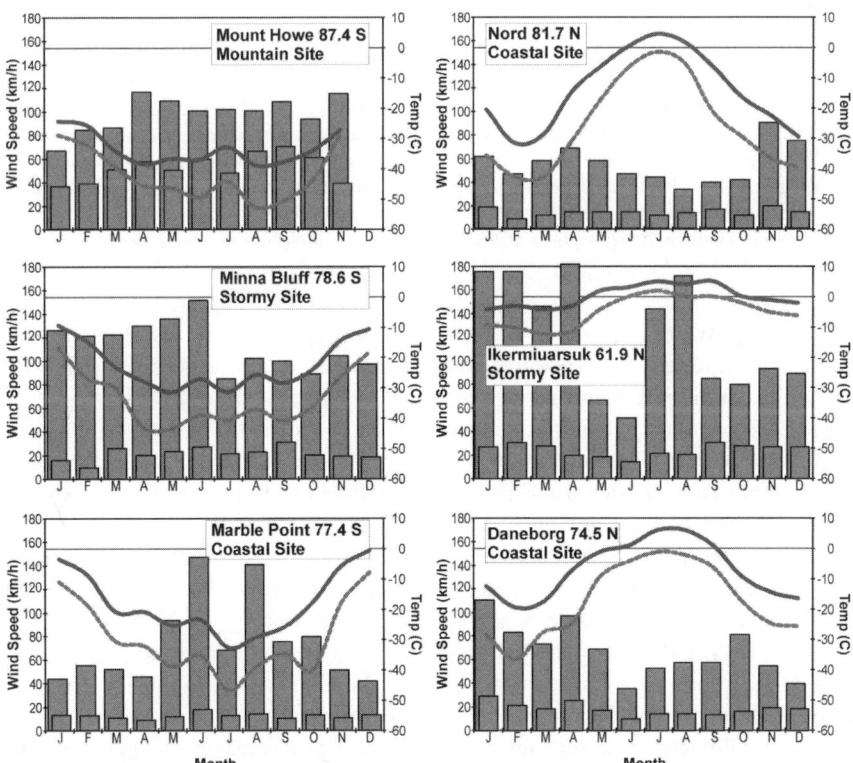

Fig. 5 Temperature and wind speeds for six polar sites. Thick red line is average monthly temperature in Celsius, Blue dashed line is minimum monthly temperature in Celsius, Dark grey columns are peak monthly wind speed in km/h, orange columns are average monthly wind speed in km/h. The plots show Antarctic sites (on the *left*) are typically much colder than Greenlandic sites (on the *right*). Extreme peak wind speed are measured during winter storms in both regions. Winter average wind speeds are typically higher in both regions

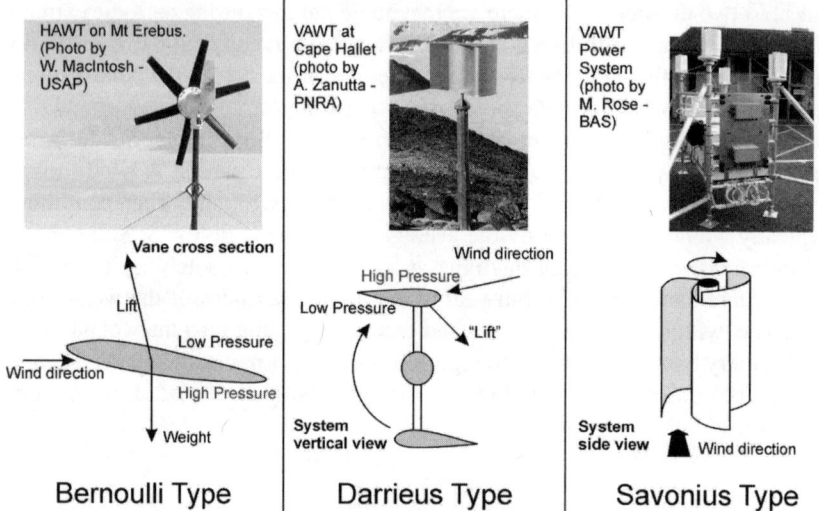

Fig. 6 Small wind turbine types. (*left*) Bernoulli type horizontal axis wind generator (HAWT) use the aerodynamic principle of lift to rotate the blades. Blade tips can achieve speeds greater than wind speed. (*center*) Ropatec hybrid Darrieus and Savonius type vertical axis wind turbine (VAWT). The sides of the turbine are airfoils that also use lift, this time oriented in a horizontal plane. The blades of this turbine can also turn at speeds faster than the wind speed. The middle of the turbine is a Savonius type drag system. (*right*) Savonius type drag driven wind turbine. Drag force on the face of the blade turns the turbine. The leading edge of the turbine cannot spin at rates faster than the wind speed. This makes a less efficient design, but promotes durability

Figure 6 provides schematics of three different types of small wind generators that have been tested in the polar regions. Horizontal axis wind turbines (HAWT) use the Bernoulli principle of aerodynamic lift to spin the turbine airfoils to generate power. Some vertical axis wind turbines (VAWT) use the Darrieus design, which has two or more aerodynamic airfoils aligned vertically and directs "lift" in the horizontal plane. Turbines that use aerodynamic airfoils can turn at very high rotation speeds and produce large amounts of power. The high rotor speeds, where rotation can be faster than the wind speed, require the systems to be finely balanced. Darrieus type turbines are not self starting and those used in Antarctica have incorporated a small drag style turbine in their design to initialize rotation.

Several VAWT designs operate on the Savonius principle using the drag force of the wind to turn an S-shaped vane (Fig. 6). Vertical axis Savonius systems are inherently less efficient than horizontal axis or Darrieus systems, as the energy of the wind is split between a force turning the turbine and a force on the opposing side of the turbine resisting the rotation. System using this principle cannot have rotation speeds that exceed the wind speed, an advantage in storm-prone polar regions.

Tests of three systems fielded in Antarctica are described here. The Italian National Antarctic Program (PRNA) has installed several models of Ropatec wind turbines at Cape Hallett and Mario Zucchelli Station (F. Mancini, 2006, Personal Communication). Larger Ropatec turbines have been installed in the northern hemisphere by Tele-Greenland. The PRNA Ropatec systems produce about 30 watts of power at a wind speed of 36 km/h and require fairly extensive bracing and balancing as the vane weight is upwards of 30 kg. The generator has a total weight of 120 kg, or about the same as four large batteries. The foundation for the system must also be substantial. Ropatecs are hybrid Darrieus-Savonius turbines that can potentially produce significant amounts of power at high rotation rates, but so far the remote systems have yet to survive the Antarctic winters of Northern Victoria Land (F. Mancini, 2006, Personal Communication).

The Jet Propulsion Laboratory and the Finnish National Antarctic Program (FINNARP) have successfully used vertical axis Savonius systems fabricated by the Finnish Windside company. The Finnish Aboa base in Antarctica is partially powered by a suite of six Windside turbines (J. Mäkinen, 2006, Personal Communication). JPL used two Windsides on a remote GNNS site at Mount Coats during 2000. The turbines were then removed for further evaluation at sites around McMurdo Station and on Mount Erebus. During 2000, the Windside turbines provided enough power to bring a substantial bank of batteries up to a level where they could run a receiver with a 15 W draw for a limited amount of time (C. Raymond, 2006, Personal Communication). These turbines are of similar performance to the Ropatec models, producing about 40 watts at a wind speed of 36 km/h. The Windside system vane weighs only 2 kg, but also must be braced well to survive high wind events. The total weight of a Windside generator minus bracing is approximately 80 kg or about the same as two 100 Amp-Hour SLA batteries. Both the Ropatec and Windside generators are built with extremely strong, heavy construction techniques. The amount of bracing and engineering that these types of generator require make them challenging to deploy in the deep field.

The British Antarctic Survey has been using eleven smaller wind turbines of a Savonius design for the last three years. The units they use, Forgen 500® turbines, are very lightweight (3.65 kg) and relatively cheap. The 500 model can be provided with cold weather bearings and has been reliably operated to −65°C in Antarctica (M. Rose, 2006, personal communication.) Larger Forgen 1000® models have been used before in Antarctica (Anderson et al. 2005), but the bearings on these model have shown decreasing levels of reliability through time (M. Rose, 2006, Personal communication). UNAVCO has started field testing the smaller Forgen 500 units in Antarctica and has placed them at a site which experiences extreme wind speeds in an effort to further test the reliability of the turbines (Minna Bluff, Fig. 5). The Forgen 500 produces a very low amount of power, about 4 watts at a wind speed of 28 km/h (B. Johns, 2006, Personal Communication), but the light weight of the units and their affordability means multiple units can be installed at relatively low cost, improving both the power generation and redundancy of the system. Like all wind generators, Forgens need to be leveled precisely before their use, but a great advantage of the lightweight turbines is that they do not require the bracing that must be built for larger systems.

For further information concerning wind generator installation and operation in polar regions the reader is referred to http://www.polarpower.org/technologies/wind_power.

3.3 Fuel Cells

Fuel cells operate by converting chemical energy directly into electrical power. There are several types of fuel cell available on the open market, some of which are more applicable to remote GNSS sites than others. The main parameters to consider when evaluating the feasibility of deploying a fuel cell into the deep field are the efficiency of the conversion process, the characteristics of the fuel, and the operating temperature range for the conversion process. There are currently five main types of fuel cells stacks, although more are under development. The current deployable types are defined by the type of chemical energy sourced used; alkaline, phosphoric acid, molten carbonate, solid oxide and proton exchange membrane (PEM) fuel cells.

Most fuel cell technology will not work at remote polar GNSS stations. Alkaline fuel cells are extremely efficient but require pure oxygen and pure hydrogen fuels. Phosphoric acid fuel cells operate at internal temperatures of between 150°C and 200°C and must be kept at temperatures of above 40°C in order to start operation. Molten carbonate fuel cells are industrial sized units that operate at temperatures above 650°C. Solid oxide fuel cells also typically work at high temperatures that are not sustainable at remote sites in polar regions. The remaining technology is PEM fuel cells.

There are three main types of PEM fuel cell – those that use pure hydrogen, those that use hydrogen that is stripped off other hydrogen bearing compounds (reformed hydrogen), and those that use methanol directly. A PEM hydrogen fuel cell stack has been deployed at Beaver Lake, in the Lambert Graben of East Antarctica since 2000 (Tregoning et al. 2000), but its performance has been poor. There was difficulty transporting the required number of compressed hydrogen gas canisters out to the site and the parasitic maintenance load while the fuel cell was in idle, was higher than the energy produced by the system (Tregoning, 2007, personal communication.)

Of the PEM technologies in consideration, the reformed hydrogen and direct methanol (or ethanol – in development) systems seem to offer the most promise. The US military has shown interest in reformed hydrogen fuel cells (Sifer and Gardner 2004) but details on their capabilities at low temperature are unavailable at present. Direct Methanol Fuel Cells (DMFC) operate at relatively low efficiency compared to other fuel cells (Srinivasan et al. 1999), but they work at much lower internal temperatures of between 10°C and 100°C than other fuel cell types (Srinivasan et al. 1999). In addition, the methanol fuel has a high energy density, has a freezing point close to $-100°C$ and as a liquid is relatively simple to procure and transport. Fuel cell systems would work in the same manner as a hybrid wind/solar system, the fuel cell providing energy when a specific battery voltage threshold is passed. Fuel cells will need maintenance visits to recharge the fuel supply. The waste products of a DMFC are heat, water vapor and carbon dioxide. The heat can be used to keep batteries warm, allowing the batteries to store more charge, but the water vapor is a difficult by-product to manage in the polar environment see (Datta et al. 2002). Several projects in Antarctica in the

1990s used thermal electric generators to power geophysical instruments. The by-product of the propane-fuelled generator was water vapor and carbon dioxide. The generator often shut down over the polar winter when the exhaust pipe, used to remove the waste products away from the generator, froze over and, on several occasions, ice formed almost all the way into the body of the generator (see reports at http://space.augsburg.edu/ago/index.html). For this reason, the placement of DMFCs should be carefully considered. A small EFOY® DMFC fuel cell manufactured by Smart Fuel Cell will be deployed in northern Greenland to power science instrumentation in 2007 (R. Forsberg, 2007, Personal Communication).

3.4 Power Management Systems

The sophistication of the power management system at a remote GNSS site is dependant on the sources of power available and the number of loads to power. A simple passive array of photovoltaic panels, feeding a single battery bank composed of SLA batteries and powering a single GNSS system and its communications, only requires a reliable temperature-compensated charge controller and a low voltage disconnect. The temperature sensor for the charge controller should, ideally, be attached to a battery post so as to sense the internal temperature of the battery bank. Sophistication of the power management system must be increased when dealing with batteries of varying chemistries and/or with hybrid power systems. Complexity can increase even further when using management systems that gradually shut parts of the system down in a tiered approach until only the GNSS receiver is working. For example, as power availability drops any state-of-health instruments could be put into hibernation, followed by the communications systems. This would drop the total load on the battery banks and ensure that year-round data recording was achieved. As energy is returned to the batteries when power generation restarts, the hibernating systems can be restored to power.

In general, keeping things simple has been the most reliable path, but as systems are being deployed at distances further and further away from logistical hubs, the redundancy of hybrid power systems becomes more attractive. Several groups have built and run power management systems of various configurations. The PCON system described by Tregoning et al. (2000) has the typical characteristics of a hybrid control system that monitors both the thermal state and loads of the GNSS system and intelligently controls the power supply. When diversifying the power sources (and especially if using wind power), the power management system needs to be able to divert excess load to a dump load that can either be used to warm the system or shed excess heat to the outside environment. When using diversified storage systems, such as a mix of single use battery banks and rechargeable battery banks, the power controller needs to select which energy storage system will be used preferentially.

Power generation components that generate very small amounts of power, such as the Forgen wind turbines mentioned above, need to be considered carefully within a hybrid system. The output of very small wind turbines is so low that their effect on very large battery banks is unclear, as the power provided is usually less than the draw of the GNSS system and the bank will not charge. UNAVCO is building a test site in Antarctica to see if the addition of small wind generators significantly flatten the battery drawdown curve by providing a trickle charge to the battery bank during the polar night when the photovoltaic panels no longer produce usable current (B. Johns, 2007, Personal Communication). A second suggestion is to link the turbines to a small subsidiary battery of limited capacity. The wind generator may then through time actually be able to charge this separate battery to a level useable by the GNSS receiver. If using a super-insulated enclosure that traps heat very efficiently, a third alternative is to direct the output of the small wind generator to a very small dump load that can be used to heat the battery bank, improving the capacitance of the batteries. Care must be taken not to over heat the GNSS system in a super-insulated enclosure.

3.5 Batteries

Sealed Lead Acid (SLA) batteries useful for polar deployment are found in two styles, Absorbed Glass Mat (AGM) and gel-cells. AGM batteries are able to handle larger charging currents than gel-cells, which is an advantage when considering supplementing power with wind turbines in the field, but gel-cells are faster to recover from deep discharge and the effects of the cold. They are also more appropriate when using low charging rates such as those provided by PV arrays and fuel cells (Dahl 2006).

During the course of the TAMDEF project we have used gel-cell batteries and they have been extremely reliable. The very small draws that we place on the battery banks allow them to run with surprising efficiency to very low temperatures, also observed by Häring and Giess (2003). The capacity of our battery banks is severely degraded when the internal temperature within the battery enclosures reaches $-50°C$, but the system recovers when the temperature rises. Some of our batteries have been in use since 1999 and have shown no signs of permanent degradation. The major drawback with using SLA batteries is their relatively large weight and resultant low energy densities.

Lithium ion (Li-ion) batteries for both primary (non rechargeable) and secondary (rechargeable) systems merit further investigation. The Li-ion chemistries can be tailored to cold weather operations (Smart et al. 2003) and have shown excellent reliability in the cold conditions of Mars (Ratnakumar et al. 2003a,b; Smart et al. 2004). The JPL-operated Mars expedition rovers, which are still running at the time of writing, are powered by high-efficiency Gallium Arsenide solar panels and two 8 Amp-hour rechargeable Li batteries. The batteries are warmed and have shown exceptional durability.

The NSF has funded a project to tailor the use of primary off-the-shelf Li-ion batteries for use at remote sites (http://www.nsf.gov/awardsearch/showAward.do?

AwardNumber=0619708). The great advantage of Li-ion batteries are their light weight, high energy densities and strong performance in the cold. For example a remote station with a draw of 5 W at a latitude of 75°, if equipped with a 430 W PV system, needs a 2000 Amp hour bank of batteries to run year round. A typical gel-cell battery bank of this size weighs 626 kg. A 2000 Amp hour primary Li battery system would reduce this weight by about a factor of four to 160 kg. The system would be much lighter, but would require replenishment of the Li-ion battery bank each year. Due to the logistical expense of visiting remote sites, this mode of operation is not ideal for stations designed to run more than two years, where rechargeable systems would provide the greatest benefit.

A second type of power storage system under investigation by the Japanese Antarctic Research Expedition is super capacitors. The systems are very efficient at storing large amounts of charge and are being used to power a remote GNSS system in Enderby Land, Antarctica (Shibuya et al. 2006).

4 Environmental Systems

A remote GNSS station installed in polar regions will face the most extreme climate on the planet, so it has to be tolerant of very low temperatures and robust enough to survive extreme wind speeds. At the same time, the GNSS system has to be easily transportable and lightweight. Batteries react in two different ways as temperatures drop below 0°C. The lifespan of the battery improves, but the capacitance of the battery drops and the batteries operate at ever-decreasing efficiencies. The environmental systems that follow will discuss enclosure types for batteries and instrumentation, the thermal management issues within those enclosures, and measures that can be taken to improve system survival in extreme environments.

4.1 Enclosures

Several levels of sophistication are available when considering enclosures. First and foremost, the enclosure must protect the instruments and batteries from spindrift, which is light, fast-moving snow that enters even the smallest apertures. If the station is to be placed in a region that experiences melting or rainfall, then the enclosure must also be waterproof. During TAMDEF, we used highly insulated vacuum-panel boxes to protect the receivers and communications equipment from the Antarctic climate. These enclosures were first tried by JPL during projects in the Transantarctic Mountains and Marie Byrd Land in Antarctica. Vacuum-panel boxes are custom made with an enclosed foam-based vacuum sandwiched between the inner and outer walls of the box. Blueboard is a cheap extruded Styrofoam product commonly used to insulate equipment boxes and buildings in the USAP.

A comparative study between an expensive super-insulated vacuum-panel box (Fig. 7 left) and a plastic shipping case (Fig. 7 right) lined with "blueboard" insulation was undertaken at McMurdo by UNAVCO. In this test two identical GPS systems were deployed side by side. One was placed in a vacuum-panel enclosure, the other in a blueboard insulated enclosure. Each system consisted of a Trimble 5700® receiver, a freewave radio modem and 6 batteries. The systems shared the power from a solar array. A Campbell datalogger recorded external temperature, battery temperatures, receiver temperatures, battery voltages, solar panel voltage, receiver voltages and system draw. Temperature results shown in Fig. 8 indicate that the vacuum-panel box successfully captured the excess heat produced by the GPS receivers more effectively than the plastic case. The interior of the vacuum-panel system stays at an average of 9°C above ambient temperature, while providing on average a 4°C warmer environment than the shipping case. Two other temperature records collected at a maritime site (a "warm" polar site) and polar plateau site (a "cold" polar site) by the TAMDEF project show that the inside of the vacuum-panel box stayed 11°C above ambient outside temperature at the maritime site and 7°C above ambient at the plateau site. At the plateau site, a 1 cm-thick, plywood-walled box packed with 5 cm of blueboard kept external batteries at only 1°C above ambient.

Vacuum-panel technology is well established, but there are newer "super" insulating materials that are being tested for their ability to capture the residual heat produced by the GNSS system. Aerogel, an extremely low density silica-based solid with exceptional thermal insulation properties (Schmidt and Schwerfeger 1998; Smith et al. 1998) is being used to line an experimental enclosure housing a seismometer over the winter period at the South Pole as part of the MRI testing program. The enclosure will be monitored in order to see if the gains through the use of this material outweigh the expense. Results of this and the previous test will be posted to http://facility.unavco.org/project_support/polar/remote/remote.html.

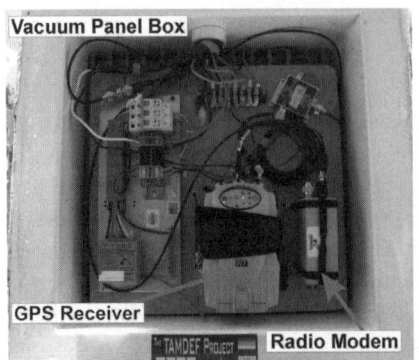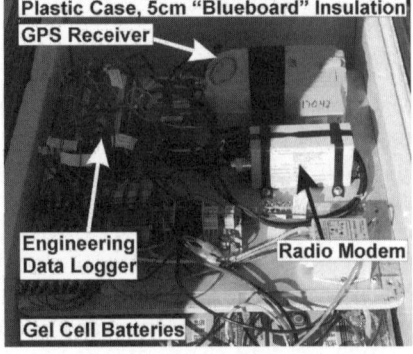

Fig. 7 Comparative test of insulating properties of two different kinds of enclosure at McMurdo. (*left*) Vacuum-panel insulated enclosure. Beneath the electronics board are six 100 amp hour gel-cell batteries. (*right*) Blueboard insulated enclosure, showing batteries and engineering data logger

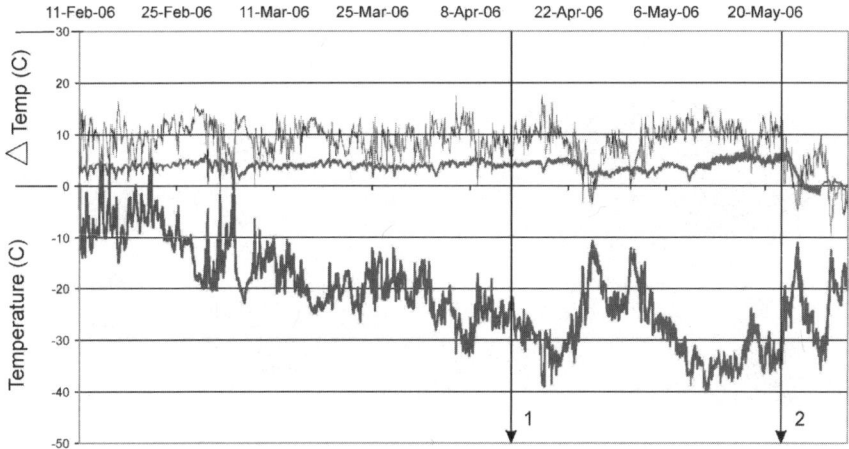

Fig. 8 Performance of a super-insulated enclosure and an adjacent blueboard insulated enclosure at temporary site in McMurdo. Both enclosures contain identical systems and share output from an 80 W solar panel. Red line represents ambient temperature, Blue line is temperature inside the vacuum-insulated enclosure minus ambient, green line is temperature inside vacuum-insulated minus temperature in blueboard insulated enclosure. Data shows that the super-insulated enclosure stays at approximately 10°C above ambient outside temperature while the system is powered. The super-insulated box stays on average 4°C warmer than the blueboard box, until the power system is exhausted. At this point the temperature in the enclosures equalize. Data courtesy of UNAVCO

4.2 Thermal Management

Several problems can occur when using super-insulated enclosures. The first is that constant but small heat sources can, through time, heat the interior of these enclosures to temperatures above operating specifications. Secondly, if battery banks are separated, with some in super-insulated enclosures and some in exterior non super-insulated enclosures, then the temperatures experienced by the batteries can be quite different. In these situations there should be separate temperature-compensated charging systems. In the TAMDEF project, we avoided using active heating in our stations. This is feasible for areas where the station-interior temperature will typically stay above −40°C, which from our calculations should be areas where the ambient temperature does not stay below −50°C for extended periods. This suggests active heating is not needed for GNSS stations around Greenland, but that it will be needed in the deep interior of Antarctica.

4.3 Environmental Hardening

The main hazards to equipment in the polar regions are high winds and extreme cold, the same hazards that are faced by installation crews. The TAMDEF project

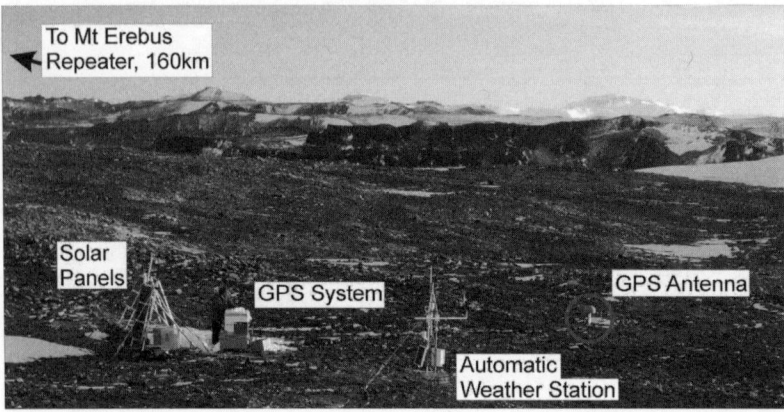

Fig. 9 Mount Fleming TAMDEF GPS site (FLM5) showing typical site configuration. Photo courtesy of Seth White (UNAVCO)

provided experience with sites in katabatic wind regimes and showed that over-engineering is essential in these environments. Two brief examples follow. At Mount Fleming, Fig. 9, an automatic weather station has been placed within 20 m of the GPS installation. This weather station has measured wind speeds greater than 150 km/h, but the GNSS station has shown no evidence of damage. The solar panels are on two metal frames. The larger array of 260 watts was built in 2004 and has a large surface area. It has been heavily anchored, which is prudent but not feasible at deep field sites. The second array of four 40 W panels is on an A-frame simply weighed down with rocks. This smaller array has been on site since 2001 and shows no sign of damage, nor have the solar panels required maintenance. The A-frame design has a substantial gap between the base of the solar panels and the ground. This allows wind to blow snow away from around the rocks that hold the frame down and stops the panels from drifting over (see Fig. 3 for an identical frame set up at Westhaven Nunatak).

At Lonewolf Nunatak (81°20′S, 152°43′E) the solar frame components were repeatedly damaged. Lonewolf is the most westerly bedrock outcrop in the region and sits directly in the katabatic wind regime of the Byrd Glacier, the largest outlet glacier that passes through the Transantarctic Mountains, and experiences extreme wind speeds. Solar panels have been destroyed on two occasions at Lonewolf. In the first instance a small 0.5×0.5 m 40 W panel was torn from the "swing set" leaving only parts of the metal frame surrounding the PV attached. To counter this, we lowered the profile of the solar panels from ~70° to ~30° and reinforced the edges of the solar panel frame so that it was attached to the "swing set" at many points, not just at the standard corner bolts. We also doubled the power of the solar panels to make up for the reduced efficiency of the tilted panels. A solar panel was again ripped from its frame by the wind.

New tests will include affixing a thin aluminum plate to the frame on the back of solar panels in order to reduce some of the wind load on the solar cells. The metal on the rear of the panels will be left with an air gap between the wood and the panel, so that the panel back remains air cooled but the aluminium can flex to a small extent. The aim is not to block the wind, but to deflect it. A small internal brace will ensure the assembly is rigid. Armored plexiglass over the front of the panels may also help the rigidity of the structures, although tests for UV tolerance must be undertaken. Plexiglass and especially UV degraded, cloudy plexiglass will cut the efficiency of the panels, and more batteries will be needed in compensation.

Abrasion from windborne ice and gravel particles is not evident at most of the TAMDEF sites but has been witnessed around McMurdo by this author and in Marie Byrd Land, where flying rocks damaged several solar panels deployed at GPS sites in the late 1990s (A. Donnellan, 1999, Personal Communication). The winds at Lonewolf Nunatak have also forced spin drift into our well-sealed enclosure. The site is so cold that the receiver temperature never passed above 0°C and no liquid water was in the system, however, at lower latitude or more maritime sites, water-proofing of electrical components would be prudent.

It is not known whether our GPS antenna radomes (SCIGN short domes) ice up during the winter period, although it is unlikely at sites far from a moisture source. Examination of the radomes in early spring show no evidence of riming and the time series do not indicate any obvious problems. Rime ice has been observed on unprotected choke ring antennae (Fig. 10) during the summer campaign deployments at several low-elevation, windy sites. Rime ice is known to build up on SCIGN radomes in more maritime regions such as Iceland (Fig. 10). There is no clear path to ameliorate the problem of rime ice, and it may prove a major difficulty for wind turbines if enough builds up during low wind periods to unbalance or stop the blades. At Alert, northern Ellsemere island, Canada, staff at the Canadian Forces Base remove snow and ice by hand from their unprotected antenna (M. Craymer, IGS Station Mail 1291, 2006). This solution is untenable at deep field sites.

Fig. 10 (*left*) Rime ice build up on unprotected GPS antenna at Minna Bluff, Antarctica. The site is prone to low clouds and high wind speeds. Photo by Bradish Johnson, USGS. (*right*) Rime ice build up on a GPS antenna equipped with a short SCIGN dome radome at Skrokkalda, Iceland, photo by Halldór Geirsson, modified from SCIGN website. Rime can be a difficult problem

5 Logistics

As discussed above, polar environments dictate that strong, heavy systems are required for year round operation. The logistical realities of the regions, however, demand lightweight systems. Remote sites are, by definition, far removed from logistical hubs, which consist of towns, bases and military installations in the Arctic and science bases in the Antarctic. Sites must usually be deployed by fixed wing aircraft or helicopter, both of which have limited range and cargo capacities. Fixed wing work additionally requires a flat enough expanse of land or ice to land upon safely. For sites that are a long way from a logistical hub there is a trade-off between the range of the aircraft, the number of people on board and the amount of equipment that can be taken out to the GNSS site. There are also trade-offs in terms of the amount of time required for an installation. Pilots can only be on active duty for limited amounts of hours, therefore the amount time taken to reach a site and the amount of time taken on site both have to be taken in to consideration. In polar regions, the costs of air transport to very remote sites, including the infrastructure underpinning the air transport system, far exceeds typical deployment costs elsewhere in the world. All of these factors have lead to goal of deploying GNSS stations that can record year-round, but operate for a minimum of 2 years without the need for a site visit.

5.1 Weight

The environmental requirements discussed above call for very strong, well-insulated systems. For deep field sites, the balance between the amount of fuel needed to reach the site and the amount of equipment must be examined. Costs can be greatly reduced by minimizing the number of trips to the site. Lowering the science payload weight allows more fuel to be carried, meaning more distant sites can be reached more easily. Any avenues that can be taken to reduce the science payload weight whilst increasing strength of the system are very important. One example would be using fiberglass or plastic instead of thick metal for enclosures. Figure 11 shows an extrapolation of the battery capacity in Amp hours required to run a GNSS system year-round at varying latitudes. This study is based upon measurements made at TAMDEF sites between 76°S and 81°S and at testing rigs in McMurdo. This diagram highlights that, at coastal sites around Antarctica (approximately 70°S for East Antarctica and about 75° for West Antarctica), the power requirement will be anywhere between 800 and 2000 Amp hours of battery storage capacity, depending on system draw. This equates to a weight of between 250 kg and 1000 kg just for the battery bank, if using SLA batteries. The logistical cost to move this amount of weight and volume of batteries is high. A de Havilland Twin Otter carries approximately 1000 kg of payload including passengers, this means that multiple trips must be made for a site requiring 1000 kg of batteries. It can be seen from Fig. 11 that, at almost all latitudes, a reduction in system draw of 0.5 W reduces the battery

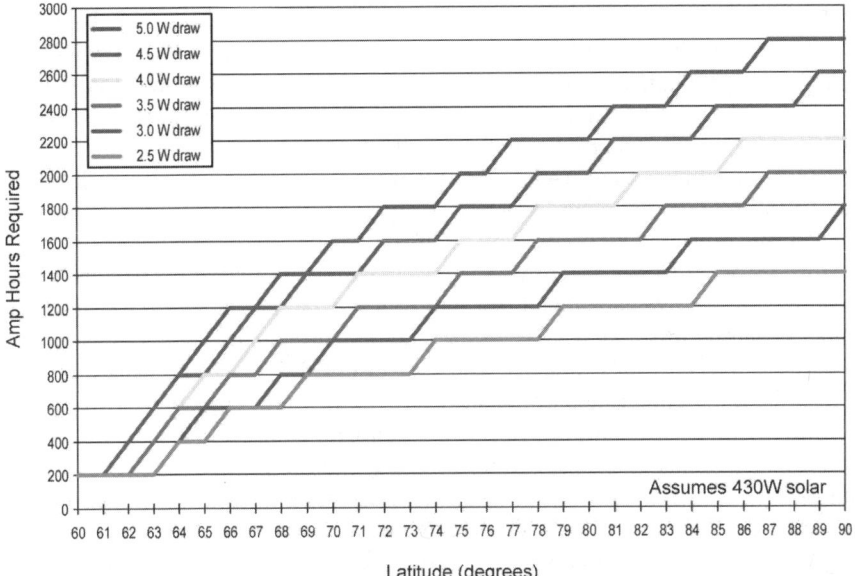

Fig. 11 Number of Amp Hours of battery capacity needed to last one full year of data collection with varying system draw and latitude, Assumes a large array of solar panels (400 W+) and that instrument temperatures are always warmer than −40°C. Temperature effects cannot be accounted for at colder sites, especially at latitudes greater than 80°

capacity required by about 200 amp hours, equivalent to 65 kg weight of SLA batteries. Therefore, the more power-efficient the GNSS and communications system, the smaller the total weight of the system. A similar amount of weight saving can be made by increasing the power output of the solar panels. Larger panels provide more power through more of the year and reduce the number of batteries needed. Keeping the system warm through increased insulation, which by its nature is lightweight, increases the capacitance of the battery bank, again making the system lighter. Another promising method of reducing the logistical cost of remote GNSS systems will be the introduction of lithium ion batteries, which can cut the total weights involved by a factor of three or four.

5.2 Time On-Site

During TAMDEF, the major time sinks once in the field were moving equipment from the landing site to the construction site, constructing the solar panel frames, installing the geodetic monument and wiring the components of the system together.

Reducing the weight of the system makes moving materiel around in the field easier. Having a solar panel frame set that simply unfolds, locks into place and then is anchored easily can reduce the time on site dramatically. This system has to be flexible enough that it can be reconfigured quickly on site, with the minimal amount of tools and with the reduced dexterity that comes from working with heavy gloves. Having pre-built connectors with cable pass-throughs and built-in monitoring systems reduces wiring to a few twists of military-style connectors. This leaves the majority of time to be focused on selecting the best bedrock and sky view for the monument installation. The TAMDEF-style monuments are relatively quick to install, often less than two hours to emplace an extremely stable, leveled aluminum pillar anchored 50 cm into bedrock by four epoxy coated expansion bolts.

More people reduces on-site times, but increases the weight of the payload to be carried to the site. We have found that four or five experienced people can install a remote system in less than 5 h. Using the technologies touched upon in this article we have reduced both these numbers and have installed 20+ PV and wind-powered remote GNSS systems around Greenland using two or three people and between two and four hours of ground time. If the flight to a field site, the installation, and the flight back from the field site does not exceed the flight-hour limit for aircraft pilots, then the site can be installed in one day. If an installation team is required to stay for more than one day in the field, then additional safety equipment and tents are required and flight numbers escalate. A strong lightweight site that can be quickly installed by two or three people has a much lower logistical cost.

6 Conclusions

This work provides some details of new technological developments suitable for the installation of remote GNSS stations in the polar environment. Some best practices are derived from the TAMDEF project experience. These are:

- Robust, low power, reliable GNSS receivers that have been thoroughly tested are the single most important piece of equipment at a remote GNSS station. The reliability and performance of the receiver is absolutely paramount.
- Communications systems are important to reduce the number of costly visits to the remote site either for maintenance or data collection.
- Diagnostic monitoring helps understand any problems that occur and provides an extremely valuable data set for developing reliable technologies.
- Passive photovoltaic arrays are the simplest, best way to generate reliable power during the daytime at high latitudes.
- Sealed lead acid batteries provide a reliable, albeit heavy, storage capability that works to about $-50°C$.
- Wind generators are still problematic in the polar regions and should be used with caution.
- Highly insulated enclosures, though expensive, provide batteries and equipment with an extended operating season.
- Any means to reduce the weight of systems without compromising their structural strength and capability should be explored for future remote deployments.

The blending of best practices with emerging technologies will provide new capabilities to deploy continuous GNSS stations at remote sites. Measurements at a large number of remote stations across the breadth of the polar regions will revolutionize our understanding of the interplay between ice sheets, the solid Earth and sea-level. GNSS deployments during the International Polar Year will provide a legacy of long-term observations and observatories that will be useful for many years to come.

Acknowledgments Many thanks to my advisor, Terry Wilson, PI of the TAMDEF project. My colleagues at the Byrd Polar Research Center are also due thanks for help deploying these autonomous systems in Antarctica. Larry Hothem, Bob Glover and others of the USGS and Graeme Blick and others of Land Information New Zealand were also central to the design and installation of TAMDEF GPS sites. Thanks to Bjorn Johns, Seth White and Thomas Nylen at UNAVCO for running the enclosure tests at McMurdo and doing the calculations for solar panel performance at low temperatures. Finally, my thanks to the organizers of "GPS in the IPY: The POLENET project" in Dresden that was the driver for this publication. The Greenland GPS Network – GNET has been partially installed using technologies and practices outlined in this paper. This material is based upon work supported by the National Science Foundation under Grant No. ANT-0230285. Any opinions, findings, and conclusions or recommendations expressed in this material are those of the author and do not necessarily reflect the views of the National Science Foundation.

References

Abdalati W, Krabill WB, Frederick E, Manizade S, Martin C, Sonntag J, Swift R, Thomas RH, Wright W and Yungel J (2001) Outlet glacier and margin elevation changes: Near-coastal thinning of the Greenland ice sheet. J Geophys Res, vol 106 (D24): 33729–33741

Akhmatov V, Galster G and Larsen E (1998) Questionable effects of antireflective coatings on inefficiently cooled solar cells. Solar Energy Materials and Solar Cells, vol 56: 17–28

Anandakrishnan S, Voigt D, Burkett PG, Long B and Henry R (2000) Deployment of a broadband seismic network in West Antarctica. Geophys Res Lett, vol 27 (14): 2053–2056

Anderson PS, Ladkin RS and Renfrew IA (2005) An autonomous doppler sodar wind profiling system. J Atmos Oceanic Technol, vol 22: 1309–1325

Chen JL, Wilson CR, Blankenship DD and Tapley BD (2006a) Antarctic mass rates from GRACE. Geophys Res Lett, vol 33: doi: L11502, 10.1029/2006GL026369

Chen JL, Wilson CR and Tapley BD (2006b) Satellite gravity measurements confirm accelerated melting of Greenland Ice Sheet. Science, vol 313: 1958–1960 doi: 10.1126/science.1129007

Dahl T (2006) "Wind Power Systems." PolarPower.org, 5th April 2006, p. 38 http://www.polarpower.org/static/docs/WindPower05Apr06.pdf

Datta BK, Velaytham G and Goud AP (2002) Fuel cell power source for a cold region. J Power Sources, vol 106: 370–376

Deb SK (1998) Recent developments in high efficiency photovoltaic cells. Renewable Energy, vol 15: 467–472

Delahoy AE, Chen L, Akhtar M, Sang B and Guo S (2004) New technologies for CIGS photovoltaics. Solar Energy, vol 77: 785–793 doi:10.1016/j.solener.2004.08.012

Donnellan A and Luyendyk BP (2004) GPS evidence for a coherent Antarctic plate and for postglacial rebound in Marie Byrd Land. Global and Planetary Change, vol 42: 305–311 doi:10.1016/j.gloplacha.2004.02.006

Dow JM, Neilan RE and Gendt G (2005) The International GPS Service (IGS): Celebrating the 10th Anniversary and Looking to the Next Decade. Adv SpaceRes, vol 36 (3): 320–326 doi:10.1016/j.asr.2005.05.125

Fleming K and Lambeck K (2004) Constraints on the Greenland Ice Sheet since the Last Glacial Maximum from sea-level observations and glacial-rebound models. QSR, vol 23: 1053–1077 doi:10.1016/j.quascirev.2003.11.001

Häring P and Giess H (2003) Performance of a VRLA battery in the Arctic environment. J Power Sources, vol 116: 257–262 doi:10.1016/S0378-7753(02)00699-7

Ivins ER and James TS (2005) Antarctic glacial isostatic adjustment: a new assessment. Antarctic Sci, vol 17 (4): 541–553 doi: 10.1017/S0954102005002968

James TS and Ivins ER (1998) Predictions of Antarctic crustal motions driven by present-day ice sheet evolution and by isostatic memory of the Last Glacial Maximum. J Geophys Res, vol 103 (B3): 4993–5017

Joughin I, Rignot E, Rosanova CE and Lucchitta BK (2003) Timing of recent accelerations of Pine Island Glacier, Antarctica. J Geophys Res, vol 30 (13): doi: 10.1029/2003GL017609, 2003

Kyle PR (2007) "The Edge of Discovery." The Antarctic Sun, 14th January 2007. http://antarcticsun.usap.gov/2006-2007/documents/01-14-2007_antarcticsun.pdf

Llubes M, Lemoine J-M and Remy F (2007) Antarctica seasonal mass variations detected by GRACE. Earth Plan Scie Lett, vol 260: 127–136 doi: 10.1016/j.epsl.2007.05.022

Luthcke SB, Zwally J, Abdalati W, Rowlands DD, Ray RD, Nerem RS, Lemoine FG, McCarthy JJ and Chinn DS (2006) Recent Greenland Ice Mass Loss by drainage system from satellite gravity observations. Science, vol 314: 1286–1289 doi: 10.1126/science.1130776

Moore A, W. (2007). "The International GNSS Service: Any Questions?" GPS World 18 (1): 58–62

Nishioka K, Hatayama T, Uraoka Y, Fuyuki T, Haguhara R and Watanabe M (2003) Field-test analysis of PV system output characteristics focusing on module temperature. Solar Energy Materials and Solar Cells, vol 75: 665–671

Prescott WH, Anderson K, Johns B and Simpson D (2006) "Collaborative Research: Development of a Power and Communication System for Remote Autonomous GPS and Seismic Stations in Antarctica." National Science Foundation, http://www.nsf.gov/awardsearch/showAward.do?AwardNumber=0619908 & http://www.nsf.gov/awardsearch/showAward.do?AwardNumber=0619708

Rachelson W (2006) "AGO Field Service Summary, December 2005–February 2006." Augsburg College, p. 15 http://space.augsburg.edu/ago/05_06sum.pdf

Rannels J (2000) The DOE Office of Solar Energy Technologies' Vision for Advancing Solar Technologies in the New Millennium. Solar Energy, vol 69 (5): 363–368

Ratnakumar BV, Smart MC, Kindler A, Frank H, Ewell RC and Surampudi S (2003a) Lithium batteries for aerospace applications: 2003 Mars Exploration Rover. J Power Sources, vol 119–121: 906–910 doi:10.1016/S0378-7753(03)00220-9

Ratnakumar BV, Smart MC, Whitcanack LD, Knight J, Ewell RC, Surampudi R, Pugliam F and Curran T (2003b). "Lithium Ion Batteries for Mars Exploration Rovers". Space Power Workshop, Redondo Beach, CA. April 21–24th

Rignot E, Jezek KC and Sohn HG (1995) Ice Flow Dynamics of the Greenland Ice Sheet from SAR Interferometry. Geophys Res Lett, vol 22 (5): 575–578

Ryu K, Rhee J-G, Park K-M and Kim J (2006) Concept and design of modular Fresnel lenses for concentration solar PV system. Solar Energy, vol 80: 1580–1587 doi:10.1016/j.solener.2005.12.006

Schmidt M and Schwerfeger F (1998) Applications for silica aerogel products. J Non-Crystalline Solids, vol 225: 364–368

Shibuya K, Doi K, Nogi Y, Aoyama Y, Ohzono M, Morita K and Egawa K (2006). "Recent GPS Application for geodynamic research during the Japanese Antarctic Research Expeditions". GPS in the IPY: The POLENET Project, Dresden, Germany. October 4–6

Sifer N and Gardner K (2004) An analysis of hydrogen production from ammonia hydride hydrogen generators for use in military fuel cell environments. J Power Sources, vol 132: 135–138 doi:10.1016/j.jpowsour.2003.09.076

Skone S and de Jong M (2000) The impact of geomagnetic substorms on GPS receiver performance. Earth Planets Space, vol 52: 1067–1071

Skone SH (2001) The impact of magnetic storms on GPS receiver performance. J Geodesy, vol 75 (9–10): 457–468 doi:10.1007/s001900100198

Smart MC, Bugga R, Ewell RC, Whitcanack LD, Chin KB and Surampudi R (2004). "Validation of Lithium-Ion Cell Technology for JPL's 2003 Mars Exploration Rover Mission." 2nd International Energy Conversion Engineering Conference, Providence, Rhode Island. August 16–19, 2004

Smart MC, Ratnakumar BV, Whitcanack LD, Chin KB, Surampudi S, Croft H, Tice D and Staniewicz R (2003) Improved low-temperature performance of lithium-ion cells with quaternary carbonate-based electrolytes. J Power Sources, vol 119–121: 349–358 doi:10.1016/S0378-7753(03)00154-X

Smith DM, Maskara A and Boes U (1998) Aerogel-based thermal insulation. J Non-Crystalline Solids, vol 225: 254–259

Srinivasan S, Mosdale R, Stevens P and Yang C (1999) FUEL CELLS: Reaching the Era of Clean and Efficient Power Generation in the Twenty-First Century. Ann Rev Energy Environ, vol 24. pp. 281–328

Sterling R (2005) "AGO Field Service, December 2004–January 2005." Augsburg College, p.12 http://space.augsburg.edu/ago/04_05sum.pdf

Thomas RH, Frederick E, Krabill WB, Manizade S and Martin C (2006) Progressive increase in ice loss from Greenland. Geophys Res Lett, vol 33 (L10503) doi: 10.1029/2006GL026075

Tregoning P, Welsh A, McQueen H and Lambeck K (2000) The search for postglacial rebound near the Lambert Glacier, Antarctica. Earth Planets Space, vol 52: 1037–1041

Willis MJ, Wilson TJ and James TS (2006) Bedrock Motions From a Decade of GPS Measurements in Southern Victoria Land, Antarctica. Eos Trans AGU, vol 87 (52): Abstract G33B-0059

Yoshioka K, Hasegawa J, Saitoh T and Yatabe S (2002). "Performance analysis of a PV array installed on building walls in a snowy country." Photovoltaic Specialists Conference, 2002 Conference Record of the Twenty-Ninth IEEE

Zwartz D and Helsen M (2002) "GPS Observations for Ice Sheet History (GOFISH)." Institute for Marine and Atmospheric Research, Utrech University, Dec 2001–Jan 2002, p. 22 http://www.phys.uu.nl/%7Ewwwimau/research/ice_climate/gofish.pdf

VLNDEF Project for Geodetic Infrastructure Definition of Northern Victoria Land, Antarctica

A. Capra, M. Dubbini, A. Galeandro, L. Gusella, A. Zanutta, G. Casula, M. Negusini, L. Vittuari, P. Sarti, F. Mancini, S. Gandolfi, M. Montaguti and G. Bitelli

Abstract Scientific investigations in Antarctica are, for many different reasons, a challenging and fascinating task. Measurements, observations and field operations must be carefully planned well in advance and the capacity of successfully meeting the goals of a scientific project is often related to the capacity of forecasting and anticipating the many different potential mishaps. In order to do that, experience and

---

A. Capra
DIMeC, Universita' di Modena e Reggio Emilia

M. Dubbini
DIMeC, Universita' di Modena e Reggio Emilia

A. Galeandro
DIASS, Politecnico di Bari

L. Gusella
DISTART, Universita' di Bologna

A. Zanutta
DISTART, Universita' di Bologna

G. Casula
INGV, Roma

M. Negusini
IRA – INAF, Bologna

L. Vittuari
DISTART, Universita' di Bologna

P. Sarti
IRA – INAF, Bologna

F. Mancini
DAU, Politecnico di Bari

S. Gandolfi
DISTART, Universita' di Bologna

M. Montaguti
DISTART, Universita' di Bologna, IRA – INAF, Bologna

G. Bitelli
DISTART, Universita' di Bologna

logistic support are crucial. On the scientific side, the team must be aware of its tasks and be prepared to carry out observations in a hostile environment: both technology and human resources have to be suitably selected, prepared, tested and trained. On the logistic side, nations, institutions and any other organisation involved in the expeditions must ensure the proper amount of competence and practical support.

The history of modern Italian Antarctic expeditions dates back to the middle 80's when the first infrastructures of "Mario Zucchelli Station", formerly Terra Nova Bay Station, were settled at Terra Nova Bay, Northern Victoria Land. Only a few years later, the first geodetic infrastructures were planned and built. Italian geodetic facilities and activities were, ever since, being constantly maintained and developed. Nowadays, the most remarkable geodetic infrastructures are the permanent Global Positioning System (GPS) station (TNB1) installed at Mario Zucchelli and the GPS geodetic network Victoria Land Network for DEFormation control (VLNDEF) entirely deployed on an area extending between 71°S and 76°S and 160°E and 170°E.

These facilities do not only allow carrying out utmost geodetic investigations but also posses interesting capacities on the international multidisciplinary scientific scenario.

In order to fully exploit their potentiality, management and maintenance of the infrastructure are crucial; nevertheless, in order to perform high quality scientific research, these abilities must be coupled with the knowledge concerning a proper use and a correct processing of the information that these infrastructures can provide.

This work focuses on the different methods that can be applied to process the observations that are performed with GPS technique in Northern Victoria Land, aiming at reaching the highest accuracy of results and assuring the larger significance and versatility of the processing outcomes. Three software were used for the analysis, namely: Bernese v.5.0, Gipsy/Oasis II and Gamit/Globk. The working data sets are (i) the permanent GPS station TNB1 observations continuously performed since 1998 and (ii) the five episodic campaigns performed on the sites of VLNDEF.

The two infrastructures can be regarded as neat examples of standard geodetic installation in Antarctica. Therefore, the technological solutions that were adopted and applied for establishing the GPS permanent station and the VLNDEF geodetic network as well as the data processing strategies and the data analysis procedures that were tested on their observation will be illustrated in detail. The results will be presented, compared and discussed. Furthermore, their potentials and role in geodetic research will be carefully described; their versatility will also be highlighted in the foreground of a multidisciplinary Antarctic international scientific activity.

1 Introduction and Motivation

The success of Antarctic scientific investigations is often related to the capacity of optimising the quality and quantity of the observations that can actually be performed. Difficulties of different nature may suddenly arise during an observing session or a field campaign. The ability of facing unforeseen occurrences is therefore mandatory in immediate terms. On the other hand, an equally large amount of difficulties may arise when long term or permanent observations are needed. Complications eventually increase when long time series of observations have to be recorded and securely stored at isolated and unmanned sites, where weather conditions as well as power supply can often place a serious limit on the capacity of confidently perform a set of observations.

The importance of existing (reliable and accurate) scientific records is therefore central to Antarctic research. Indeed, observations performed in Antarctica are often central to global mathematical models that describe the state and the evolution of the entire Earth system.

If possible, the number of data and observations should be constantly increased, keeping in mind that maximal versatility of observations is obtained through international actions aimed at coordinating the efforts, the financial investments and the needs of the entire scientific community. Sect. 2 is devoted to a description of the most important actions and of the recent developments that were undertaken in the field of Antarctic geodesy and geosciences.

Section 3 describes the geodetic Italian infrastructures in Antarctica and their role within the abovementioned international scenario. A detailed description of the practical solutions that were adopted for exploiting and maintaining these two geodetic facilities is given: they were designed, planned and established during the several Italian Antarctic scientific expeditions. They both are GPS based facilities which require different efforts and managing strategies; their contribution to geosciences and geodetic research is, in principle, different but, undoubtedly, complementary. Section 4 identifies the main scientific contributions that can derive from a proper management of these infrastructures.

In order to perform good scientific investigations and infer from results, the ability of properly process the available data sets is crucial: the procedures and the most efficient analysis strategies were identified and tested and are presented in Sect. 5. In particular, this latter Section presents the different scientific software that were used for processing (i) the seven years of continuous GPS observations performed by the permanent GPS station TNB1, installed at Mario Zucchelli station and (ii) the GPS data sets acquired by more than 28 stations episodically occupied during the five VLNDEF campaigns. Software's characteristics and their basic processing approaches are shortly summarized. The analysis strategies that were adopted for processing the two data sets are described in detail.

Section 6 is devoted to a critical discussion concerning the efficiency of each analysis strategy and to a detailed comparison of results.

2 Historical Background of Geodetic Infrastructure in Antarctica

The SCAR (Scientific Committee on Antarctic Research) GIANT (Geodetic Infrastructure of Antarctica) was established in 1992 for providing a common geodetic framework over Antarctica, as the basis for recordings of positional related science.

Some large scale GPS surveys over the Antarctic continent were planned within the SCAR WG-GGI (Working Group – Geodesy and Geographic Information) aiming at determining the rates of crustal deformation within the Antarctic tectonic plate and the relative motion between the Antarctic plate and the surrounding tectonic plates and micro-plates. The measurements were carried out in three campaigns: 1989–90, 1990–91 and 1991–92.

At the XXII SCAR meeting held in San Carlos de Bariloche, Argentina, 8–19 June 1992, the results of the SCAR GPS Antarctic Project 90–92 were assessed. It was also decided to further extend the GPS project to develop co-locations with other instruments and other observing techniques, such as VLBI, Absolute Gravity, DORIS and tide gauges. This was collectively identified as the Geodetic Infrastructure for Antarctica (GIANT), the coordinating program for Geodesy, initially chaired by Mr. J. Manning from Australia (Manning, 2005).

GIANT program objectives are:

– To provide a common geodetic and geographic reference system for all Antarctic scientists and operators;
– To contribute to global geodesy for the study of the physical processes of the earth and the maintenance of the precise terrestrial reference frame;
– To provide information for monitoring the horizontal and vertical motion of the Antarctic.

Since 1992 the GIANT program, has been revised and endorsed at each major SCAR conferences on a two years basis.

Several successful application of GPS surveys were performed in Antarctica; in particular GPS Epoch Campaigns based on series of summer GPS acquisitions started in 1995, within GIANT program (see e.g. http://www.tu-dresden.de/ipg/ FGHGIPG/Aktuell-Dienste/scargps/database.html).

The SCAR GPS Epoch campaigns aim at establishing and maintaining an Antarctic GPS geodetic network framed within the International Terrestrial Reference Frame (ITRF); the densification of the IGS network established by the permanent GPS observatories (Fig. 1) is another operational goal. The observations acquired during the episodic campaigns are used for geodynamic as well as geodetic research (geodynamics, crustal deformation, reference frame definition).

SCAR GPS Epoch Campaigns are coordinated by Prof. R. Dietrich from TU-Dresden, Germany. The whole set of data comprises observations performed at numerous sites (Fig. 2) and is archived at the University of Dresden as ongoing collection devoted to scientific investigation; for the moment being, the entire set of data has been processed and re-processed several times (Dietrich et al., 2001, 2004; Steigenberger et al., 2006).

GIANT is giving a significant contribution to the work of other Antarctic earth scientists such as the newly formed ANTEC (Antarctic NeoTECtonics) group of specialists; its main research task is related to the development of a better understanding of the crustal dynamics process undergoing in Antarctica. In order to meet the ever increasing need of accurate observations for studying Antarctic geodynamic movements, the scientific community has planned to expand the geodetic network to provide a very stable Antarctic reference frame for geodynamics.

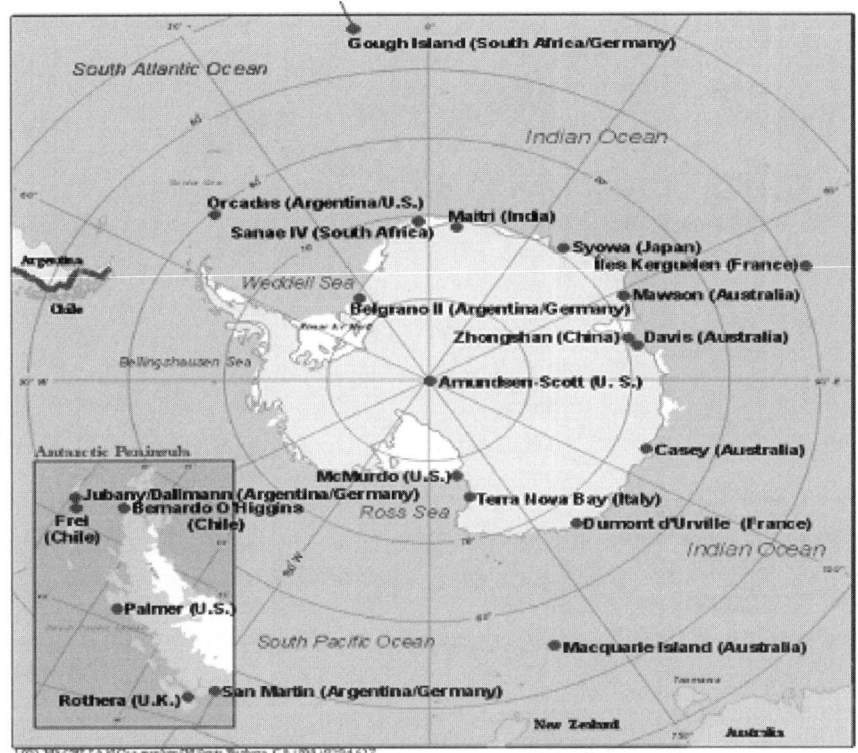

Fig. 1 GPS permanent trackers in Antarctica (2005), courtesy of SCAR website

A very interesting proposal which has been endorsed by the International Polar Year (IPY) 2007 Committee is POLENET (POLar Earth observing NETwork). This programme is an example of the most recent efforts that are undertaken by the scientific polar community to widen the perspective of operational scientific investigations towards the realization of multidisciplinary coordinated facilities and observations (Fig. 3).

POLENET aims at "[...] investigating the system-scale interactions within the polar earth system and polar geodynamics by deploying autonomous remote observatories [...]". The technical and scientific challenges that are going to be addressed are numerous. On the scientific side, the primary co-location of GPS systems (Fig. 4) and seismometers, possibly completed with meteorological sensors, geomagnetic observatories, tide gauges and bottom pressure gauges (Fig. 5), will realize a step towards the acquisition of coordinated geophysical observations of the Earth system and its processes.

POLENET can be regarded as a step toward a practical realization of IGOS (Integrated Global Observing System) guidelines (http://www.igospartners.org/over.htm) and as an effective contribution to GGOS (Global Geodetic Observing System) (see e.g. Altamimi et al., 2005; Rummel et al., 2005; Woodworth et al., 2005).

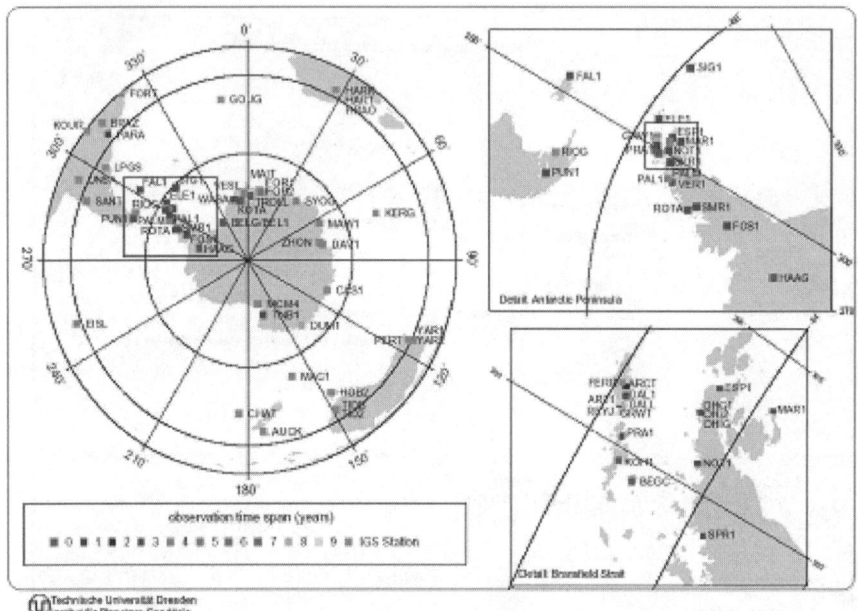

Fig. 2 SCAR GPS Epoch campaigns sites, courtesy of SCAR website

In particular, a continuous geodetic monitoring, exploited through GPS permanent stations, would accurately determine the 3-D motion of the crust. Besides all geodynamical applications, an accurate monitoring of vertical displacements will undoubtedly benefit the investigations related to cryosphere stability and ice mass balance, providing robust constraints on ongoing processes. Sea level change and post-glacial rebound are directly affected by modifications of the ice sheet; these investigations are particularly important as the modifications seem to occur at an unpredicted high rate of change. On the operational side, the efforts that must be made in optimising and ensuring highly accurate standardised performances of remote unmanned observing stations will be maximised by a coordinated effort in designing and planning the installations and the related technical solutions.

Several permanent GPS receivers have already been installed around the Antarctic continent; the global data set is thus continuously increasing. Although, not all observing sites are designed so as to provide observations' transmission in real time or with latency below one day. Furthermore, at remote sites the only suitable way to frequently retrieve experimental data is realized through a satellite link. The scientific instruments and all the hardware remotely installed need to be assisted by technological solutions capable of continuously supplying electric power. This is surely the case for GPS equipments at unattended and remote Antarctic sites. The lack of sunlight and the very low temperature characterising the dark Antarctic winters limit the use of solar panels and/or batteries. Technical solutions based on wind turbines and electric power generators are currently being investigated and developed:

VLNDEF Project

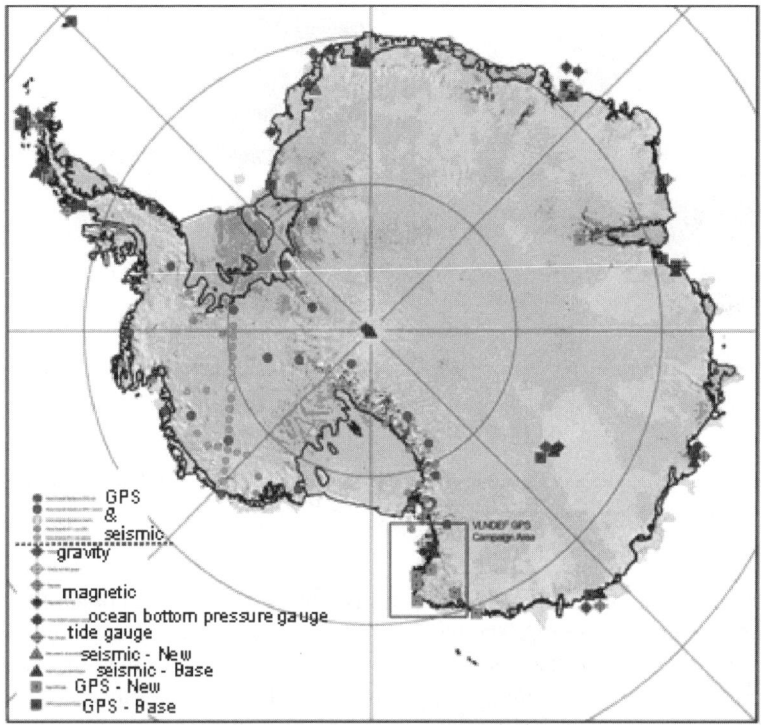

Fig. 3 Geodetic and Geophysical observatories in POLENET, courtesy of SCAR website

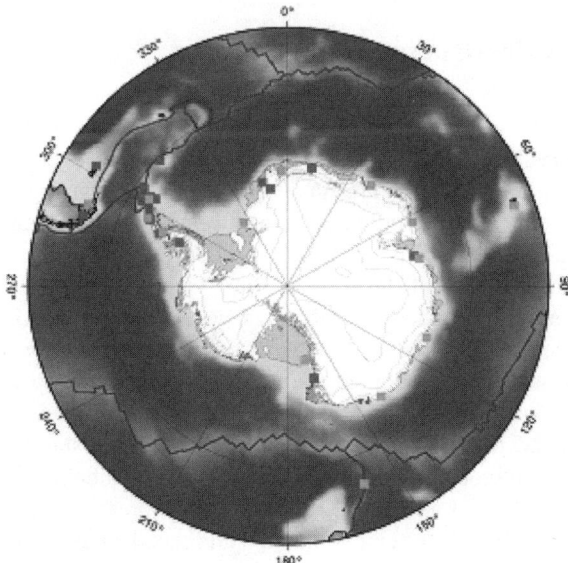

Fig. 4 GPS Observatories in POLENET, courtesy of SCAR website

Fig. 5 Observation Techniques and related instrumentation to be co-located at remote polar observatories according to POLENET guidelines, courtesy of SCAR website

the different stages of the process are characterized by varying degrees of success. Ideally, such a remote observing station should be equipped with an inexpensive and reliable remote power unit and a satellite connection for data transmission. This remote operation technology is not quite proven and needs further development and testing to be ready for the International Polar Year in 2007, IPY07.

3 The Italian Geodetic Infrastructures in Antarctica

Geodetic applications of GPS technique in Northern Victoria Land (NVL), Antarctica, started during the 1988–1989 Italian scientific expedition. A local geodetic network was established aiming at creating an experimental GPS geodetic infrastructure that could be used as local geodetic reference frame as well as control network for local crustal motion and deformation detection purposes. In order to do that, 12 sites around the area of Terra Nova Bay (Victoria Land, East Antarctica) were carefully selected: geodetic benchmarks were monumented in suitable locations characterized by exposed and stable bedrock, thus creating a new local reference frame that could be adopted for geodetic, photogrammetric and geological surveys. Moreover, with the purpose of investigating deformations possibly occurring in the area nearby the Melbourne volcano, eight more stations were monumented over stable outcrops around the volcanic edifice (Gubellini & Postpischl 1991; Gubellini et al., 1994; Capra et al., 1996). The map shown in Fig. 6 depicts the location of the benchmarks belonging to the first GPS geodetic network and its densification around the Mt. Melbourne.

In 1998, a permanent GPS station (ITRF code: TNB1) was installed on a granite hill close to Mario Zucchelli Station at, Terra Nova Bay (see Fig. 6). The receiver

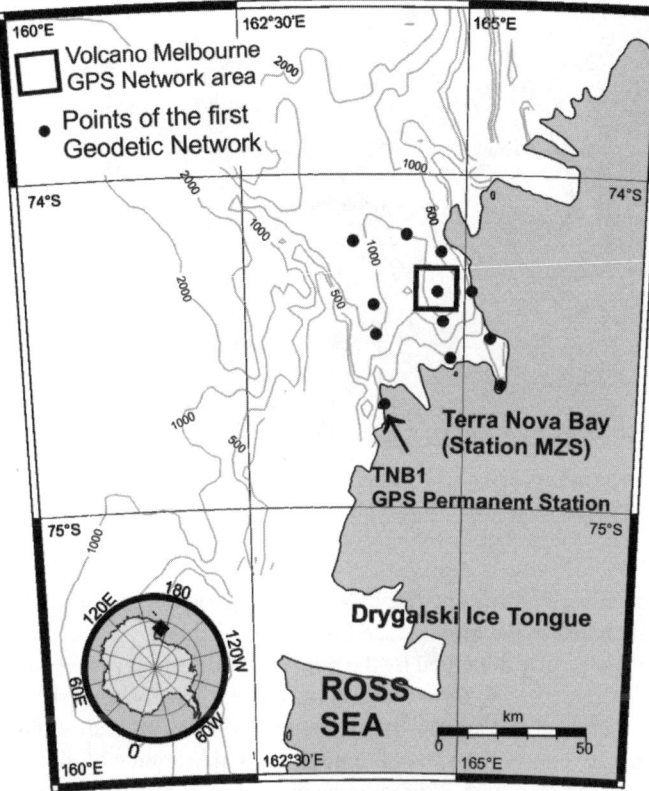

Fig. 6 Map showing the area around Terra Nova Bay: the sites that formed the first local geodetic network are represented with dark dots. The densification network around the Mount Melbourne volcano is situated within the dark square. The dark arrow indicates the location of the permanent GPS station TNB1

has, since, been continuously operating (its location with respect to Mario Zucchelli station and its monument and concrete pillar are shown in Fig. 7) (Capra et al., 2004).

In order to provide accurate estimates of site's positions, a stable monumentation of the geodetic benchmark is needed. The monument of the permanent GPS station TNB1 guarantees excellent technical solutions as well a very high stability. A concrete pillar was built on a well-preserved granite outcrop; it was materialized in 1988 (Gubellini & Postpischl, 1991) and its design and technical characteristics aim at ensuring high-precision measurements. The ITRF tracking point (DOMES number: 66036M001) is unambiguously and accurately identified by means of a forced centring system installed on top of the pillar. Three ex-centre markers were materialized close to the main ITRF tracking point; their positions were surveyed and estimated for maintaining and preserving the original location of the main tracking point in case of damages or inconveniences that might incidentally occur.

The GPS system that was chosen for and installed at Terra Nova Bay is composed by an ASHTECH Z-XII receiver and an ASH700936 Dorne Margolin antenna with

Fig. 7 The monument of TNB1 GPS permanent station. In the small upper-right inset, a black circle highlights the location of the GPS site in the vicinity of the Mario Zucchelli Station (MZS).

SNOW radome. The receiver and the ancillary electronic equipment are safely located in a box firmly anchored to the ground (see Fig. 7). The technology eventually installed and implemented greatly benefited by the numerous GPS surveys that were repeatedly performed in the area of Terra Nova Bay: this experience remarkably helped in selecting the receivers and antenna best suited to work in Antarctic conditions. TNB1 is permanently connected to a continuous and stable source of electric power: its location nearby MZS allows a direct connection to the engines that provide electricity to all the systems, instruments and equipments that must keep on working when the base is closed. Unfortunately, a lack of a cheap satellite connection does not only prevent a daily download of the observations performed during the closing period of the base: it also prevents the use of TNB1 observations to their full extent. They are stored on a personal computer and are downloaded every year, when geodesists arrive at MZS. Therefore, despite its extraordinary geodetic characteristics and its scientific potential, TNB1 cannot yet be part of the IGS network.

A successive fundamental step for Italian geodetic GPS-based activities in Antarctica is related to the monumentation of VLNDEF (Victoria Land Network for DEFormation control) network. This project aimed at planning, designing, establishing and maintaining a dense GPS network, deployed on solid bedrock, with the purpose of measuring horizontal and vertical crustal displacements. The network is nowadays formed by 28 sites (Fig. 8) which were monumented during the 1999–2000 and 2000–2001 field campaigns.

The benchmarks are stainless steel rods with a 5/8 inch screw thread that realize a 3-D forced centring set up and orientation of the GPS antenna (Mancini et al., 2004). Information of different origin had to be merged for properly designing the geometry of the network. Since one of the main scientific purposes was the realization of a geodynamic model of Northern Victoria Land (NVL), the main tectonic features of the area were taken into account and derived by the work of Salvini et al. (1997;

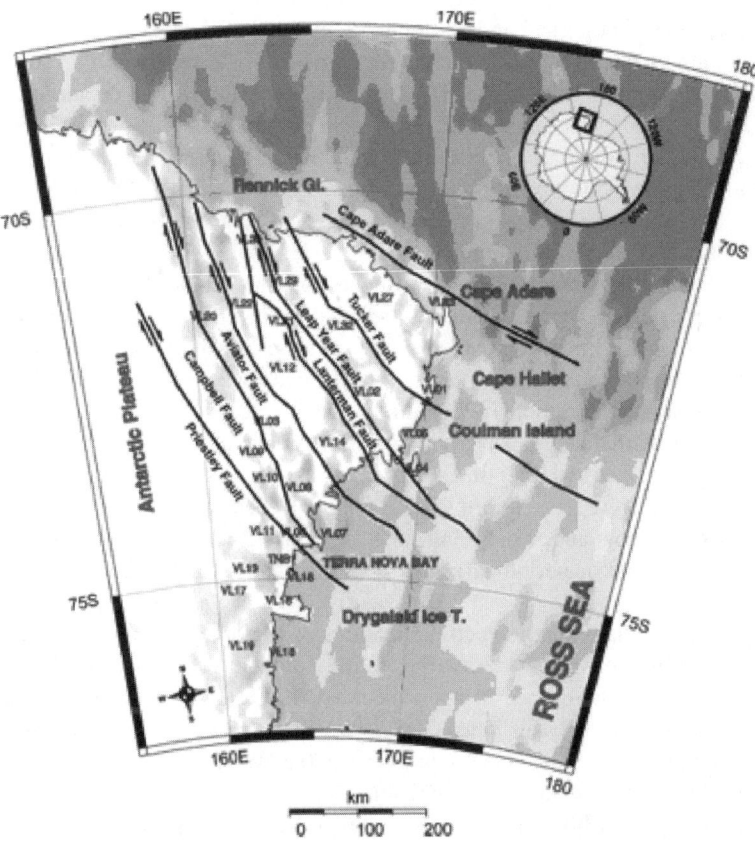

Fig. 8 VLNDEF sites geometry and proposed major strike-slip fault system as derived from the Cenozoic tectonic map of the Ross Sea region proposed by Salvini et al. (1997; see Sect. 4).

see Fig. 8). This led to the identification of the overall shape and extension of the network. In addition to these latter geometric aspects, derived by neotectonic hypothesis, several other requirements had to be evaluated in the planning phase; in particular, the geologic as well as rheologic properties of the bedrock where the monumentation could potentially be realized. Furthermore, the location of the sites had to be suitable for performing GPS measurements: this was a challenging task since the candidate spots needed to be characterized by further stringent properties. In practice, once the overall geometry and the potential area had been identified on a map, the site had to be surveyed by helicopter before taking a final decision and proceed with an immediate materialization of the benchmark. As a matter of fact, stable outcrops, characterized by solid and compact portions of rock are not very common in NVL. Furthermore, a wide open field of view, free from obstructions, obstacles and other limiting factors which might restrict the visibility of satellites during the GPS observations is an aspect that was also taken into account. All sites were described in log-files where all relevant information was summarized, such as position, main morphological and geological features, helicopter approach pathway, suggestions for a correct and quicker set up of the GPS receiver and related equipment (solar panel, battery pack). The first complete campaign-style survey of VLNDEF (1999–2000) was carried out, with limited memory storage capabilities, during the XV Italian Antarctic expedition, harmonising the monumentation of the network with the surveying operations. The following campaigns (2002–2003 and 2005–2006) led to the availability of a larger number of receivers which were also equipped with bulker memory devices (up to 1 Gb). VLNDEF sites are remote and not easily accessible; the observations can only be performed assuring a continuous source of electric power, which is usually provided by means of solar panels and batteries. Technological improvements of a few sites are currently under investigation, in order to meet the requirements that are needed for creating self-standing, unmanned, permanent and remote GPS stations (see Sect. 2).

The increasing redundancy of observations offers new opportunities concerning data processing strategies and long time series analysis. Results obtained by the last-generation GPS receivers have improved the ability to detect crustal deformation from GPS measurements and, when long time series of data are being acquired, to measure local (seasonal or periodical) non-geodetic effects which could affect the investigation.

It has to be highlighted that neotectonic features of Victoria Land are still under investigation both in the Northern as well as in the Southern part, where VLNDEF partially overlaps whit the United States project TAMDEF (TransAntarctic Mountains DEFormation network); this latter GPS network is managed by the Ohio State University and the United States Geological Survey (see Willis et al., 2006).

TNB1 and VLNDEF realize a remarkable Antarctic geodetic infrastructure, which can be view as a whole or as two separate facilities that can serve the whole international geoscience community. The scientific potentials are different as scientific applications are strictly related to the quantity and quality of observations, their duration and their main geodetic characteristic, principally, permanent vs. episodic observations. Maintenance, management, development, optimisation and full exploitation of these facilities imply the capability to approach similar and connected, though different, practical and theoretical aspects.

On one hand, TNB1 can be episodically considered as part of VLNDEF and a subset of its continuous observations are regularly processed along with the episodic observations performed during VLNDEF campaigns. The entire set of VLNDEF observations was processed applying different analysis strategies and different software with the purpose to achieve the most accurate and reliable results. Section 5.2 presents in detail the procedures that were applied while Sect. 4 focuses on the geodynamical potential of the network and its possible contributions to an understanding of the neotectonic features of the area. It is worth mentioning that a selected subset of TNB1 observations was used in the framework of the SCAR GPS epoch campaigns (Dietrich et al., 2001, 2004) (see Sect. 2) and in the densification of the ITRF2000 in the Antarctic and peri-Antarctic regions (Altamimi et al., 2002).

On the other hand, TNB1 is an Antarctic permanent GPS station and its continuous observations, despite the 12 months latency, can improve and support scientific investigations in a wide variety of fields, not only geodetic but multidisciplinary (see e.g. Sarti et al., 2008). The data processing approach, when dealing with TNB1 continuous observations, must be optimised searching, testing and applying the proper methods and the correct strategies to be adopted for continuous permanent GPS observations. Section 5.1 illustrates the detail of the processing strategies and solutions that were tested on seven years of TNB1 observations.

4 Geological and Geodynamical Settings of Northern Victoria Land

VLNDEF network extends on a wide portion of NVL, between 71°S and 76°S and 160°E and 170°E (Fig. 8). Its design (see Sect. 3) was planned according to the most recent information concerning the rheology, neotectonics and geodynamics of the area.

Most of the authors involved in offshore and onshore geology investigations of the Ross Sea region (Salvini et al., 1997; Salvini et al., 1998; Salvini, 1999; Storti et al., 2001) suggest that the tectonic framework revealed in the NVL region is dominated by a major right lateral strike-slip motion and a faults system with a dominant NW-SE direction. It represents the onshore side of the widest dextral transform shear which characterizes the Tasman Fracture Zone and the Balleny Fracture Zone in the Southern Ocean. Three main terrane that follow the NW-SE striking fault zone have been initially described by the GANOVEX TEAM (1987). From NE to SW they are: Robertson Bay Terrane, Bowers Terrane and Wilson Terrane. The Leap Year Fault divides the Robertson Bay from the Bowers Terrane whereas the Lanterman Fault separates the Bowers from the Wilson Terrane. Within the major NW-SE fault system five more faults have been mapped. From NE to SW they are the Cape Adare Faults, Tucker Faults, Aviator Faults, Campbell Faults and the Priestley Faults. Figure 8 shows the distribution of the 28 stations belonging to VLNDEF together with the deformation pattern along the NW-SE faults; it can be clearly noticed that the design of the network aims at detecting the deformations due to neotectonic phenomena and at confirming the strike-slip kinematics which is believed to control the Victoria Land basin.

All previous aspects will be further discussed in the next Sections after a short review of the plate kinematic and Glacial Isostatic Adjustment (GIA) in NVL as presented by several authors involved in the field of Antarctic geosciences.

4.1 Plate Kinematics in NVL

Since 1990, information about the plate kinematics of Antarctica was obtained from the No-Net-Rotation NUVEL 1 model (Argus & Gordon, 1991; DeMets et al., 1990). At the beginning of the 90's the IGS stations installed around the

Antarctic coast started to provide the first measurements of displacements derived from GPS observations. Since middle 90's, the SCAR GPS Epoch project (see Sect. 2), through a set of data episodically acquired by more than 30 IGS and non-IGS permanent stations, has been providing an accurate regional solution for sites' positions and velocities. The results obtained by SCAR GPS Epoch campaigns have highlighted a major clockwise motion of Antarctica with a magnitude somewhere greater than 1 cm/yr (Dietrich et al., 2001) whereas the relative motion occurring between the Antarctic Peninsula area and the eastern Antarctica is not larger than 1–2 mm/yr (Dietrich et al., 2004). These results have been recently confirmed by Negusini et al. (2005) processing a subset of GPS data recorded at 15 IGS Antarctic and peri-Antarctic sites to validate the reference frame stability. These displacements are consistent with the idea of a minimal amount of recent relative motion between East and West Antarctica. Besides that, the vertical rates of displacements are currently under considerations, in terms of accuracy and reliability, as a constraint in the available Glacial Isostatic Adjustment models. However, the analysis of vertical rates has to be combined with the study of other geophysical signals which could potentially affect the observations. It must be highlighted that vertical rates are less accurately estimated when compared to horizontal ones; nevertheless, values up to 10 mm/year were detected in the Northern Antarctic Peninsula (Dietrich et al., 2004) while smaller values were estimated elsewhere in the Antarctic continent.

Several GIA models predict uplift rates in Antarctica. ICE-3G (1991), ICE-4G (1994), ICE-5G (1998) and D91 (1998) show different scenarios of glacial history (Ivins et al., 2003, 2005; Kaufmann et al., 2005; Peltier, 1994; Peltier, 1998; Raymond et al., 2004). Even if this paper does not focus on a detailed discussion of GIA vertical rates as derived by GPS measurements, a first attempt to compare vertical rates (GIA vs GPS) will be discussed in the final part of this paper. The next Sections are devoted to an illustration of the different software and processing strategies applied to TNB1 and VLNDEF data and to a detailed description of the results.

5 GPS Data Sets and Related Analysis Approaches

Nowadays, the continuous observations performed with TNB1 form a 7-years data set; the VLNDEF data refer, instead, to observations performed and stored during five episodic GPS campaigns. In order to find a suitable processing strategy for these specific sets of data, different approaches and software had to be tested; the different tests aimed at obtaining the most accurate results out of the geodetic observations.

Taking into consideration the peculiar orographic and geographic features of Antarctica and its location with respect to the other continents, which distinguish Antarctica as a very remote and isolated site, the baselines that may be possibly formed with extra-Antarctic IGS network stations are characterized by very large moduli. Therefore, the effects of such a stringent constraint was investigated with (i) a classical double difference approach that was applied using Bernese GPS software, (ii) a classical distributed processing applied by means of Gamit/Globk data

Table 1 Some relevant information concerning the software packages and the parameterisation eventually applied to the GPS data sets

Software	Bernese 5.0	Gamit/Globk 10.2	Gipsy Oasis II 4.04
Producer	Astronomical Institute University of Bern (AIUB)	Department of Earth Atmospheric and Planetary Sciences, MIT	Jet Propulsion Laboratory
Orbits	IGS precise	Colombo (1986), Beutler et al. (1994)	FLINN
Antenna Phase Center Variation	IGS_05 – absolute	Antex.dat – relative	ant_info.003 – NGS – Relative
Satellite Phase Center Variation	Yes	No	No
Pole Tide	McCarthy (1996), McCarthy & Petit (2004)	McCarthy (1996)	Yoder (in Webb & Zumberge, 1995)
Solid Tide	McCarthy 1996, McCarthy & Petit (2004)	McCarthy (1996)	IERS2003
Ocean Tide	Matsumoto et al. (2000) Schwiderski (1980)	Scherneck (1991)	No
Tropospheric Mapping	Niell (1996)	Niell (1996)	Niell (1996)
Troposphere Function	Saastamoinen (1972)	Saastamoinen (1972)	Saastamoinen (1972)
Ionosphere correction	LC	LC	LC
Solution	Double difference	Double difference	Precise Point Positioning
Adjustment	Least Square	Least Square (Gamit) Kalman Filter (Globk)	Free net fiducial and 7 parameter transformation to ITRF00

processing package and (iii) Precise Point Positioning (PPP) applied by means of Gipsy/Oasis II. Table 1 resumes some relevant characteristics of each software as well as the specific models that were selected and applied in the analysis.

5.1 TNB1 Data Processing Approaches

In order to achieve an accurate estimate of TNB1 global position and velocity, eventually framed into ITRF2000, two different approaches were tested: a double difference approach, based on the analysis of a network formed by selected IGS permanent stations, by means of Bernese V.5.0 software and an undifferenced approach, based on the Precise Point Positioning strategy, applied with Gipsy/Oasis II software. The observations performed at TNB1 span, almost continuously, a seven-year period: from 1999 till 2006.

Sections 5.1.1 and 5.1.2 contain a short summary of the peculiarities, which characterise each solution; results are summarised in Sect. (5.1.3).

5.1.1 Bernese Carrier Phase Differenced Approach

Taking into consideration the peculiar orographic and geographic features of Antarctica, a relative positioning of TNB1 with respect to other IGS stations can be achieved using observations acquired at GPS sites located in Antarctica, on peri-Antarctic islands or in the nearest continents. In fact, this approach ensures a very high redundancy of observations: it is possible to select those IGS stations that have homogeneously and continuously acquired since 1998, being this latter the year when observations started at Terra Nova Bay (see Sect. 1). A network of GPS stations was therefore selected among the permanent IGS observing sites located both on Antarctica Plate (CAS1, DAV1, DUM1, KERG, MAW1, MCM4, OHIG/OHI2, SYOG e VESL) as well as on other tectonic plates (CHAT, GOUG, HOB2, HRAO, MAC1, PERT e RIOG, which constitute the so called peri-Antarctic network). The stations were chosen on the basis of several criteria: stability, availability of continuous observation series, data quality, availability of ITRF2000 positions and velocities as well as the geometrical shape of the network.

A traditional double difference approach was adopted in order to perform the analysis. Bernese GPS Software v5.0 (Dach et al., 2007) and the suite of scripts and programs named Bernese Processing Engine (BPE) were the processing tools. The analysis started from the observations files in RINEX (Receiver INdependent EXchange) format and aimed at producing and storing daily solutions in the SINEX V.1.0 (Software INdependent EXchange) format as well as Normal EQuation (NEQs).

A data sampling rate of 30 s, standard parameters such as precise orbits and Earth rotation parameters (ERP) provided by IGS were used. An elevation cut-off angle of 10° was adopted since it was considered the best compromise between quantity and quality of data available at the latitudes of the network.

A priori information about tropospheric delay, computed from a standard atmosphere (Berg, 1948), was estimated using Dry Niell mapping function for the dry part, while continuous piecewise linear troposphere parameters were estimated at 1-h intervals using the Wet Niell mapping function, in order to obtain the total zenith tropospheric delay.

The first order term of the ionospheric refraction was eliminated by forming the ionosphere-free linear combination (LC) of the L1 and L2 measurements.

Absolute phase antenna center variations (PCVs) for receivers (Menge et al., 1998) as well as block-specific values for the Block II/IIA and Block IIR satellites (Schmid & Rothacher, 2003) were used.

The ambiguity resolution on all baselines was performed adopting the Quasi-Ionosphere Free (QIF) strategy (Mervart, 1995) along with Global Ionosphere Models (GIMs), provided by CODE that were used as a priori information. The geodetic datum was defined by a No-Net Rotation (NNR) condition with a minimal constraint approach, fixing 6 peri-Antarctic stations which were considered as those having the most reliable a-priori coordinates into ITRF2000.

5.1.2 Undifferenced Precise Point Positioning Gipsy/Oasis II Approach

The Precise Point Positioning (PPP) approach (Zumberge et al., 1997), implemented in Gipsy/Oasis II V.4.04 developed at JPL (Jet Propulsion Laboratory), allows GPS data undifferenced processing of code and carrier phases observables acquired by one single receiver. PPP was run on a Linux Red Hat 9 platform using accurate orbits information and accurate satellite clock data as provided by the IGS or JPL.

PPP represents a major step towards the realization of high accuracy positioning based on stand-alone receivers. From a theoretical point of view, differentiate and undifferentiated approaches differ in clock bias and ambiguity modelling. When observations are differentiate, single, double and triple differences are formed and clock bias and ambiguity are thus removed. With the undifferentiated approach, clock bias and ambiguity are considered as unknowns and are estimated using a proper statistic model by a sequential filter: Square Root Information Filter (SRIF) (Blewitt, 1993). The SRIF filter is a modified Kalman filter, developed at JPL by Gerald Bierman. This filter includes the capability to assess effects from mismodelling by process noise. It allows parameters to have a stochastic behaviour: this is particularly useful for clock bias and tropospheric delay. The stochastic models that have been implemented are mainly a time-uncorrelated behaviour (white noise model) or a time dependent behaviour (random walk model); also unvarying model can be applied. A random walk model is explicitly used for evaluating wet tropospheric delay (Zumberge et al., 1998; Kouba, 2000); a white noise error model is then used for clock bias, since GIPSY doesn't solve double difference. The geodetic precision of PPP solution was demonstrated with the 'March 1985 High Precision Baseline Test' (Davidson et al., 1985; Beutler et al., 1986; Gouldman et al., 1986; Parrot et al., 1986).

The PPP strategy was also tested in remote regions, concluding that GPS daily solutions accuracy can be compared with traditional, differenced GPS positioning. The computation of orbits and satellite clocks solutions comes from the FLINN (Fiducial Laboratories for International Natural science Network) global network (it is a sub-network of the IGS): many of the FLINN stations are equipped with a hydrogen maser or a good quality rubidium or caesium clock. Thus a very stable time reference is available at the receiver site, crucial when estimating high-rate satellite clock corrections.

The orbits used for GIPSY processing have some peculiarities: for each day, JPL analysis center offers IGS-like fiducial orbits, for the current ITRF, and also non fiducial orbits, obtained by a fiducial free solution of the FLINN (Fiducial

Laboratories for International Natural Science Network) stations (Panafidina & Malkin, 2002; Heflin et al., 1992).

The analysis was carried out using non-fiducial orbits which were produced using poorly constrained ground stations. The relative geometry of all orbits is determined by GPS data only and the orbits will not be perturbed by any imperfect knowledge of station coordinate. After the non fiducial solution is obtained, a daily similarity transformation, also provided by JPL during the orbits downloading phase, was used to remove the uncertainty of the frame and to express the solution in ITRF2000.

All daily GPS solution were combined using a GIPSY utility (STAMRG) that use the station positions and the associate complete variance matrix, then the position and the velocity at 2003.0 epoch was calculated. The results were iteratively refined using a data snooping strategy with a 3 sigma tolerance. Default configuration and IGS antenna calibration parameters were also implemented. All the time series were processed using C-shell scripts in order to automate the processing procedure.

5.1.3 TNB1 Data Processing Results

The local geodetic velocity of TNB1 (North, East and Up components), as estimated using both the undifferenced (Gipsy/Oasis II) and the differenced (Bernese V.5.0) approaches, are shown in Fig. 9.

The results were obtained applying the software and the related analysis approach described in Sects. 5.1.1 and 5.1.2; Fig. 10 shows a flow diagram were the processing steps related to these two solutions are highlighted in yellow. Table 2 schematically shows the same results along with new estimates of TNB1 local geodetic velocity components that were obtained by means of Gamit/Globk software package (details can be found in Sect. 5.2.1). The working data set processed with Gamit/Globk differs from the one which was processed with Bernese V.5.0 and Gipsy/Oasis II: it is a subset of the whole 7-year observations set which was formed extracting the TNB1 data that overlap the VLNDEF episodic observations.

5.2 VLNDEF Data Processing Approaches

This Section illustrates the different tests that were performed on the data acquired during the VLNDEF campaigns. More than one strategy and combination of models are investigated, aiming at producing the most accurate final estimation and, at the same time, obtaining a comparable, homogeneous analysis strategy from each software.

The observations performed at VLNDEF sites were acquired during episodic campaigns (see Sect. 3). The large distances between the sites of the network and the external IGS stations as well as the limited duration of the observations on each site (lasting from a minimum of a few days to several weeks) prevent a straightforward application of a processing approach similar to e.g. the one described

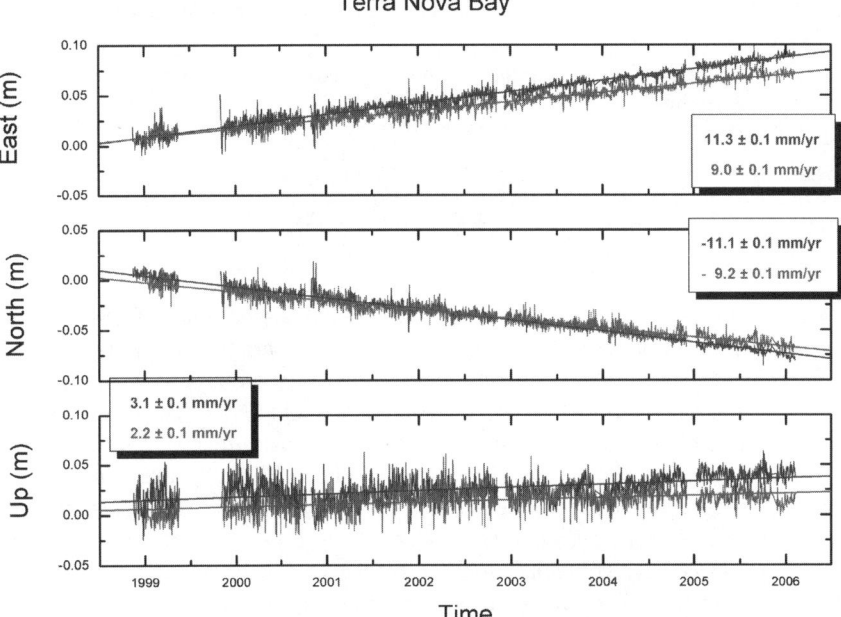

Fig. 9 TNB1 local geodetic coordinates time series derived with Bernese V.5.0 (*red*) and Gipsy/Oasis II (*blue*); the graph also shows the linear fit computed on the time series and the corresponding linear velocities

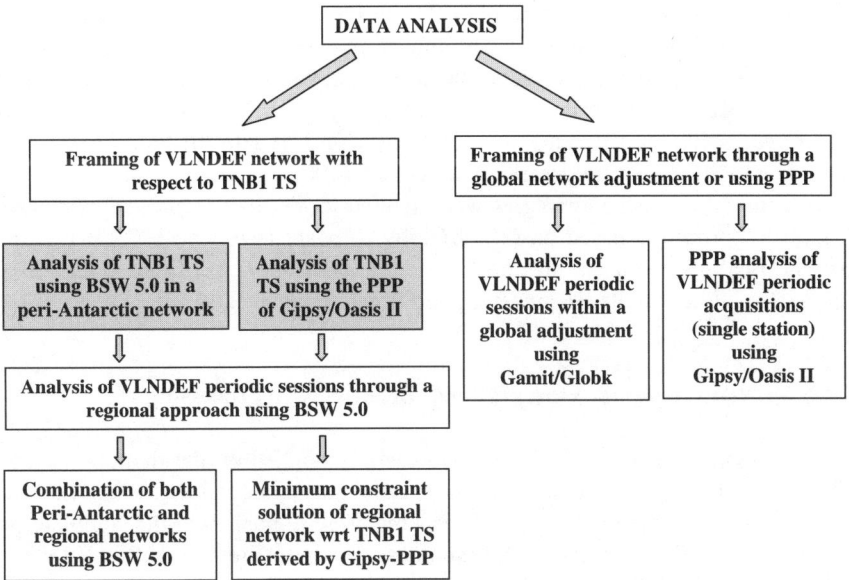

Fig. 10 Schematic flow diagram offering an overview of the strategies and the software packages that were adopted within each different analysis procedure

Table 2 TNB1 GPS permanent station velocities and errors expressed in the local geodetic reference frame

	East componentr (mm/yr)	North component (mm/yr)	Up component (mm/yr)
Bernese v. 5.0	9.0 ± 0.1	−9.2 ± 0.1	2.2 ± 0.2
Gipsy/Oasis II	11.3 ± 0.1	−11.1 ± 0.1	3.1 ± 0.2
Gamit/Globk	9.8 ± 0.2	−11.4 ± 0.2	1.6 ± 0.4

in Sect. (5.1.1). Nevertheless, the good quality and the duration of observations performed at each network's site as well as the their limited reciprocal distance (approximately 100 km) are sufficient to ensure an accurate relative positioning of the points of the network with respect to each others and TNB1. There is no unique analysis approach; therefore, in order to achieve accurate estimations of sites' absolute positions and velocities, eventually framed into ITRF2000, several processing strategies were applied and evaluated. The flow diagram of Fig. 10 resumes the different analysis approaches that were adopted and tested for estimating the most accurate positions' and velocities' solution for the sites of VLNDEF network.

The first approach is based on an accurate estimation of TNB1 coordinates and velocities within ITRF2000: both Bernese V.5.0 and Gipsy/Oasis II were used to estimate these parameters according to the approaches described in Sects. (5.1.1) and (5.1.2), respectively. The first analysis approach of VLNDEF observations was performed with Bernese V.5.0 in order to produce a set of Normal Equations (NEQ) that were successively combined with those produced and stored according to Sect. (5.1.1) procedure: an absolute framing of VLNDEF sites was therefore achieved (see Sect. 5.2.1). A second approach, which is also based on Bernese V.5.0, was adopted and evaluated (see Sect. 5.2.2): VLNDEF data processing was performed, tightly constraining the coordinates of TNB1 to the values estimated according to the procedure described in Sect. (5.1.2).

Two more processing strategies were applied to VLNDEF data and evaluated. A global approach is based on Gamit/Globk software (see Sect. 5.2.3) while the last and final approach that has been tested is based on Gipsy/Oasis II PPP (see Sect. 5.2.4). Final results are presented and discussed in Sect. (5.2.5).

5.2.1 VLNDEF Episodic Observations: Bernese V.5.0 Processing

This processing procedure had been successfully applied on the first and second VLNDEF campaigns to compute a set of coordinates and velocities for the sites of the network (Capra et al., 2007). It was therefore re-applied to the complete set of four VLNDEF campaigns, using the latest version of Bernese software (V.5.0) with similar strategy but a different approach for network framing. In order to take into account the effects of ocean tide on site coordinates, the model NAO.99b was applied at each VLNDEF station as computed by H. G. Scherneck, Onsala Space

Observatory. This ocean tide model is based on the same hydrodynamics as the Schwiderski model but uses TOPEX/Poseidon data given on a 0.5 by 0.5 degree grid (Matsumoto et al., 2000). In particular, station-specific amplitudes and phase of the eleven largest tidal terms for the vertical as well as for the horizontal station components were used; the IERS "standard" format was adopted, accordingly. A consistent set of coefficients was used for all VLNDEF stations.

Daily solutions were performed using the whole set of observations and NEQs were stored.

The final solution for VLNDEF network sites' coordinates (Table 3), properly expressed into ITRF2000, was produced by means of Bernese V.5.0 ADDNEQ routine: VLNDEF NEQs were combined with those produced and stored for TNB1 (5.1.1). Six peri-Antarctic stations (CHAT, KERG, MAC1, HOB2, HRAO, PERT) were used to define the geodetic datum of the site coordinates imposing a No-net translation condition.

Site velocities, estimated with the same ADDNEQ run, are shown in Table 4.

Table 3 VLNDEF sites positions expressed into ITRF2000 (epoch 2003.0) computed with Bernese/peri-Antarctic approach

	X (m)	Y (m)	Z (m)
TNB1 66036M001	−1623858.4319 ± 0.0001	462478.1171 ± 0.0001	−6130048.9861 ± 0.0001
VL01	−1898355.0792 ± 0.0001	344131.5796 ± 0.0001	−6059589.5961 ± 0.0002
VL02	−1871176.6926 ± 0.0001	419007.1067 ± 0.0001	−6064822.2174 ± 0.0002
VL03	−1793824.6435 ± 0.0001	550947.6384 ± 0.0001	−6077986.2946 ± 0.0002
VL04	−1786676.3872 ± 0.0001	323127.1023 ± 0.0001	−6095659.5976 ± 0.0004
VL05	−1833378.6460 ± 0.0001	336084.8236 ± 0.0001	−6079754.0536 ± 0.0002
VL06	−1665381.7116 ± 0.0001	455889.0590 ± 0.0001	−6122163.3470 ± 0.0004
VL07	−1731871.0228 ± 0.0001	451786.7821 ± 0.0001	−6103456.3734 ± 0.0002
VL08	−1717947.9199 ± 0.0001	501076.6043 ± 0.0001	−6104184.8502 ± 0.0003
VL09	−1747656.6838 ± 0.0001	562141.0904 ± 0.0001	−6090108.2408 ± 0.0003
VL10	−1716963.1572 ± 0.0001	532520.4138 ± 0.0001	−6101777.8056 ± 0.0002
VL11	−1644838.3637 ± 0.0001	517300.7456 ± 0.0001	−6122511.0323 ± 0.0004
VL12	−1870330.3720 ± 0.0001	545966.3540 ± 0.0001	−6054921.0795 ± 0.0002
VL13	−1592717.3243 ± 0.0001	511213.2531 ± 0.0001	−6135756.5247 ± 0.0003
VL14	−1791154.8186 ± 0.0001	449717.6339 ± 0.0001	−6086636.3236 ± 0.0002
VL15	−1596324.0004 ± 0.0001	466324.0209 ± 0.0001	−6136835.4752 ± 0.0003
VL16	−1555858.1899 ± 0.0001	489202.5719 ± 0.0001	−6145734.3337 ± 0.0003
VL17	−1561179.2635 ± 0.0001	521189.6915 ± 0.0001	−6142166.0340 ± 0.0002
VL18	−1487512.5303 ± 0.0001	466337.4687 ± 0.0001	−6164019.9253 ± 0.0004
VL19	−1490548.0456 ± 0.0001	490597.3284 ± 0.0001	−6162196.0181 ± 0.0004
VL20	−1909089.5266 ± 0.0002	677740.0492 ± 0.0001	−6029108.8684 ± 0.0006
VL21	−1932100.1091 ± 0.0001	563780.2301 ± 0.0001	−6033962.4198 ± 0.0004
VL22	−1938968.3244 ± 0.0001	628496.4975 ± 0.0001	−6023703.1621 ± 0.0003
VL23	−2017366.1617 ± 0.0001	344665.3680 ± 0.0001	−6021793.9214 ± 0.0001
VL27	−1999402.0838 ± 0.0002	432212.5222 ± 0.0001	−6022817.4305 ± 0.0006
VL29	−1985915.1837 ± 0.0001	573344.5455 ± 0.0001	−6015394.1881 ± 0.0004
VL30	−2027440.4130 ± 0.0002	638271.5344 ± 0.0001	−5994964.3005 ± 0.0004
VL32	−1947613.4119 ± 0.0001	479657.0732 ± 0.0001	−6036111.4515 ± 0.0003

Table 4 VLNDEF sites velocities expressed into ITRF2000 computed with Bernese/peri-Antarctic approach

	V_x (mm/yr)	V_y (mm/yr)	V_z (mm/yr)
TNB1 66036M001	5.9 ± 0.1	-11.1 ± 0.1	-5.1 ± 0.1
VL01	6.3 ± 0.1	-12.3 ± 0.1	-2.0 ± 0.1
VL02	6.7 ± 0.1	-10.7 ± 0.1	-2.5 ± 0.1
VL03	5.9 ± 0.1	-11.6 ± 0.1	-4.4 ± 0.1
VL04	8.0 ± 0.1	-10.3 ± 0.1	-1.1 ± 0.3
VL05	6.5 ± 0.1	-10.5 ± 0.1	-3.5 ± 0.1
VL06	5.5 ± 0.1	-10.5 ± 0.1	-3.1 ± 0.2
VL07	6.7 ± 0.1	-9.7 ± 0.1	-5.3 ± 0.1
VL08	5.1 ± 0.1	-11.2 ± 0.1	-5.0 ± 0.1
VL09	5.6 ± 0.1	-11.4 ± 0.1	-5.1 ± 0.1
VL10	5.4 ± 0.1	-11.2 ± 0.1	-4.7 ± 0.1
VL11	5.6 ± 0.1	-10.7 ± 0.1	-2.6 ± 0.1
VL12	5.6 ± 0.1	-10.9 ± 0.1	-6.0 ± 0.1
VL13	6.0 ± 0.1	-9.8 ± 0.1	-3.2 ± 0.1
VL14	6.7 ± 0.1	-11.8 ± 0.1	-2.1 ± 0.1
VL15	7.0 ± 0.1	-8.1 ± 0.1	-4.3 ± 0.1
VL16	5.3 ± 0.1	-11.8 ± 0.1	-3.6 ± 0.1
VL17	6.3 ± 0.1	-8.3 ± 0.1	-4.1 ± 0.1
VL18	6.9 ± 0.1	-8.9 ± 0.1	-4.8 ± 0.2
VL19	4.7 ± 0.1	-8.4 ± 0.1	-3.7 ± 0.2
VL20	6.7 ± 0.2	-13.8 ± 0.1	-6.6 ± 0.6
VL21	7.2 ± 0.1	-4.1 ± 0.1	-5.7 ± 0.2
VL22	6.5 ± 0.1	-12.5 ± 0.1	-4.5 ± 0.1
VL23	6.0 ± 0.1	-12.1 ± 0.1	-4.9 ± 0.2
VL27	11.3 ± 0.2	-11.4 ± 0.1	-6.3 ± 0.6
VL29	4.5 ± 0.1	-12.3 ± 0.1	-8.7 ± 0.2
VL30	4.0 ± 0.1	-11.8 ± 0.1	-6.5 ± 0.2
VL32	6.3 ± 0.1	-12.7 ± 0.1	-4.9 ± 0.1

Figure 11 shows the horizontal components of the absolute velocities of Table 4, properly transformed and expressed into the local geodetic frame (NEU).

5.2.2 Gipsy/Oasis II PPP and Bernese GPS Minimum Constraint Joint Solution

A second VLNDEF data processing strategy, based on a minimal constrained solution, was tested. The TNB1 coordinates time series, as estimated with the procedure described in Sect. (5.1.2) by means of Gipsy/Oasis II, was used in order to frame, into ITRF2000, the VLNDEF double-differences solutions produced according to Sect. (5.2.1). VLNDEF sites' positions, expressed into ITRF2000 at epoch 2003.0, are shown in Table 5.

Absolute velocities (expressed into ITRF2000) obtained with the same estimation process are shown in Table 6.

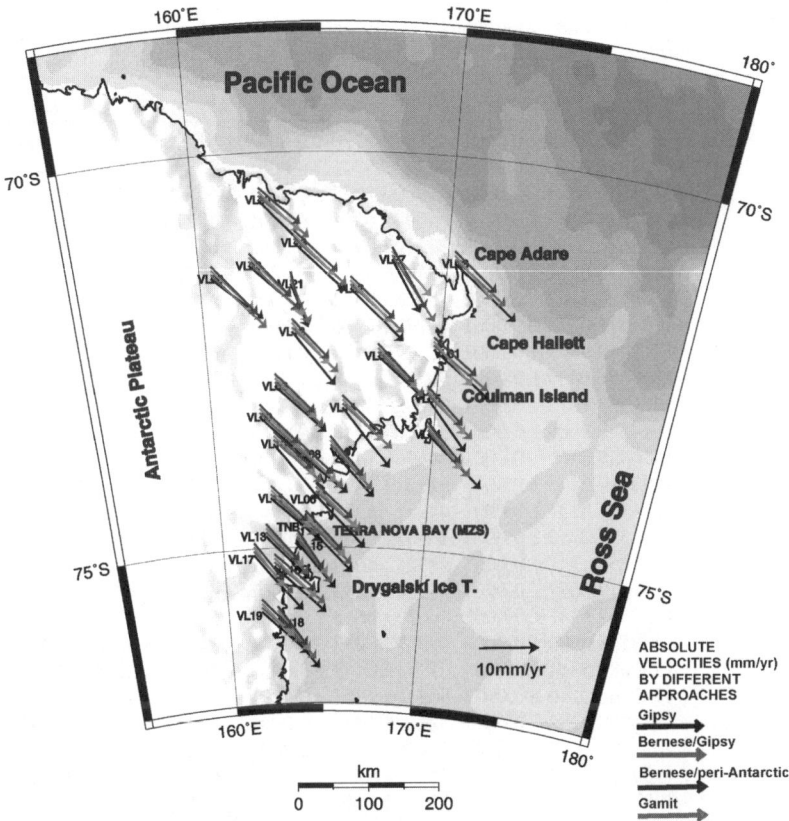

Fig. 11 Absolute velocities of the VLNDEF GPS sites in ITRF2000 computed by means of the three software and four solutions: Gipsy/Oasis (*black arrows*), Bernese/Gipsy (*red arrows*), Bernese/peri-Antarctic (*dark blue arrows*) and Gamit/Globk (*green arrows*)

This method, compared to a traditional approach based on larger networks of reference stations where positions are over-constrained (or loosely "over-constrained") at ITRF values, represents an interesting approach to deformation control: when accurate position(s) of the reference station(s) are available, the intrinsic geometry and orientation of the network is preserved, yielding datum definition while simultaneously maintaining the original characteristic of the regional solution.

5.2.3 Gamit/Globk Global Approach

The third processing strategy that was applied to VLNDEF data relies on the use of Gamit/Globk software package. The statistical model implemented in Gamit is based on weighted least squares. Gamit produces two or more solutions: a pre-fit solution is run in order to obtain a priori values of relevant parameters. Afterwards,

Table 5 ITRF2000 VLNDEF sites' coordinates (epoch 2003.0) estimated with a minimal constrained solution performed combining Bernese V.5.0 and Gipsy/Oasis II software.

	X (m)	Y (m)	Z (m)
TNB1 66036M001	−1623858.4374 ± 0.0001	462478.1120 ± 0.0001	−6130048.9968 ± 0.0001
VL01	−1898355.0840 ± 0.0001	344131.5745 ± 0.0001	−6059589.6055 ± 0.0003
VL02	−1871176.6972 ± 0.0001	419007.1022 ± 0.0001	−6064822.2267 ± 0.0004
VL03	−1793824.6482 ± 0.0001	550947.6334 ± 0.0001	−6077986.3041 ± 0.0004
VL04	−1786676.3925 ± 0.0002	323127.0975 ± 0.0002	−6095659.6071 ± 0.0009
VL05	−1833378.6511 ± 0.0001	336084.8188 ± 0.0001	−6079754.0631 ± 0.0003
VL06	−1665381.7163 ± 0.0002	455889.0542 ± 0.0001	−6122163.3562 ± 0.0007
VL07	−1731871.0270 ± 0.0001	451786.7776 ± 0.0001	−6103456.3816 ± 0.0004
VL08	−1717947.9245 ± 0.0001	501076.5994 ± 0.0001	−6104184.8594 ± 0.0005
VL09	−1747656.6889 ± 0.0001	562141.0851 ± 0.0001	−6090108.2502 ± 0.0004
VL10	−1716963.1612 ± 0.0001	532520.4087 ± 0.0001	−6101777.8129 ± 0.0003
VL11	−1644838.3685 ± 0.0001	517300.7410 ± 0.0001	−6122511.0410 ± 0.0005
VL12	−1870330.3769 ± 0.0001	545966.3488 ± 0.0001	−6054921.0892 ± 0.0003
VL13	−1592717.3293 ± 0.0001	511213.2496 ± 0.0001	−6135756.5343 ± 0.0005
VL14	−1791154.8233 ± 0.0001	449717.6289 ± 0.0001	−6086636.3328 ± 0.0003
VL15	−1596324.0050 ± 0.0001	466324.0166 ± 0.0001	−6136835.4843 ± 0.0006
VL16	−1555858.1942 ± 0.0001	489202.5669 ± 0.0001	−6145734.3419 ± 0.0005
VL17	−1561179.2677 ± 0.0001	521189.6883 ± 0.0001	−6142166.0423 ± 0.0005
VL18	−1487512.5347 ± 0.0002	466337.4643 ± 0.0001	−6164019.9351 ± 0.0006
VL19	−1490548.0500 ± 0.0002	490597.3237 ± 0.0001	−6162196.0278 ± 0.0006
VL20	−1909089.5314 ± 0.0005	677740.0443 ± 0.0003	−6029108.8780 ± 0.0010
VL21	−1932100.1140 ± 0.0002	563780.2254 ± 0.0001	−6033962.4295 ± 0.0006
VL22	−1938968.3293 ± 0.0002	628496.4928 ± 0.0001	−6023703.1707 ± 0.0005
VL23	−2017366.1668 ± 0.0003	344665.3634 ± 0.0001	−6021793.9310 ± 0.0008
VL27	−1999402.0887 ± 0.0005	432212.5175 ± 0.0002	−6022817.4404 ± 0.0014
VL29	−1985915.1887 ± 0.0002	573344.5406 ± 0.0001	−6015394.1972 ± 0.0006
VL30	−2027440.4181 ± 0.0002	638271.5296 ± 0.0001	−5994964.3101 ± 0.0007
VL32	−1947613.4171 ± 0.0002	479657.0683 ± 0.0001	−6036111.4609 ± 0.0005

post-fit solutions may be iterated until post-fit residuals are minimized in a least squares sense (King & Bock, 2000; Herring et al., 2006).

GPS phase observations equations are double differenced (Shaffrin, 1988; Serpelloni et al., 2006) and loosely constrained daily solutions are computed. These latter are stored in h-files and are based on a simultaneous estimation of the following parameters: coordinates, satellite orbits parameters, atmospheric corrections, integer-cycle ambiguities and variance-covariance matrices.

A short resume of the parameterisation adopted in the analysis follows (see Table 1). g-files precise orbit computed at Scripps Orbit and Permanent Array Center (SOPAC) and downloaded from the ftp site "ftp://lox.ucsd.edu" were used. An atmospheric zenith gradient was estimated for each station together with the zenith delay parameter; both atmospheric parameters are modelled by means of a piecewise-linear function (Chen & Herring, 1997). The default models that have been adopted to compute the hydrostatic (or dry) and wet, a priori, zenith delays are those described by Saastamoinen (1972), King and Block (2000), Herring et al. (2006). The dynamical models that were used in the initial conditions integration are taken

Table 6 Absolute velocities of VLNDEF sites expressed into ITRF2000, estimated with a minimal constrained solution performed combining Bernese V.5.0 and Gipsy/Oasis II software

	V_x (mm/yr)	V_y (mm/yr)	V_z (mm/yr)
TNB1 66036M001	6.4±0.1	−13.6±0.1	−5.9±0.1
VL01	6.6±0.1	−15.0±0.1	−3.3±0.1
VL02	6.9±0.1	−13.5±0.1	−4.0±0.1
VL03	6.2±0.1	−14.2±0.1	−5.8±0.1
VL04	8.2±0.1	−12.9±0.1	−2.4±0.1
VL05	6.8±0.1	−13.2±0.1	−4.9±0.1
VL06	5.8±0.1	−13.2±0.1	−4.5±0.1
VL07	6.7±0.1	−12.5±0.1	−7.1±0.1
VL08	5.3±0.1	−13.8±0.1	−6.6±0.1
VL09	5.8±0.1	−14.1±0.1	−6.6±0.1
VL10	5.5±0.1	−13.8±0.1	−7.1±0.1
VL11	5.8±0.1	−13.3±0.1	−4.3±0.1
VL12	5.9±0.1	−13.6±0.1	−7.4±0.1
VL13	6.2±0.1	−12.6±0.1	−4.8±0.1
VL14	7.0±0.1	−14.5±0.1	−3.6±0.1
VL15	7.2±0.1	−11.0±0.1	−5.8±0.1
VL16	5.4±0.1	−14.5±0.1	−5.5±0.1
VL17	6.2±0.1	−11.8±0.1	−6.3±0.1
VL18	6.9±0.1	−11.7±0.1	−6.2±0.1
VL19	4.6±0.1	−11.3±0.1	−5.0±0.1
VL20	7.1±0.1	−16.5±0.1	−8.0±0.3
VL21	7.6±0.1	−6.7±0.1	−7.1±0.1
VL22	6.9±0.1	−15.4±0.1	−6.1±0.1
VL23	6.3±0.1	−14.7±0.1	−6.2±0.1
VL27	11.6±0.1	−14.0±0.1	−7.6±0.2
VL29	4.9±0.1	−15.0±0.1	−10.0±0.1
VL30	4.4±0.1	−14.5±0.1	−7.9±0.1
VL32	6.6±0.1	−15.3±0.1	−6.2±0.1

from IGS/IERS 1992 standards (McCarthy, 1996). In particular, the tidal model that was used to take into account displacements of station coordinates originated by tides is the IERS/IGS standard model for diurnal, semi-diurnal and ter-diurnal earth tides (McCarthy, 1996). Finally, concerning pole tide corrections, the IERS standard model was adopted (McCarthy, 1996); pole tides affects station coordinates and pole tide model was also used during the GLOBK run (King & Bock, 2000; Herring et al., 2006).

Post-fit residuals were screened by means of an AUTomatic CLeaNing tool (AUTCLN) which is applied with the purpose to repair cycle-slips and eliminate outliers.

5.2.3.1 Data Combination, Stabilization and Computation of Velocity Field

VLNDEF solutions described in Sect. (5.2.3) were combined with the global solutions computed by SOPAC for the IGS global network stations using the GLRED tool of GLOBK package, treating them as quasi-observations (Dong et al., 1998).

A distributed session approach by means of sequential Kalman filtering procedures was followed, in order to preserve the uniformity of the reference frame definition and to reduce CPU usage (Hudnut et al., 1996; Dong et al., 1998; Mazzotti et al., 2003; Serpelloni et al., 2006). During the quasi-observation combination process, the goodness of fit of the model was estimated with the reduced Chi-Square estimator, whose values were always smaller than 2.

Adjusted coordinates and velocities were translated into ITRF2000, constraining positions and velocities of 54 stations belonging to the IGS Global Permanent Network and provided by SOPAC ftp site (ftp://lox.ucsd.edu) and obtaining the results shown in Tables 7 and 8, respectively.

Horizontal absolute velocities expressed in the local geodetic frame are shown in Fig. 11.

The overall Weighted Root Mean Square (WRMS) of the estimated stabilized station positions is between 1 and 2 mm for horizontal coordinates and 4–5 mm for

Table 7 ITRF2000 adjusted coordinates of VLNDEF network sites (epoch 2003.0) obtained with Globk, after the stabilisation process of 1999–2006 GPS episodic campaigns

	X (m)	Y (m)	Z (m)
TNB1 66036M001	−1623858.4330 ± 0.0004	462478.1236 ± 0.0004	−6130048.9910 ± 0.0010
VL01	−1898355.0800 ± 0.0007	344131.5863 ± 0.0006	−6059589.6060 ± 0.0016
VL02	−1871176.6890 ± 0.0008	419007.1132 ± 0.0007	−6064822.2280 ± 0.0020
VL03	−1793824.6430 ± 0.0007	550947.6454 ± 0.0006	−6077986.3050 ± 0.0018
VL04	−1786676.3870 ± 0.0010	323127.1064 ± 0.0008	−6095659.6060 ± 0.0026
VL05	−1833378.6500 ± 0.0004	336084.8323 ± 0.0004	−6079754.0680 ± 0.0010
VL06	−1665381.7120 ± 0.0006	455889.0620 ± 0.0005	−6122163.3580 ± 0.0015
VL07	−1731871.0240 + 0.0006	451786.7872 ± 0.0005	−6103456.3830 ± 0.0017
VL08	−1717947.9210 ± 0.0008	501076.6108 ± 0.0006	−6104184.8650 ± 0.0022
VL09	−1747656.6850 ± 0.0008	562141.0966 ± 0.0006	−6090108.2550 ± 0.0024
VL10	−1716963.1570 ± 0.0006	532520.4202 ± 0.0004	−6101777.8110 ± 0.0012
VL11	−1644838.3660 ± 0.0006	517300.7526 ± 0.0005	−6122511.0520 ± 0.0019
VL12	−1870330.3750 ± 0.0005	545966.3580 ± 0.0004	−6054921.0880 ± 0.0011
VL13	−1592717.3230 ± 0.0009	511213.2602 ± 0.0007	−6135756.5370 ± 0.0029
VL14	−1791154.8190 ± 0.0005	449717.6398 ± 0.0004	−6086636.3350 ± 0.0011
VL15	−1596323.9990 ± 0.0008	466324.0257 ± 0.0007	−6136835.4870 ± 0.0026
VL16	−1555858.1900 ± 0.0006	489202.5783 ± 0.0005	−6145734.3440 ± 0.0018
VL17	−1561179.2630 ± 0.0007	521189.6953 ± 0.0006	−6142166.0410 ± 0.0018
VL18	−1487512.5310 ± 0.0007	466337.4738 ± 0.0006	−6164019.9360 ± 0.0024
VL19	−1490548.0460 ± 0.0006	490597.3336 ± 0.0006	−6162196.0300 ± 0.0018
VL20	−1909089.5270 ± 0.0026	677740.0542 ± 0.0023	−6029108.8750 ± 0.0052
VL21	−1932100.1100 ± 0.0009	563780.2380 ± 0.0007	−6033962.4280 ± 0.0023
VL22	−1938968.3240 ± 0.0009	628496.5050 ± 0.0007	−6023703.1720 ± 0.0023
VL23	−2017366.1650 ± 0.0010	344665.3744 ± 0.0008	−6021793.9340 ± 0.0020
VL27	−1999402.0870 ± 0.0029	432212.5283 ± 0.0024	−6022817.4420 ± 0.0062
VL29	−1985915.1840 ± 0.0009	573344.5532 ± 0.0006	−6015394.1980 ± 0.0021
VL30	−2027440.4130 ± 0.0012	638271.5411 ± 0.0009	−5994964.3110 ± 0.0028
VL32	−1947613.4120 ± 0.0006	479657.0799 ± 0.0005	−6036111.4620 ± 0.0016

Table 8 Absolute velocity components of VLNDEF sites expressed into ITRF2000, estimated with Globk

	V_x (mm/yr)	V_y (mm/yr)	V_z (mm/yr)
TNB1 66036M001	7.3 ± 0.5	−12.5 ± 05	−4.5 ± 0.7
VL01	7.6 ± 0.7	−13.8 ± 0.6	−2.5 ± 1.1
VL02	6.5 ± 0.6	−12.2 ± 0.6	−2.4 ± 1.2
VL03	6.0 ± 0.6	−13.1 ± 0.6	−4.3 ± 1.1
VL04	7.8 ± 1.0	−10.9 ± 0.8	−2.8 ± 2.1
VL05	8.3 ± 0.7	−12.7 ± 0.6	−0.8 ± 1.0
VL06	5.3 ± 0.7	−11.0 ± 0.7	−5.0 ± 1.5
VL07	8.5 ± 0.6	−11.6 ± 0.5	−1.1 ± 1.1
VL08	6.0 ± 0.6	−12.8 ± 0.6	−4.2 ± 1.4
VL09	6.2 ± 0.6	−12.8 ± 0.6	−4.8 ± 1.3
VL10	5.6 ± 0.6	−12.9 ± 0.6	−4.8 ± 1.1
VL11	6.1 ± 0.7	−12.2 ± 0.6	−1.0 ± 1.4
VL12	7.4 ± 0.6	−11.9. ± 0.6	−4.9 ± 1.0
VL13	5.7 ± 0.7	−11.6 ± 0.6	−3.9 ± 1.6
VL14	7.6 ± 0.6	−13.4 ± 0.5	−0.4 ± 1.0
VL15	6.8 ± 0.7	−10.0 ± 0.6	−4.0 ± 1.5
VL16	5.7 ± 0.6	−13.2 ± 0.6	−4.4 ± 1.3
VL17	6.3 ± 0.6	−9.8 ± 0.6	−6.3 ± 1.3
VL18	7.6 ± 0.7	−10.8 ± 0.6	−5.2 ± 1.6
VL19	4.8 ± 0.7	−10.0 ± 0.6	−4.1 ± 1.5
VL20	7.5 ± 2.0	−15.1 ± 1.5	−6.6 ± 5.1
VL21	8.4 ± 0.7	−5.5 ± 0.6	−6.0 ± 1.6
VL22	6.3 ± 0.7	−13.9 ± 0.6	−5.1 ± 1.3
VL23	7.4 ± 1.0	−13.8 ± 0.9	−4.5 ± 1.8
VL27	12.2 ± 2.6	−14.0 ± 1.6	−2.0 ± 6.3
VL29	5.3 ± 0.8	−13.7 ± 0.6	−8.7 ± 1.6
VL30	5.0 ± 0.8	−12.8 ± 0.6	−5.2 ± 1.6
VL32	7.2 ± 0.7	−14.1 ± 0.6	−4.3 ± 1.4

heights, depending on the number of daily sessions (varying between 5 and 10) and their duration.

GPS data error spectra are spatially correlated due to common orbital, earth rotational and regional atmospheric errors (Shen et al., 2000). In order to evaluate these effects, coordinates' time series were interactively edited using Tom Herring Matlab Utilities (Herring et al., 2006); outliers and discontinuities were evaluated along with time series common mode errors, seasonal signals and site per site Weighted Root Mean Square (WRMS). GPS time series' noise can be represented as a combination of white, flicker and random walk functions (Mao et al., 1999; Dixon et al., 2000; Mazzotti et al., 2003). Provided that a sufficiently large amount of data is given, the WRMS of the site coordinate time series is well correlated with the amount of white and colored noise. VLNDEF quasi-observations were re-weighted taking into account the WRMS of site coordinates time series.

5.2.4 Estimation of VLNDEF Sites' Positions Using the Gipsy/Oasis II PPP Approach

VLNDEF observations were processed with the PPP approach described in Sect. (5.2.1), with a c-shell scripting especially realized to achieve automatic data processing.

GPS daily sessions were processed independently; every solution was treated as part of a time series, from which positions (Table 9) and velocities (Table 10) were estimated.

The absolute velocities shown in Table 10, transformed and expressed into the local geodetic frame, are shown in Fig. 11; this latter figure also shows the estimated horizontal velocities obtained in Sects. (5.2.1) and (5.2.3).

Table 9 Positions of VLNDEF sites estimated with Gipsy/Oasis II software and expressed into ITRF2000 (epoch 2003.0)

	X (m)	Y (m)	Z (m)
TNB1 66036M001	−1623858.4378 ± 0.0002	462478.1111 ± 0.0002	−6130048.9959 ± 0.0005
VL01	−1898355.0857 ± 0.0003	344131.5745 ± 0.0003	−6059589.5989 ± 0.0008
VL02	−1871176.7022 ± 0.0004	419007.1021 ± 0.0003	−6064822.2633 ± 0.0009
VL03	−1793824.6495 ± 0.0005	550947.6313 ± 0.0004	−6077986.3151 ± 0.0011
VL04	−1786676.3936 ± 0.0011	323127.1006 ± 0.0009	−6095659.5932 ± 0.0026
VL05	−1833378.6580 ± 0.0003	336084.8160 ± 0.0003	−6079754.0701 ± 0.0008
VL06	−1665381.7196 ± 0.0007	455889.0539 ± 0.0006	−6122163.3506 ± 0.0017
VL07	−1731871.0292 ± 0.0005	451786.7774 ± 0.0004	−6103456.3973 ± 0.0012
VL08	−1717947.9246 ± 0.0006	501076.5988 ± 0.0006	−6104184.8703 ± 0.0016
VL09	−1747656.6869 ± 0.0005	562141.0847 ± 0.0005	−6090108.2575 ± 0.0012
VL10	−1716963.1666 ± 0.0007	532520.4080 ± 0.0006	−6101777.8046 ± 0.0016
VL11	−1644838.3713 ± 0.0007	517300.7426 ± 0.0006	−6122511.0562 ± 0.0018
VL12	−1870330.3822 ± 0.0003	545966.3461 ± 0.0003	−6054921.0899 ± 0.0008
VL13	−1592717.3281 ± 0.0006	511213.2483 ± 0.0005	−6135756.5400 ± 0.0015
VL14	−1791154.8304 ± 0.0005	449717.6299 ± 0.0005	−6086636.3306 ± 0.0013
VL15	−1596324.0039 ± 0.0006	466324.0145 ± 0.0005	−6136835.4869 ± 0.0015
VL16	−1555858.1967 ± 0.0005	489202.5655 ± 0.0005	−6145734.3474 ± 0.0015
VL17	−1561179.2733 ± 0.0006	521189.6925 ± 0.0005	−6142166.0424 ± 0.0015
VL18	−1487512.5365 ± 0.0006	466337.4638 ± 0.0006	−6164019.9389 ± 0.0017
VL19	−1490548.0516 ± 0.0007	490597.3276 ± 0.0006	−6162196.0308 ± 0.0019
VL20	−1909089.5387 ± 0.0015	677740.0379 ± 0.0014	−6029108.8760 ± 0.0035
VL21	−1932100.1119 ± 0.0008	563780.2237 ± 0.0008	−6033962.4288 ± 0.0019
VL22	−1938968.3420 ± 0.0006	628496.4951 ± 0.0006	−6023703.2235 ± 0.0014
VL23	−2017366.1762 ± 0.0014	344665.3665 ± 0.0012	−6021793.9336 ± 0.0032
VL27	−1999402.0914 ± 0.0011	432212.5133 ± 0.0010	−6022817.4345 ± 0.0025
VL29	−1985915.1915 ± 0.0008	573344.5385 ± 0.0007	−6015394.2060 ± 0.0018
VL30	−2027440.4214 ± 0.0009	638271.5284 ± 0.0008	−5994964.3107 ± 0.0020
VL32	−1947613.4191 ± 0.0006	479657.0635 ± 0.0006	−6036111.4637 ± 0.0015

Table 10 Components of absolute velocities of VLNDEF sites expressed into ITRF2000, estimated with Gipsy/Oasis

	V_x (mm/yr)	V_y (mm/yr)	V_z (mm/yr)
TNB1 66036M001	7.1 ± 0.1	-14.0 ± 0.1	-14.0 ± 0.2
VL01	6.7 ± 0.2	-16.1 ± 0.2	-16.1 ± 0.5
VL02	7.5 ± 0.2	-14.1 ± 0.2	-14.1 ± 0.4
VL03	6.2 ± 0.2	-14.9 ± 0.2	-14.9 ± 0.4
VL04	7.6 ± 0.6	-15.7 ± 0.5	-15.7 ± 1.4
VL05	10.0 ± 0.1	-12.4 ± 0.1	-12.4 ± 0.4
VL06	6.5 ± 0.3	-14.3 ± 0.3	-14.3 ± 0.8
VL07	7.9 ± 0.2	-13.1 ± 0.2	-13.1 ± 0.5
VL08	4.6 ± 0.2	-14.9 ± 0.2	-14.9 ± 0.6
VL09	5.3 ± 0.2	-14.3 ± 0.2	-14.3 ± 0.5
VL10	8.1 ± 0.3	-13.9 ± 0.2	-13.9 ± 0.7
VL11	5.9 ± 0.2	-14.4 ± 0.2	-14.4 ± 0.6
VL12	8.2 ± 0.1	-13.4 ± 0.1	-13.4 ± 0.4
VL13	5.8 ± 0.2	-12.8 ± 0.2	-12.8 ± 0.6
VL14	10.4 ± 0.2	-14.8 ± 0.2	-14.8 ± 0.6
VL15	6.6 ± 0.2	-11.7 ± 0.2	-11.7 ± 0.7
VL16	5.9 ± 0.2	-15.0 ± 0.2	-15.0 ± 0.6
VL17	7.6 ± 0.3	-14.7 ± 0.2	-14.7 ± 0.7
VL18	7.8 ± 0.2	-12.9 ± 0.2	-12.9 ± 0.7
VL19	4.7 ± 0.3	-13.4 ± 0.3	-13.4 ± 0.9
VL20	6.2 ± 0.9	-15.6 ± 0.9	-15.6 ± 2.0
VL21	7.5 ± 0.3	-6.6 ± 0.3	-6.6 ± 0.7
VL22	9.1 ± 0.2	-16.1 ± 0.2	-16.1 ± 0.6
VL23	8.0 ± 0.7	-17.8 ± 0.6	-17.8 ± 1.6
VL27	14.7 ± 1.0	-10.9 ± 1.0	-10.9 ± 2.4
VL29	6.6 ± 0.3	-15.2 ± 0.3	-15.2 ± 0.7
VL30	7.0 ± 0.4	-14.5 ± 0.3	-14.5 ± 0.8
VL32	7.7 ± 0.2	-15.6 ± 0.2	-15.6 ± 0.6

6 Combination of Results and Discussion

In order to compare the consistency of the solutions that were estimated adopting the four different approaches described in Sect. 5.2, a Helmert transformation between the results of Table 5 (derived with Bernese in Sect. 5.2.2) and the results of Table 3, Table 7 and Table 9 was performed. The transformation's residuals are shown in Table 11.

All the approaches that were adopted give similar results, both in terms of absolute and relative values. The comparison of the different solutions highlights biases that affect only a few points.

- Bernese (Sect. 5.2.2) vs. Gipsy: East and North components are similar while larger differences can be observed in the Up component. Two anomalous values

Table 11 Helmert transformation's residuals between the different solutions of Sect. 5.2.

	Gipsy/Oasis II (cf. Table 9)			Gamit/Globk (cf. Table 7)			Bernese (Sect. 5.2.1) (cf. Table 3)		
	ΔN (mm)	ΔE (mm)	ΔU (mm)	ΔN (mm)	ΔE (mm)	ΔU (mm)	ΔN (mm)	ΔE (mm)	ΔU (mm)
TNB1	−0.9	−1.2	0.8	−1.0	12.6	5.9	0.4	6.2	11.5
VL01	−4.0	−0.4	6.1	1.9	11.5	0.6	0.8	5.8	10.7
VL02	6.1	−1.1	−35.8	5.5	12.5	0.7	−0.3	5.8	10.4
VL03	2.9	−2.2	−10.6	1.5	12.9	−0.6	−0.4	6.0	9.3
VL04	−5.4	2.8	13.0	4.0	8.9	2.2	2.4	5.0	10.1
VL05	−4.0	−4.2	−8.5	0.2	13.0	−5.2	1.2	5.8	9.8
VL06	−5.1	−1.1	3.8	2.2	8.8	−1.5	0.0	5.9	9.4
VL07	2.6	−1.5	−14.9	0.9	9.5	−0.8	0.2	4.9	9.4
VL08	3.1	0.0	−10.6	2.1	12.6	−5.6	0.7	6.2	9.6
VL09	4.1	0.6	−7.1	1.6	12.6	−4.8	0.5	6.3	9.5
VL10	−7.5	−2.7	6.2	0.0	11.7	2.1	0.3	6.0	7.4
VL11	1.6	1.3	−15.1	2.3	12.3	10.8	0.7	6.3	9.5
VL12	−3.5	−4.3	−2.2	−0.9	9.2	0.8	0.5	6.2	9.6
VL13	2.8	−0.6	−5.5	3.1	12.3	−2.3	1.1	5.3	9.6
VL14	−7.3	−0.7	−0.1	1.7	11.6	−1.6	0.0	5.8	9.4
VL15	2.5	−2.6	−2.4	3.9	10.3	−2.1	1.2	5.2	9.6
VL16	−1.2	−1.9	−5.5	1.0	11.7	−1.8	0.2	6.0	8.3
VL17	−5.8	2.2	−1.5	2.2	8.2	1.6	0.4	5.1	8.4
VL18	0.0	−0.3	−4.1	1.0	10.7	−0.8	0.7	6.3	10.5
VL19	−2.3	3.2	−3.7	1.1	10.7	−1.8	0.0	5.1	10.3
VL20	−5.9	−8.3	0.1	−0.5	10.8	3.0	−1.2	6.1	10.1
VL21	1.8	−0.4	1.6	−0.4	13.6	2.0	0.1	6.2	10.6
VL22	4.6	−2.1	−54.4	1.3	13.0	−0.6	0.2	6.3	9.6
VL23	−8.1	2.4	−5.9	1.1	11.2	−2.8	0.7	5.8	10.8
VL27	−2.4	−5.3	4.5	0.5	10.2	−1.9	1.0	5.0	9.8
VL29	1.0	−3.7	−9.2	1.7	12.9	−0.5	0.3	6.2	9.6
VL30	−1.8	−2.8	−1.7	1.7	12.0	−0.5	0.4	5.3	9.7
VL32	0.0	−4.4	−3.2	2.2	12.8	−0.3	0.7	6.1	9.7
Mean	4.2	3.0	14.3	2.1	11.7	3.3	0.8	5.9	9.9

are found for a couple of points (VL02, VL22); these discrepancies must be understood and investigated further;
- Bernese (Sect. 5.2.2) vs. Gamit: the two solutions are in good agreement with respect to North and Up components; an exception is found for VL11. A systematic bias of one cm is found in the East component;
- Bernese (Sect. 5.2.2) vs. Bernese (Sect. 5.2.1): a constant bias of about one cm is found in the Up component, while a bias of half this value is found in the East component. These biases are probably originated by the initial values of the stations velocities used to constrain the final adjustment.

A formal inversion procedure was applied to the absolute velocity field obtained by means of the processing approaches described in Sects. (5.2.2) and (5.2.3) aiming at deriving the site relative velocities in a robust stable reference frame (Antarctic

Plate) (Argus & Gordon, 1991; DeMets et al., 1994; Altamimi et al., 2002). The relevant Euler Vector of the Nuvel-1A NNR global plate model relating stable Antarctic Plate ANTA to ITRF2000 (Altamimi et al., 2002) was applied; this procedure enabled a removal of the inherently rigid body rotation generally associated with a geodetic reference frame. This approach permitted to reduce the residual horizontal velocities of TNB1 and MCM4 sites to about some tenth of millimetres: this result highlights the consistency between the global reference frame and frame defined within the approaches described in Sects. (5.2.2) and (5.2.3) and adopted for solving the position and velocity of all VLNDEF sites.

A serious limiting factor that prevents the most accurate determination of the movements undergoing in NVL is originated by the lack of permanent remote GPS stations within VLNDEF. As it has been clearly stated before, all deformation results are obtained from episodic campaigns of ever increasing duration, with performances of increasing precision but not comparable to those achievable with permanent GPS stations. An optimal exploitation of the network is therefore tightly connected to the establishment of new remote observing permanent sites.

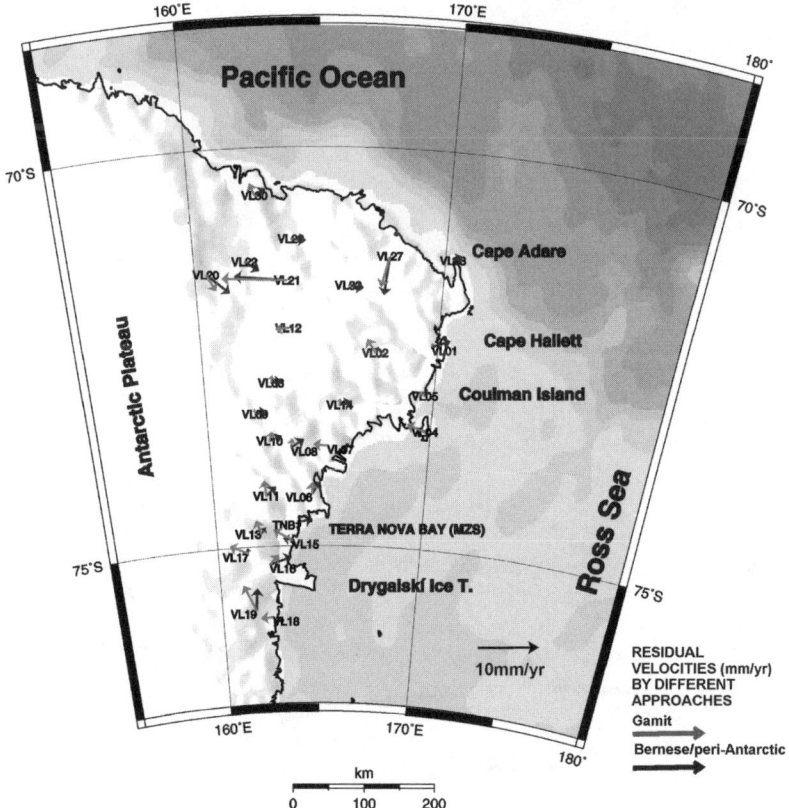

Fig. 12 Residual velocities obtained with Gamit/Globk (*green arrows*) and Bernese (*dark blue arrows*) after removal of the rigid clockwise Antarctic Plate rotation associated with the ITRF2000 reference frame

Relative horizontal velocities shown in Fig. 12 are very small and, within their own 95% confidence interval, do not highlight evident movements. The only exception is represented by the velocity of VL21, whose origin and magnitude are uncertain and should be further investigated.

Vertical velocities shown in Fig. 13 highlight a generalized positive trend of 2/3 mm/yr (1–2 in the Globk solution); this behaviour is particularly evident for the continental stations, being the estimated velocity increasing from the coast towards the Antarctic Plateau; a similar pattern of such vertical rates was already obtained by the processing of GPS data acquired up to the year 2003 (Capra et al., 2007).

The magnitude of the uplift is almost entirely consistent with most of the GIA models that are nowadays proposed to describe the vertical movements of NVL (Ivins et al., 2003, 2005); nevertheless, the uncertainties associated with our geodetic estimates are too large to draw any quantitative and final conclusion on the matter. According to most authors (Nakada et al., 2000; Hamilton et al., 2001;

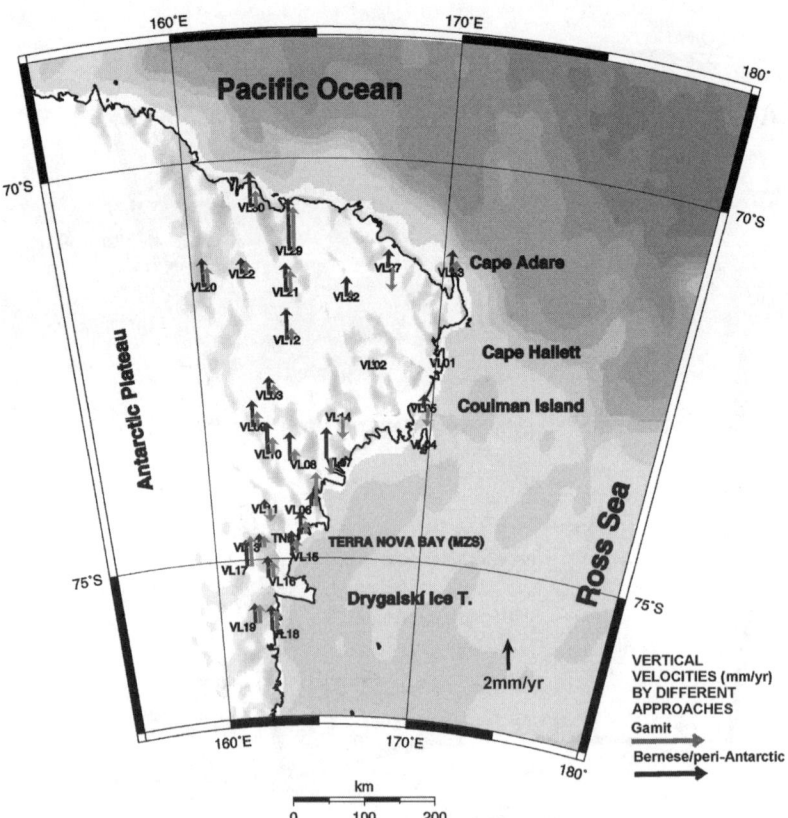

Fig. 13 Vertical velocities obtained with Gamit/Globk (*green arrows*) and Bernese (*dark blue arrows*)

Donnellan, 2004; Raymond et al., 2004; Ohzono et al., 2006), the Antarctic plate behaves as a coherent and rigid plate and our results, even if on a smaller area, are consistent with these conclusions.

This work demonstrates how geodetic Antarctic facilities, such as TNB1 and VLNDEF, can contribute to geosciences investigations in remote regions. Each facility needs special handling, management and maintenance and suitable processing strategies have to be identified, applied and evaluated.

Geodetic results are obtained testing different analysis approaches and processing tools and are framed within a wider scientific scenario, where geodynamics as well as neotectonics are the evaluating background and touchstone.

Therefore, the Italian geodetic infrastructures can be regarded as useful tools for the entire Antarctic scientific community as well as for global geosciences and their maintenance and their exploitation has to be planned, realized and performed as part of a wide, international scientific action.

Acknowledgments Research carried out within the activities of the Italian National Program for Antarctic Research (Consorzio per l'attuazione del PNRA, Rome, Italy).

References

Altamimi Z., P. Sillard, C. Boucher (2002), ITRF2000 A new release of the international terrestrial reference frame for earth science applications, *J. Geophys. Res.*, **107(B10)**, 2214, doi:10.1029/2001JB000561.

Altamimi Z., C. Boucher, P. Willis (2005), Terrestrial reference requirements within GGOS perspective, *J. Geodyn.*, **40**, 363–374.

Argus D. & R. Gordon (1991), No-Net Rotational model of current plate velocities incorporating plate motion model NUVEL-1, *Geophys. Res. Lett.*, **18(11)**, 2039–2042.

Berg H. (1948), Allgemeine Meteorologie, Dummler's Verlag, Bonn, Germany.

Beutler G., E. Brockmann, W. Gurtner, U. Hugentobler, L. Mervart, M. Rothacher (1994), Extended orbit modeling techniques at the CODE Processing Center of the International GPS Service for Geodynamics (IGS): theory and initial results, *Manuscripta Geodetica*, **19**, 367–386.

Beutler G., W. Gurtner, M. Rothacher, T. Schildknecht, I. Bauersima (1986), Evaluation of the March 1985: HPBL test: Fiducial point concept versus free network solutions. Presented at: AGU Fall Meeting, San Francisco, CA, 13.

Blewitt J (1993), Advances in global positioning system technology for geodynamics investigations: 1978–1992, AGU Crustal Dynamics Monogr. *Contributions of Space Geodesy to Geodynamics: Technol.*, **25**, 195–213

Capra, A., A. Gubellini, F. Radicioni, L. Vittuari (1996), Italian Geodetic activities in Antarctica, *In* MELONI, A & MORELLI, A., eds. *Italian Geophysical Observatories in Antarctica*, 2–20.

Capra, A., S. Gandolfi, F. Mancini, M. Negusini, P. Sarti, L. Vittuari (2004), Terra Nova Bay GPS permanent station, *In* BRANCOLINI, G., GHEZZO, C. & MORELLI, A., eds. *Terra Antartica Reports*, **9**, 21–24.

Capra A., F. Mancini, M. Negusini (2007), GPS a geodetic tool for geodynamics in northern Victoria Land, Antarctica, *Antarctic Sci.*, **19(1)**, 107–114.

Chen G & T. A. Herring (1997), Effects of atmospheric azimuthal asymmetry on the analysis of space geodetic data, *J. Geoph. Res.*, **102(B9)**, 20489–20502.

Colombo O. L. (1986), Ephemeris errors of GPS satellites, *Bullettin Geodesique*, **60**, 64–84.

Dach R., U. Hugentobler, P. Fridez and M. Meindl (2007), Bernese GPS Software Version 5.0, Astronomical Institute of University of Berne, 640 pp.
Davidson J. M., C. L. Thornton, C. J. Vegos, L. E. Young, T. P. Yunck (1985), The March 1985 demonstration of the fiducial network concept for GPS geodesy: A preliminary report, *Proceedings of the First International Symposium on Precise Positioning with the Global Positioning System, Rockville, MD*, **2**, 603–612.
DeMets C., R. Gordon, D. Argus (1990), Current plate motions, *Geophys. J. Int.*, **101**, 425–478.
DeMets C., R. G. Gordon, D. F. Argus, S. Stein (1994), Effect of recent revisions to the geomagnetic reversal time scale on estimates of current plate motions, *Geophysical Research Letters*, **21(20)**, 2191–2194.
Dietrich R., R. Dach, G. Engelhard, J. Ihde, W. Korth, H. J. Kutterer, K. Lindner, M. Mayer, F. Menge, H. Miller, C. Muller, W. Niemeier, J. Perlt, M. Pohl, H. Salbach, H. W. Schenke, T. Schone, G. Seeber, A. Veit, C. Volksen, (2001), ITRF coordinates and plate velocities from repeated GPS campaigns in Antarctica-an analysis based on different individual solutions, *J. Geod.*, **74**, 756–766.
Dietrich, R., A. Rülke, J. Ihde, K. Lindner, H. Miller, W. Niemeier, H. W. Schenke, G. Seeber (2004), Plate kinematics and deformation status of the Antarctic Peninsula based on GPS, *Gl. Plan. Change*, **42(1–4)**, 313–321.
Dixon T. H., M. Miller, F. Farina, H. Wang, D. Johnson (2000), Present-day motion of the Sierra Nevada block and some tectonic implications for the Basin and Range province, North American Cordillera, *Tectonics*, **19**, 1–24.
Dong. D. N., T. A. Herring, R. W. King (1998), Estimating regional deformation from a combination of space and terrestrial geodetic data, *J. Geod.*, **72**, 200–214.
Donnellan A. & B. P. Luyendyk (2004), GPS evidence for a coherent Antarctic plate and for postglacial rebound in Marie Byrd Land, *Global and Planetary Change*, **42**, 305–311.
GANOVEX TEAM (1987), Geological map of the Northern Victoria Land, Antarctica, 1:500000. Explanatory notes. *Geol. Jarbh.*, **B66**, 779.
Gouldman M. W., B. R. Hermann, E. R. Swift (1986), Absolute station position solutions for sites involved in the Spring 1985 GPS precision baseline test, *Proceedings of the Fourth International Geodetic Symposium on Satellite Positioning, Austin, Tex.*, **2**, 1045–1058.
Gubellini A., M. Marsella, D. Postpischl, F. Radicioni, L. Vittuari (1994), The Italian Geodetic Network in Antarctica, *Terra Antartica*, **1**, 173–177.
Gubellini A. & D. Postpischl (1991), The Mount Melbourne (Antarctica) geodetic network, *Memorie della Società Geologica Italiana*, **46**, 595–610.
Hamilton R. J., B. P. Luyendik, C. C. Sorlien (2001), Cenozoic tectonics of the Cape Roberts Rift Basin and Transantarctic Mountains Front, Southwestern Ross Sea, Antarctica, *Tectonics*, **20**, 325–342.
Heflin M. B., W. I. Bertiger, G. Blewitt, A. P. Freedman, K. J. Hurst, S. M. Lichten, U. J. Lindqwister, Y. Vigue, F. H. Webb, T. P. Yunck, J. F. Zumberge (1992), Global geodesy using GPS without fiducial sites, *Geophys. Res. Lett.*, **19**, 131–134.
Herring T. A., R. W. King, S. C. McClusky (2006), Gamit Reference Manual, GPS Analysis at MIT, Release 10.2, 28th September 2006.
Hudnut K. W., Z. Shen, M. Murray, S. McClusky, R. King, T. Herring, B. Hager, Y. Feng,. P. Fang, A. Donnellan, Y. Bock (1996), Co-Seismic displacements of the 1994 Northridge, California, Earthquake, *Bull. Seism. Soc. of America*, **86(1B)**, S19–S36.
Ivins E. R., T. S. James, V. Klemann (2003), Glacial isostatic stress shadowing by the Antarctic ice sheet, *J. Geophys. Res.*, **108(B12)**, 2560, 1–21.
Ivins E. R. & T. S. James (2005), Antarctic glacial isostatic adjustment: a new assessment, *Antarctic Sci.*, **17(4)**, 541–553.
Kaufmann G., P. Wu, E. R. Ivins (2005), Lateral viscosity variations beneath Antarctica and their implications on regional rebound motions and seismotectonics, *J. Geodyn.*, **39**, 165–181.
King R. W. & Y. Bock (2000), Documentation for the Gamit GPS Analysis Software, *Department of Earth and Planetary Sciences, Massachussets Institute of Technology, Cambridge, Scripps Institution of Oceanography University of California San Diego*.

Kouba J. & P. Heroux (2000), GPS Precise point positioning using IGS orbit products, *GPS solutions*, **5(2)**, 12–28.

Matsumoto K., T. Takanezawa, M. Ooe (2000), Ocean tide models developed by assimilating TOPEX/POSEIDON altimeter data into hydrodynamical model: A global model and a regional model around Japan, *J. Oceanog.*, **56**, 567–581.

Mancini F., A. Capra, S. Gandolfi, P. Sarti, L. Vittuari (2004), VLNDEF (Victoria Land Network for DEFormation control). Monumentation during the GANOVEX VIII – ItaliaAntartide XV: Survey and data processing, *Terra Antartica*, **11(1)**, 35–38.

Manning J. (2005), The evolution of the GIANT program, Report of Fifth Antractic Geodesy Symposium, SCAR Report, **23**, 1–6.

Mao A., G. Cristopher, A. Harrison, T. H. Dixon (1999), Noise in GPS coordinate time series, *J. Geophys. Res.*, **104(B4)**, 2797–2816.

Mazzotti S., H. Dragert, J. Henton, M. Schmidt, R. Hyndman, T. James, Y. Lu, M. Craymer (2003), Current tectonics of northern Cascadia from a decade of GPS measurements, *J. Geophys. Res.*, **108(B12)**, 2554, doi:10.1029/2003JB002653.

McCarthy D. D. (Ed.) (1996), IERS Conventions 1996, *IERS Tech. Note 21*, Int. Earth Rotation Serv., Obs. de Paris.

McCarthy D. D. & G. Petit (2004), IERS Conventions (2003), IERS Tech. Note 32, *Verlag des Bundesamts für Kartographie und Geodäsie, Frankfurt am Main, Germany*, **127**, ISBN 3-89888-884-3.

Menge F., G. Seeber, C. Volksen, G. Wubbena, M. Schmitz (1998), Results of absolute field calibration of GPS antenna PCV, in *International Technical Meeting of the Satellite Division of the Institute of Navigation ION GPS-98, Nashville, Tennessee, 15–18. September 1998*.

Mervart L. (1995), Ambiguity resolution techniques in geodetic and geodynamic applications of the Global Positioning System, *Geod. Geophys. Arb. Schweiz*, **53**, 155 pp.

Nakada M., R. Kimura, J. Okuno, K. Moriwaki, H. Miura, H. Maemoku (2000), Late Pleistocene and Holocene melting history of the Antarctic ice sheet derived from sea-level variations, *Marine Geology*, **167**, 85–103.

Negusini M., F. Mancini, S. Gandolfi, A. Capra (2005), Terra Nova Bay GPS permanent station (Antarctica): Data quality and first attempt in the evaluation of regional displacement, *J. Geodyn.*, **39(2)**, 81–90.

Niell, A. E. (1996), Global mapping functions for the atmosphere delay at radio wavelengths, *J. Geophys. Res.*, **101(B2)**, 3227–3246.

Ohzono M., T. Tabei, K. Doi, K. Shibuya, T. Sagiya (2006), Crustal movement of Antarctica and Syowa Station based on GPS measurements, *Earth Planets Space*, **58**, 795–804.

Panafidina N. & Z. Malkin (2002), On Computation of a homogeneous coordinate time series for the EPN network, Presented at the "Vistas for Geodesy in the New Millennium" IAG 2001 Scientific Assembly, Budapest, Hungary, 2–7 September 2001.

Parrot D., R. B. Langley, A. Kleusberg, R. Santerre, P. Vanicek, D. Wells (1986), The Spring 1985 GPS high-precision baseline test: Very preliminary results. Invited paper: GPS Technology Workshop, 25 March, Jet Propulsion Laboratory, Pasadena, CA.

Peltier W. R. (1994), Ice age paleotopography, *Science*, **265**, 195–201.

Peltier W. R. (1998), Antarctic geodetic signature of the ICE-5G model of the Late Pleistocene deglaciation, *EOS Transaction AGU*, F215.

Raymond C. A., E. R. Ivins, M. B. Heflin, T. S. James (2004), Quasi-continuous global positioning system measurements of glacial isostatic deformation in the Northern Transantarctic Mountains, *Global and Planetary Change*, **42**, 295–303.

Rummel R., M. Rothacher, G. Beutler (2005), Integrated Global Geodetic Observing System (IGGOS) – science rationale, *J. Geodyn.*, **40**, 357–362.

Saastamoinen J. (1972), Contributions to the theory of atmospheric refraction, *Bull. Geod.*, **105**, 279–298; **106**, 383–397.

Salvini F., G. Brancolini, M. Busetti, F. Storti, F. Mazzarini, F. Coren (1997), Cenozoic geodynamics of the Ross Sea region, Antarctica: crustal extension, intraplate strike-slip faulting, and tectonic inheritance, *J. Geophys. Res.*, **102(24)**, 669–696.

Salvini F. & F. Storti (1999), Cenozoic tectonic lineaments of the Terra Nova Bay region, Ross Embayment, Antarctica, *Global and Planetary Change*, **23**, 129–144.

Salvini, F., F. Storti, G. Brancolini, M. Busetti, C. De Cillia (1998), Cenozoic strike-slip induced basin inversion in the Ross Sea, Antarctica, *Terra Antartica*, **5(2)**, 209–215.

Sarti P, M. Negusini, C. Lanconelli, A. Lupi, C. Tomasi, A. Cacciari (2008), A GPS derived precipitable Water Vapour content and its relationship with 5 years of long-wave radiation measurements at "Mario Zucchelli" Station, Terra Nova Bay, Antarctica. On "Geodetic and Geophysical observations in Antarctica - An Overview in IPY Perspective" Eds A. Capra, R. Dietrich, pp. 145–178, Springer. ISBN 978-3-540-74881-6.

Scherneck H. G. (1991), A parametrised solid Earth tide model and ocean tide loading effects for global geodetic baseline measurements, *Geophys. J. Int.*, **106(3)**, 677–694.

Schwiderski E.W. (1980), Ocean Tides I, Global ocean tidal equations, *Marine Geodesy*, **3**, 161–217.

Schmid R. & M. Rothacher (2003), Estimation of elevation-dependent satellite antenna phase center variations of GPS satellites, *J. Geod.*, **77**, doi:10.1007/s00190-003-0339-0.

Serpelloni E., G. Casula, A. Galvani, M. Anzidei, P. Baldi (2006), Data analysis of permanent GPS networks in Italy and surrounding regions: applications of a distributed processing approach. *Ann. Geophys.*, 49 (4/5), 1073–1103.

Shaffrin B. & Y. Bock (1988), A unified scheme for processing GPS phase observations, *Bullettin Geodesique*, **62**, 142–160.

Shen Z., C. Zhao, A. Y. Yanxing Li, D. D. Jaxon, P. Fang, D. Dong (2000), Contemporary crustal deformation in east Asia constrained by Global Positioning System measurements. *J. Geophys. Res.*, **105(B3)**, 5721–5734.

Steigenberger P., M. Rothacher, R. Dietrich, M. Fritsche, A. Rülke, S. Vey (2006), Reprocessing of a global GPS network, *J. Geophys. Res.*, **111**, B05402.

Storti, F., F. Rossetti, Salvini, F. (2001), Structural architecture and displacement accommodation mechanisms at the termination of the Priestly Fault, northern Victoria Land, Antarctica. *Tectonophysics*, **341**, 141–161.

Webb F. H. & J. F. Zumberge (1995), An introduction to Gipsy Oasis II, Jet Propulsion Laboratory.

Wessel P. & W. H. F Smith (1998), Free software helps maps and display data, *EOS*, **79**, 579.

Willis M., T. Wilson, L. Hothem (2006), A Decade of GPS Measurements over the TAMDEF Network, Victoria Land, Antarctica. *Geophys. Res. Abstracts*, **8**, European Geosciences Union, Vienna, 2–7 April 2006.

Woodworth P. L., T. Aarup, R. Rummel (2005), IGGOS as a potential partner in IGOS, *J. Geodyn.*, **40**, 432–435.

Zumberge J. F., M. B. Helfin, D. C. Jefferson, M. M. Watkins, F. H. Webb (1997), Precise point positioning for efficient and robust analysis of GPS data from large networks. *J. Geophys. Res.*, **102**, 5005–5017.

Zumberge J. F., M. M. Watkins, F. H. Webb (1998), Characteristics and applications of precise GPS clock solutions every 30 seconds. *JPL Technical Reports*.

Communications Systems for Remote Polar GNSS Station Operation

Bjorn Johns

1 Introduction

Emerging polar geophysical networks will enable the collection of critical new data sets to address many fundamental questions about the nature and behavior of the crust and mantle beneath Antarctica and Greenland, and their relationship to ice sheet dynamics and climate. It has long been recognized that major advances in addressing many compelling questions in polar geosciences require continuous recording of Global Navigation Satellite Systems (GNSS) and seismic data at stations that can operate autonomously for several years, as addressed in the International Polar Year (IPY) Polar Earth Observation Network (POLENET) projects (e.g. Bevis et al. 2006; Wilson et al. 2006). This paper addresses communication systems for bringing remote polar network GNSS data to the Internet. While the focus is on communications related to GPS systems, the communication strategies are similar for GNSS applications also utilizing GLONASS (and future Galileo) signals.

Prior to the 2007–2009 International Polar Year, year-round continuous recording of geodetic data in the polar regions has been largely limited to a small number of stations adjacent to permanent settlements and operational bases, and more remote data collection has been almost entirely limited to summer-only episodic campaigns (Capra et al. 2007; Dietrich et al. 2001; Willis et al. 2006). To densify networks and to capture seasonal variations with continuous data, the challenge of truly autonomous station operation is amplified by the logistical constraints under which national science programs operate. Support for remote field deployments is usually restricted by the availability of aircraft flights and fuel. One could argue that some of the important science questions might be answered by gathering data over just the summer months using conventional solar and battery power systems and storing data on-site, but even so revisiting sites annually to retrieve data is logistically impractical in many cases. Robust communication systems are indispensable to minimizing site visits, an essential requirement for long-duration deployment of instruments at very remote sites. Experience has also shown the benefits of quasi-real time data communications in terms of scientific productivity, station maintenance and

---

Bjorn Johns
UNAVCO, 6350 Nautilus Dr., Boulder, Colorado, 80301, USA, e-mail: johns@unavco.org

troubleshooting, overall data availability and quality, and the streamlining of downstream data quality control and processing. In addition, there are some scientific objectives and monitoring functions of remote observatories (such as earthquake and tsunami hazards, measurements of calving glaciers and ice sheets, and GPS-meteorology) that require real-time telemetry of data.

Modern GNSS receiver technology makes it possible to use off-the-shelf units for autonomous recording in the extreme polar environment, and advances in power and communication technologies have now put us on the threshold of deploying large GNSS networks at the polar latitudes. A development effort is underway by UNAVCO and IRIS (Prescott et al. 2006) to capitalize on these advances to build reliable power and communication systems that will enable autonomous station operation in the remotest polar locations for several years. These power/communication systems are already supporting the next generation of polar researchers, and will allow the science community to achieve the first long-duration deployment of continuously-recording GNSS and seismic stations across the Antarctic continent as well as the in the high Arctic. The paper *Technology Development for Remote GNSS Stations in Extreme Environments* highlights the technical and environmental challenges of such systems, while this paper provides a detailed look at the specific challenges and options for communication systems for remote data retrieval.

Full GPS data transfer (30 s logging rate) requires up to 1 MB/day of bandwidth and is well within the capabilities of numerous communication services (Stowers and Fisher 2006). As a result, full data communication solutions are readily obtainable for geodetic GPS deployments anywhere in the world, and many such sites are operational as part of the IPY POLENET project (Bevis et al. 2006; Wilson et al. 2006) as illustrated in Figs. 1 and 2. The large volume of onboard memory on modern GNSS receivers also allows for backup data storage on the receivers, providing protection from extended communication outages. In the event of a communication failure the data can still be retrieved manually during the next maintenance visit. Since specific services change rapidly and new capabilities evolve, this paper is intended to highlight the options for polar regions communications at the commencement of IPY.

2 Communication Options and Considerations

There are often several communication options available with fundamental differences in terms of functionality, reliability, and cost, and the appropriate solution for any particular site is dependent on site specifics. For networks one must also consider the trade-offs between optimizing the communication solution for each site vs. maintaining a standardized network to simplify operation and maintenance. For example, an Antarctic network may be designed with Iridium communications at all remote stations even if there is a site within the Inmarsat service footprint, since the complexity introduced by a one-off solution may negate any service cost savings.

Fig. 1 The remote GPS station at Minna Bluff in Antarctica (MIN0) at 79°S was installed in 2007 as a prototype for autonomous GPS installations with Iridium data retrieval. The station uses wind power to reduce the number of batteries required to operate through the winter, and the site is part of the IPY POLENET project that aims to install approximately 40 new CGPS stations throughout Antarctica by 2009

Likewise, the use of radio links with no recurring costs is a reasonable choice where a point-to-point link gains access to an Internet hub, but once a repeater is required the added cost and complexity of the repeater needs to be considered vs. adopting a satellite based link. In cases where complete standardization is unrealistic, variations in site designs (GNSS receiver selection, power system, and communication method) should be kept to an absolute minimum to minimize future operational complexity and associated costs.

In addition to evaluating the various hardware and service options, the particular communication protocols also need to be considered. Standard protocols are preferred over manufacturers' proprietary communication application and allow the data communication center to develop tools and utilities that are not unique to a specific brand of product. When possible, communication protocols should conform to the Transmission Control Protocol/Internet Protocol (TCP/IP) de-facto standard for transmitting data over the Internet. While most communication hardware can be set up for standard TCP/IP and serial communication protocols it is also desirable that the end device, in this case the GNSS receiver, can be operated over standard protocols without having to use a proprietary control software. The added complexity of running a proprietary receiver communication program over third party communication links should be avoided when possible. IP based communications are preferred over serial communications since the goal is to get data to the Internet, and several GNSS receivers now support IP communication for data access and control.

Finally the trade-off between periodic data downloading vs. data streaming should be considered. For most geoscience applications the periodic data download

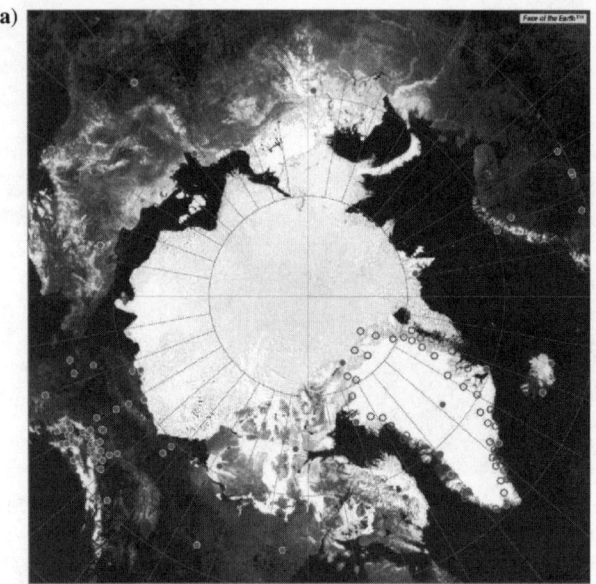

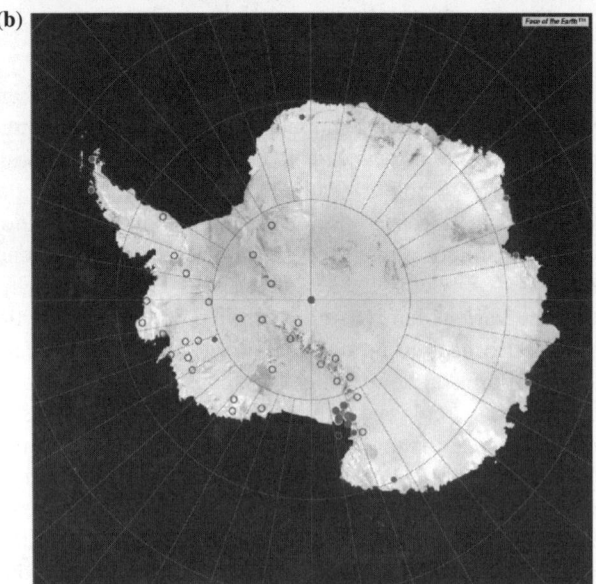

Fig. 2 New POLENET GPS stations in Antarctica (**a**) and Greenland (**b**) are shown as open circles and are pushing the limits of autonomous station operation in the extreme polar environment. The stations are being built to operate year-round with remote data retrieval

is preferred as a simple and robust method. Latency can be reduced by increasing the download frequency, and retries are straightforward in the event of partial file downloads. Advantages of streaming data include minimal data latency and the opportunity to use short-burst messaging services directly to the Internet, which may reduce operational costs and simplify the overall system. However, a means to communicate with the receiver would still need to be applied, as would a method to catch up on data recovery after eventual system outages. The most stringent latency requirements for data streaming come from real-time network applications such as GNSS virtual reference networks and other real-time differential application, and in such cases the guaranteed delivery specifications from the communications providers need to be considered.

3 Technical Characteristics

Availability, cost, performance, and reliability are the primary concerns when deciding what communication system to use. Table 1 provides a hierarchical overview of preferred communication choices, and GPS data and station performance metrics from the example stations are all available on-line from the UNAVCO data archive (facility.unavco.org). In addition to the individual communications hardware specifications, the performance of the GNSS receiver and communication hardware as a system must be well understood. System integration and testing are key activities prior to field deployment, and controlling hardware may also be used to optimize the remote operation. Custom solutions may be needed to guarantee access over the Internet when the service provider does not provide static IP addresses. Components may need to be power cycled on a regular basis to ensure robust operation, and power savings can be realized by switching the communications hardware off when not in use.

Table 1 Preferred choices for GNSS data retrieval

Situation	Solution	Concerns	Example Polar GPS station
1. The GNSS receiver can be located at an Internet (or telephone) node with available bandwidth.	Simple cable connection.	-Network security and firewall issues.	AMU2, South Pole Station, Antarctica, 90°S
2. The GNSS receiver can be located within line-of-sight to an Internet node.	Radio link between the receiver and the Internet.	-Radio and antenna reliability. -Radio power requirements. -Network security and firewall issues.	CONZ, Mt. Erebus, Antarctica, 78°S (Intuicom wireless Ethernet bridge modems)

Table 1 (continued)

Situation	Solution	Concerns	Example Polar GPS station
3. A repeater is required to establish a line-of-sight link to an Internet node.	Radio link to repeater to the Internet.	-Radio and antenna reliability. -Radio power requirements -Repeater adds complexity of another remote instrument and power system. -Network security and firewall issues.	ROB4, Cape Roberts, Antarctica, 77°S (FreeWave wireless serial modems)
4. The GNSS receiver is in the footprint of a commercial terrestrial wireless communication provider.	Connect via the appropriate network modem.	-Modem system reliability. -Modem power requirements. -Recurring costs for service.	GRNX, Healy, Alaska, 64°N (Proxicast GPRS modem)
5. The GNSS receiver is in the footprint of a commercial geosynchronous satellite wireless communication provider.	Connect via the appropriate network modem.	-Modem system reliability. -Modem power requirements. -Recurring costs for service.	SUMM, Summit Camp, Greenland, 72°N (Harris Maritime Communications Service VSAT)
6. None of the above situations apply.	Connect via the Iridium satellite constellation.	-Limited bandwidth. -Modem system reliability. -Modem power requirements. -Recurring costs for service. -Interference with GPS tracking.	FTP4, Fishtail Point, Antarctica, 79°S (NAL Research Iridium modem)

3.1 Private Radio Link

Private radio links are commonly used to address the "last mile" problem when the GNSS receiver can be located within line-of-sight to an Internet node or repeater. Many options exist including WiFi links and commercial narrowband and spread spectrum radios. Radios by FreeWave and Intuicom have been used in the Antarctic environment for several years with robust reliable performance over distances up to 100 miles between mountain peaks.

Communications Systems for Remote Polar GNSS Station Operation

Private radio link characteristics

Data rate:	100 kbps – 50 mbps typical.
Power consumption:	1–5 W typical.
Hardware cost:	$500 to $2000 typical per radio depending on specific system used.
Recurring service cost:	None.
Modes of use:	Line-of-sight links for point-to-point, point-multipoint communications. Repeaters can be used to mitigate challenging terrain. Can support two way communications, data downloads, and data streaming.
Area of availability:	Global – off the shelf hardware is available for all operating environment. Radio frequency control restriction are common and use must be coordinated with the local frequency spectrum controlling authority.
Main advantages:	Flexible and can be adapted to the specific science instrumentation needs. Owned and fully controlled by the project. No recurring costs.
Main disadvantages:	Limited to line-of-sight communications. User must set up infrastructure.

3.2 Commercial Terrestrial Wireless

Wireless terrestrial based third generation mobile communications services (G3) such as General Packet Radio Service (GPRS) are common infrastructure in populated areas. Such services can be of use in the Arctic at sites near villages with wireless Internet service. For example, GPRS service is currently available from Tele-Greenland over the Global System for Mobile Communications (GSM)

Commercial terrestrial wireless characteristics

Data rate:	15–80 kbps typical.
Power consumption:	2–6 W typical.
Hardware cost:	$1000 typical.
Recurring service cost:	$60/month typical.
Modes of use:	Provides on-site Internet access for data downloads, data streaming, and system control.
Area of availability:	Within commercial telecommunications service areas. Near larger towns and villages in the Arctic. No service in the Antarctic.
Main advantages:	Internet access at the site using existing infrastructure.
Main disadvantages:	Limited availability in remote areas. Recurring costs.

network that currently covers the larger Greenland villages. The GPRS network is an "always on", private network for data. It uses the existing GSM network to transmit and receive TCP/IP based data to and from GPRS mobile devices.

3.3 Commercial Geosynchronous Earth Orbit (GEO) Satellite Wireless

Very Small Aperture Terminal (VSAT) networks and the Inmarsat service use GEO satellites to communicate with remote sites. Various VSAT solutions are available depending on the area of interest, while Inmarsat provides a more global service. GEO satellites provide near global coverage, but they do not cover the highest latitudes and service becomes very limited above 70 degrees. Within the satellite footprints, GEO satellite based systems can be used to transfer GNSS data in near real time in areas where no ground based communication service exists.

3.3.1 VSAT Satellite Wireless

VSAT satellite space segment providers offer three types of satellite beam: spot, hemisphere, and global. Spot beams are focused on population centers and are generally high power, thus allowing smaller antenna dishes to be used at remote sites. Hemisphere and global beams have a much larger footprint and weaker signal strength and thus requires a larger antenna diameter relative to spot beam systems (Jackson et al. 2001). VSAT may provide a cost effective solution when shared with other users, but the large power consumption and large dish size makes it less desirable for standalone GNSS applications.

GEO-VSAT characteristics

Data rate:	150 kbps typical.
Power consumption:	20 W typical.
Hardware cost:	$1000 typical.
Recurring service cost:	$150/month typical.
Modes of use:	Provides on-site Internet access for data downloads, data streaming, and system control.
Area of availability:	Most populated regions and parts of the high Arctic.
Main advantages:	Direct, high bandwidth Internet access to the site.
Main disadvantages:	Limited availability in high latitude polar regions. Large power requirements. Recurring costs. Directional antenna may require set-up expertise. Large antenna dish can be problematic.

3.3.2 Inmarsat

Inmarsat's Broadband Global Area Network (BGAN) provides broadband data through a single device on a global basis, including limited coverage of the polar regions (Fig. 3). This new service supports both standard and streaming IP communications, and appears promising for near-global GNSS data telemetry. As with all GEO satellites, coverage is not available at the highest latitude locations and the planned 3rd F3 satellite is needed to cover the current service gap over the Pacific, including Alaska, eastern Siberia, and the Terre Adelie coast of Antarctica.

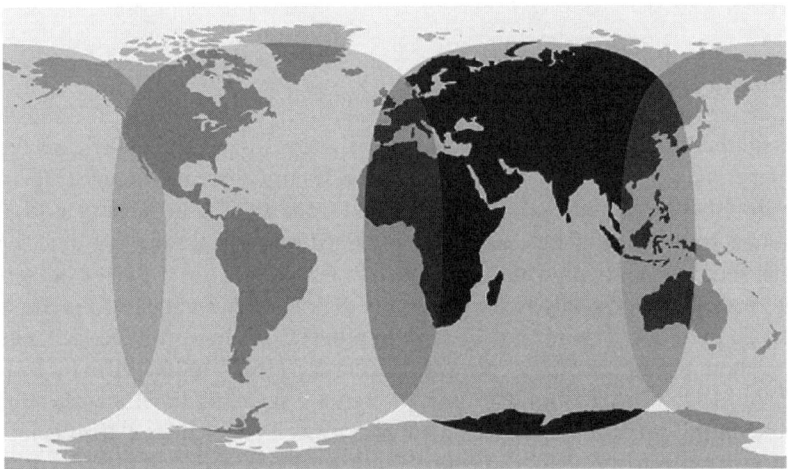

Fig. 3 Inmarsat BGAN coverage (broadband.inmarsat.com). The planned third satellite over the Pacific Ocean (*light blue footprint*) has not yet been launched but would cover Alaska and the east Siberian arctic

GEO-Inmarsat characteristics

Data rate:	60 kbps typical.
Power consumption:	1 W typical.
Hardware cost:	$200 to $2000 per radio depending on specific system used.
Recurring service cost:	$250/month for 1 MB/day data retrieval.
Modes of use:	Provides on-site Internet access for data downloads, data streaming, and system control.
Area of availability:	Global (with F3 satellite), except for highest latitudes.
Main advantages:	Reasonable cost and power consumption for polar application. IP based service direct to site.
Main disadvantages:	New service that has not yet been proven for GNSS data applications. Recurring costs. Does not cover the highest latitudes.

3.4 Commercial Low-Earth Orbit (LEO) Satellite Wireless

3.4.1 Globalstar

The Globalstar system relies on a LEO satellite constellation to provide global voice and data services. Data are routed from the user to the gateway hub via a commonly visible satellite. As a result of this "bent pipe" system, the distribution of coverage is dictated by the hub locations, and the coverage is similar to that of GEO satellite systems and sparse in the polar regions. The system provides single channel data rates of 9.6 kbps.

3.4.2 Iridium

The 66 satellite Iridium constellation is the world's largest satellite network and the only commercial communication system that provides truly global coverage (Fig. 4) suitable for GNSS data retrieval. Data are routed from satellite to satellite as necessary to reach the gateway hub. Hardware provided by NAL Research provides a traditional serial modem link with AT commands as well as a Short Burst Data service with packets sent directly to an IP address. System enhancements to increase data rates and provide IP based services are in progress (Colussy 2006). As a note of caution, both Iridium and GPS operate in the L-band (1616–1626.5 MHz for Iridium, 1227.6, 1575.42 MHz for GPS), and interference resulting in GPS cycle slips is possible during Iridium transmission. Antenna separation of at least 10 m is recommended to mitigate this. Dropped connections are also frequent occurrences with the Iridium system, and the data download system must be set up to handle this.

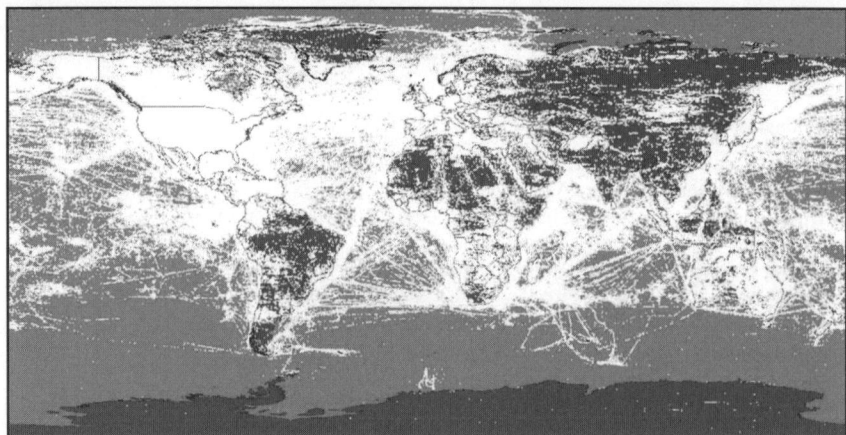

Fig. 4 Iridium voice and data activity during July 2006. Iridium is the only commercial communication system that provides truly global coverage suitable for GNSS data retrieval (Courtesy form Iridium Satellite - copyright Iridium Satellite)

LEO-Iridium characteristics

Data rate:	2.4 kbps.
Power consumption:	1 W typical.
Hardware cost:	$1500 per radio depending on specific system used.
Recurring service cost:	$400 to $3000 per month.
Modes of use:	Global stand-alone system. Supports two way communications, data downloads, data streaming, and short burst data packaging.
Area of availability:	Global – off the shelf hardware is available for all operating environments.
Main advantages:	Truly global. Low power consumption. Small omni-directional antenna for easy set-up.
Main disadvantages:	Low bandwidth. High recurring cost. Does not support 2-way IP based communications.

4 Conclusion

The convergence of reliable remote polar instrumentation, the International Polar Year, and the social relevance of the polar regions to climate change provides an exciting time to take the next "great-leap forward" in the understanding of the nature and behavior of the crust and mantle beneath Antarctica and Greenland and their relationship to ice sheet dynamics and climate. Data communications to the remote instruments promise to optimize and limit the reliance on expensive aircraft logistics, deliver data to users with very little latency, and allow for the almost immediate interpretation. As networks are built and dataflow begins, automated data analysis systems will follow allowing a level of scientific interpretation previously unobtainable. With data routed directly to data archives, they also become accessible to a much larger audience and often benefit diverse science disciplines. One of the anticipated objectives of the IPY is the establishment of enhanced infrastructure to benefit future science needs, and delivering GNSS data from the previously poorly covered remote polar regions will create a lasting legacy.

5 Web Links for Global Communication Solutions

FreeWave – www.freewave.com
Globalstar – www.globalstar.com
Inmarsat – broadband.inmarsat.com
Intuicom – www.intuicom.com
Iridium – www.iridium.com
NAL Research – www.nalresearch.com

Acknowledgments This paper describes the result of efforts by many in the geosciences research community. The author's part of this work was carried out at UNAVCO with considerable input from experts within the EarthScope Plate Boundary Observatory and Facility activities funded by the U.S. National Science Foundation Earth Sciences (NSF-EAR) and Polar Programs (NSF-OPP), and NASA Earth sciences. Special thanks to Gary Ferentchak of Raytheon Polar Services Company for assistance with developing a robust Iridium GPS data download method, Roy Stehle of SRI International for contributions regarding Arctic communications, and the POLENET and Mt. Erebus Volcano Observatoty (MEVO) teams (Larry Hothem, Mike Willis and Terry Wilson, POLENET; Rick Aster, Philip Kyle, and Bill MacIntosh, MEVO) for collaboration over the years in developing robust solutions in the Antarctic.

References

Bevis M, Csatho B, Van der Veen C, Wilson T (2006), IPY: POLENET/Greenland: Using bedrock geodesy to constrain past and present day changes in Greenland's ice mass, NSF-ARC 0632320

Capra A, Mancini F, Negusini M (2007), GPS as a geodetic tool for geodynamics in Northern Victoria Land, Antarctica, Antarctic Science, 19, pp. 107–114

Colussy D (2006), Communications in the great north: Alaska, the Arctic and global communications, Commonwealth North workshop proceedings, Anchorage, August 29

Dietrich R, Dach R, Engelhardt G, Ihde J, Korth W, Kutterer HJ, Lindner K, Mayer M, Menge F, Miller H, Muller C, Niemeier W, Perlt J, Pohl M, Salbach H, Schenke HW, Schone T, Seeber G, Veit A, Volksen C (2001), ITRF coordinates and plate velocities from repeated GPS campaigns in Antarctica-an analysis based on different individual solutions, Journal of Geodesy, 74, pp. 756–766

Jackson M, Meertens C, Ruud O, Reeder S, Gallaher W, Rocken C (2001), Real time GPS data transmission using VSAT technology, GPS Solutions, 5(4), pp. 10–19

Prescott W, Anderson K, Johns B, Simpson D (2006), Collaborative research: Development of a power and communication system for remote autonomous GPS and seismic stations in Antarctica, NSF-ANT 06199081

Stowers D, Fisher S (2006), GNSS Stations for Geodetic Applications, African Geodetic Reference Frame expert meeting proceedings, Cape Town, July 10–13

Willis M, Wilson T, Hothem L (2006), A decade of GPS measurements over the TAMDEF network, Victoria Land, Antarctica. European Geosciences Union geophysical research abstracts, 8, Vienna, April 2–7

Wilson T, Aster R, Bevis M, Dalziel I, Nyblade A, Raymond C, Smalley R, Wiens D (2006), IPY: POLENET-Antarctica: Investigating links between geodynamics and ice sheets, NSF-ANT 0632322

Current Status and Future Prospects for the Australian Antarctic Geodetic Network

Gary Johnston, Nicholas Brown and Michael Moore

Abstract The Australian Antarctic Geodetic Network consists of over 400 survey marks installed since the 1960's. This multipurpose network provides pivotal information for the International Terrestrial Reference Frame and allows us to monitor the movement of tectonic plates and the intraplate movement of the Antarctic continent. Furthermore, it supports many Australian science initiatives including post glacial rebound studies, rifting in the Amery Ice Shelf and geological mapping. Originally established using terrestrial surveying techniques, the accuracy of Network has improved over time with the help of space based techniques and in the future, as more continuous GPS sites become active, the potential science outcomes will increase significantly.

1 Background

The Australian Antarctic Geodetic Network (AAGN) (Geoscience Australia, 2001) is an essential component of the global reference frame which supports science programs such as plate tectonics, sea level rise, post-glacial rebound and geospatial activities such as mapping. Over the years, geodetic survey marks have been installed in rock outcrops and observed using both conventional surveying and space geodetic techniques (Fig. 1).

Major installation campaigns took place in the northern Prince Charles Mountains (PCM's) from 1965 to 1976, the southern PCM's in the 1970's which included Mt Komsomolskiy, the most southerly exposed rock site in the PCM's. Observations to these sites were performed using intersecting angles and loops of angle and distance observations, some of which were greater than 100 km long.

Gary Johnston
Geoscience Australia, GP0 Box 378, Canberra, ACT 2601

Nicholas Brown
Geoscience Australia, GP0 Box 378, Canberra, ACT 2601

Michael Moore
Geoscience Australia, GP0 Box 378, Canberra, ACT 2601

Fig. 1 Classical terrestrial surveying techniques (*left*) and GPS space based technique (*right*)

In more recent years, space geodetic observations, firstly Doppler NAVSTAR techniques and then the Global Positioning System (GPS) have been used to strengthen the geodetic network. GPS negated the need for stations intervisibility and consequently baseline lengths could be increased until intercontinental baselines were achievable.

Since 1990, GPS has been the primary tool for geodetic surveying in the AAGN. Originally it was used in an exclusively baseline mode over relatively short distances to improve to the accuracy of the existing network, and more recently it has allowed the establishment of new stations in a short period of time.

2 Status of Australian Antarctic Geodetic Network

The AAGN (Fig. 2) is comprised of 403 survey marks which extend from Proclamation Island (65°50′S) to Mt. Komsomolskiy (75°16′S) and from Campbell Nunatak (110°45′E) to Widows (45°27′E). The horizontal positional uncertainty (Table 1) of these survey marks varies greatly across the area due to the manner in which they have been observed, that is, by conventional techniques or by GPS.

In late 1993 Continuous GPS (CGPS) sites were established at Casey, Davis and Mawson (Fig. 3) and in the June of 1995, Macquarie Island was introduced as the fourth CGPS site in the AAGN (Geoscience Australia, 2008). Data from these CGPS sites is transferred back to Australia for distribution to the international community, in particular the International GNSS Service (IGS) (IGS, 2008).

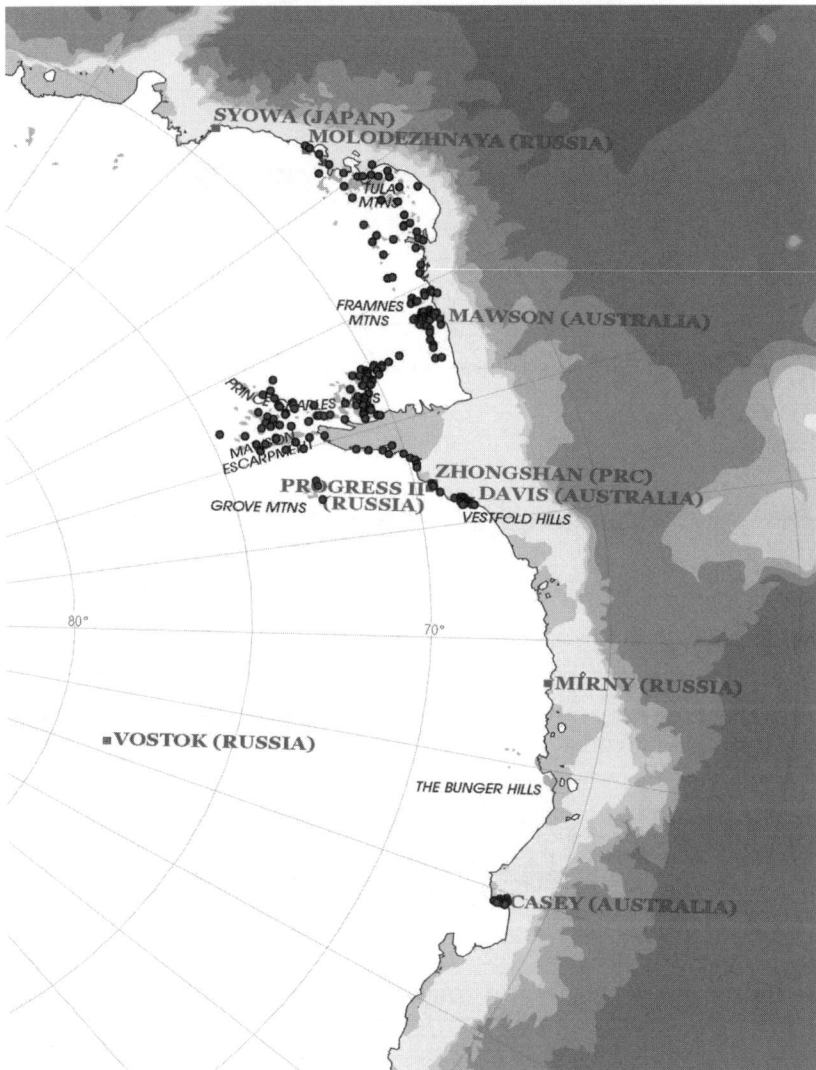

Fig. 2 The AAGN

Table 1 Summary of horizontal positional uncertainties of all AAGN sites

Horizontal Positional Uncertainty (m)	Number of Sites
< 0.01	51
0.02–0.05	29
0.06–0.10	51
0.11–0.50	86
0.51–2.00	85
> 2.00	101

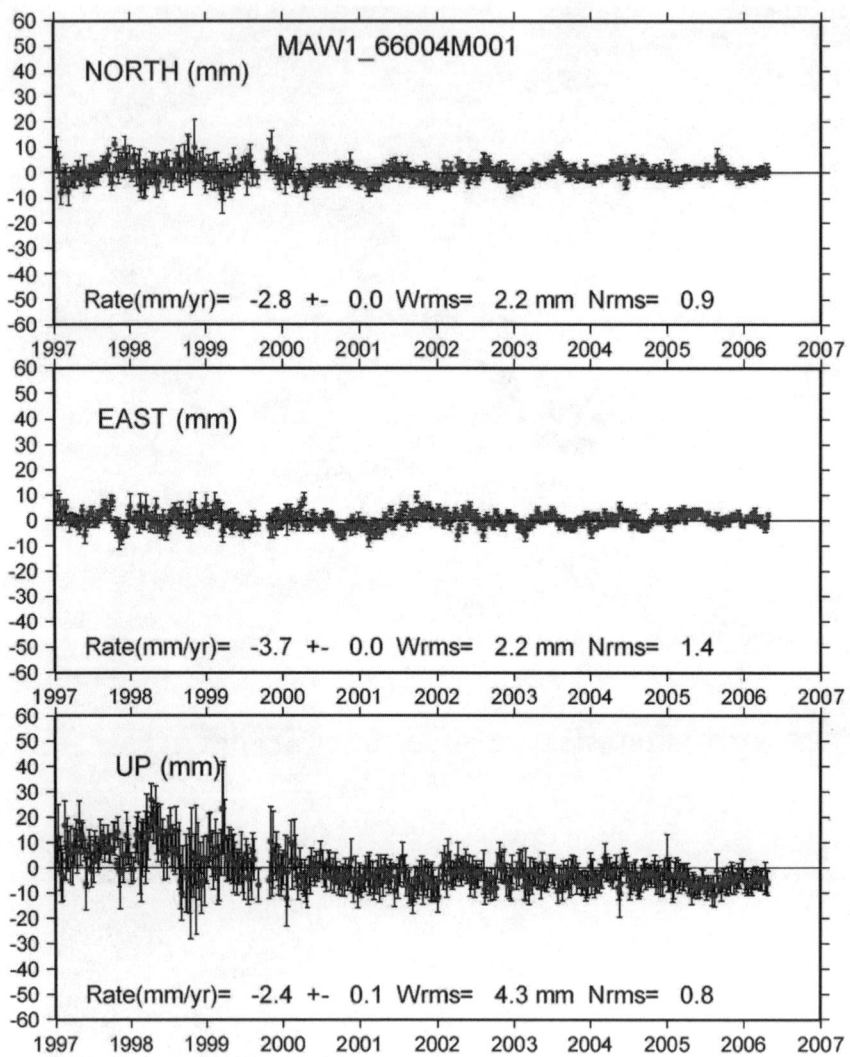

Fig. 3 GPS time series from Casey, Davis and Mawson CGPS stations from 1997 to 2007

Data from these sites is processed by the IGS analysis centres to determine coordinates for the stations in the latest International Terrestrial Reference Frame (ITRF, 2008) (currently ITRF2005). The remainder of the AAGN is constrained to the ITRF coordinates of these four stations and adjusted using Least Squares techniques, thus propagating ITRF2005 coordinates through the network.

In recent years, Geoscience Australia and The Australian National University has been attempting to augment the CGPS network with pseudo / part-time CGPS sites in the Grove Mountains (2000–2007), (Fig. 4) Wilson Bluff (2002–2007), Landing Bluff (2000–2004), Dalton Corner (2000), The Bunger Hills (2006–2007) and

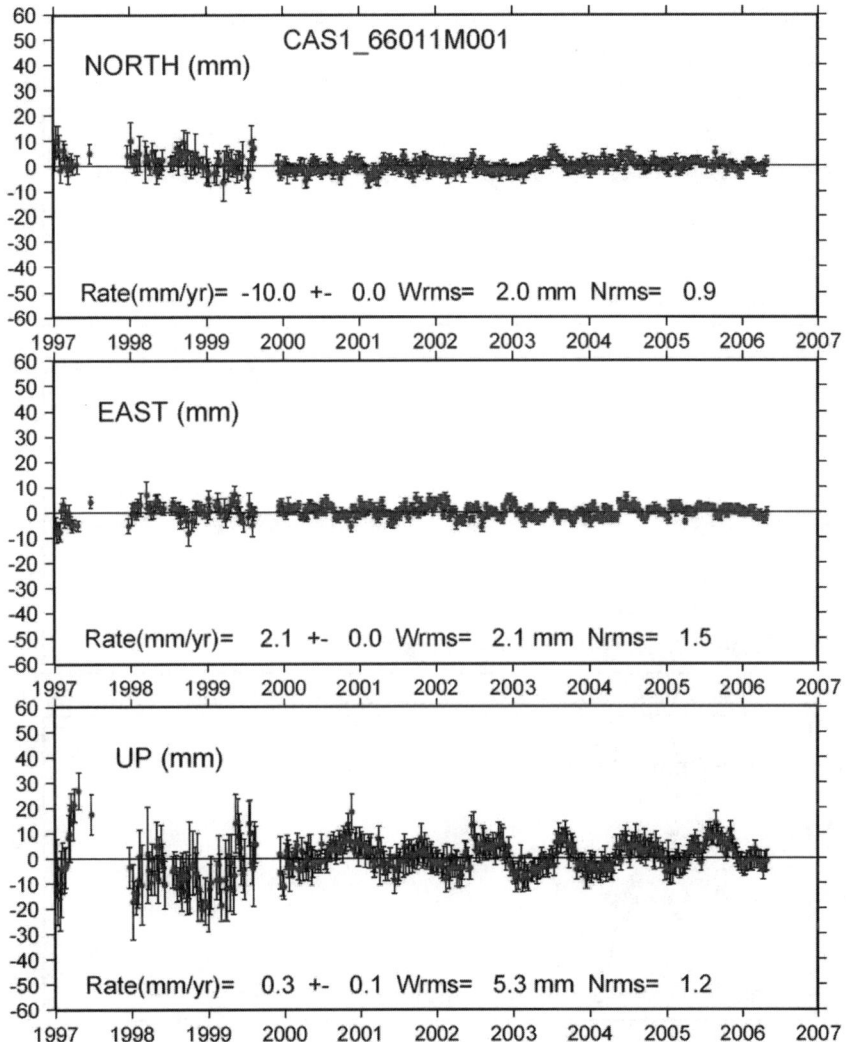

Fig. 3 (continued)

Richardson Lake (2006–2007). To date, the sites operate predominantly over the Austral summer period when solar power maintains battery voltage; however, sustaining power supply in the winter months continues to be difficult. Geoscience Australia has attempted to use wind turbines to charge the batteries with little success to date.

Since 1997, Geoscience Australia has also been undertaking GPS campaigns to upgrade the AAGN. These campaigns have concentrated on the southern PCM's

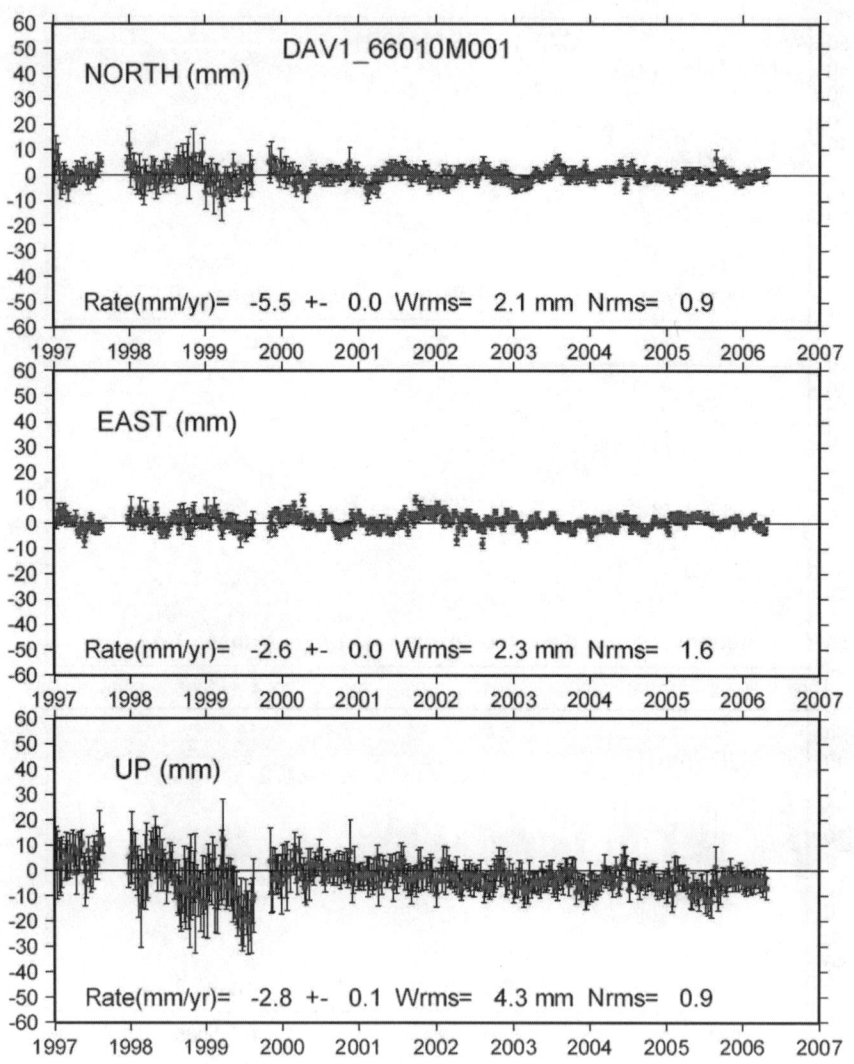

Fig. 3 (continued)

where large coordinate uncertainties exist due to cumulative propagation errors from the constrained sites at Davis and Mawson.

The biggest of these campaigns was the Prince Charles Mountains Expedition of Germany and Australia (PCMEGA). A network of twenty-one accurate geodetic/geodynamic control points was established over the southern PCM's, a number of which were placed adjacent to existing AAGN stations and connected using terrestrial observations (Fig. 5). Geodetic GPS observations from these stations have significantly improved the accuracy of the AAGN and it will allow neotectonic

Fig. 4 Pseudo/Part time continuous GPS sites at Grove Mountains and Bunger Hills

research to be undertaken once a second epoch is observed. A summary of CGPS and campaign GPS data observed in the AAGN to date is shown in Table 2.

Gravity observations were also taken at twenty-four stations during PCMEGA and connected back to the Mawson and Davis gravity network marks (Geoscience Australia, 2004). These results permit ground calibration for the airborne gravity

Table 2 Geoscience Australia's GPS data catalogue

Region	1995	1998	2000	2001	2002	2003	2004	2005	2006	2007
Amery Ice Shelf				■						
Bunger Hills									■	
Casey	■	■	■	■	■	■	■	■	■	■
Davis	■	■	■	■	■	■	■	■	■	■
Richardson Lake										■
Grove Mountains			■							
Heard Island	■					■				
Landing Bluff										
Larsemann Hills		■								
Macquarie Island	■	■	■	■	■	■	■	■	■	■
Mawson	■	■	■	■	■	■	■	■	■	■
PCM's					■	■				
Rauer Islands							■			

Fig. 5 PCMEGA GPS observations have strengthened the geodetic network

surveys as well as a densification of the gravity network which is very sparse over this part of Antarctica.

3 Vestfold Hills Geometric Geoid Determination

The Vestfold Hills is a worksite and playground for all expeditioners at Davis station. Those venturing into the Vestfold Hills from Davis station are commonly using GPS as a tool for science and navigation. Given that GPS receivers only capture the height above the ellipsoid, Fig. 6 displays the need to convert this value to the orthometric height, that is, a more practical height which approximates the height above means sea level (MSL).

The height above MSL (H) is closely approximated by the ellipsoidal height (h) (acquired by GPS) minus the geoid to ellipsoid separation (N).

Due to the sparseness of ground gravity data and geometric control over Antarctica, a local gravimetric geoid model is not readily available. Currently, the best geoid model available is the Earth Geopotential Model 1996 (EGM96) (NASA, 2004), however, it has inaccuracies due to the fact that most of the satellites utilised for data capture were not polar orbiting and geometric control of the satellite orbits is weakest in the southern latitudes. The consequence is an inaccurate gravimetric geoid model over Antarctica.

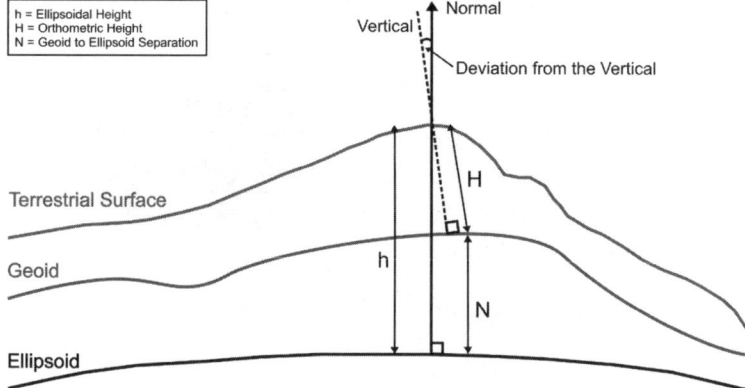

Fig. 6 Offsets between the ellipsoid, geoid and terrestrial surface

The Vestfold Hills contain many saline and freshwater lakes. To facilitate the accurate measurement of their heights 90 benchmarks were placed in the vicinity of each of these lakes through the 1980s and the majority of them have been connected to the Davis height datum using third or fourth order-levelling techniques.

These benchmarks have been used to create a geometric geoid. GPS observations were recorded on 23 of these benchmarks by Geoscience Australia and students from the University of Tasmania to compute the ellipsoidal heights. The geoid to ellipsoid separation value is derived by subtracting the orthometric height (observed by levelling) from the ellipsoidal height.

The derived geoid to ellipsoid separation values were used to develop a geometric geoid over the Vestfold Hills (Fig. 7). The benchmarks closer to the plateau (eastern side of Fig. 7) have not been connected to the Davis height datum and as a result the grid is weaker in that part of the Vestfold Hills, however, it is a more accurate representation of the geoid to ellipsoid separation values than the EGM96 (Fig. 8) which differs from the geometric model by approximately two meters near the plateau. This shift is possibly due to a lack of data in the EGM96 model.

This is only the first attempt at creating a geometric geoid for the Vestfold Hills and will undoubtedly be strengthened in future seasons as more of these lake benchmarks are coordinated using accurate GPS techniques.

4 Future Prospects

Increasingly, users of the AAGN are expecting it to be an active network rather then a passive network consisting of unoccupied survey marks. Already researchers are using GPS equipment of their own to position themselves within the framework of the ITRF as realised through the CGPS sites at Casey, Davis and Mawson. Therefore, the need to enhance the accuracy of the existing marks based control

VESTFOLD HILLS GEOMETRIC GEOID

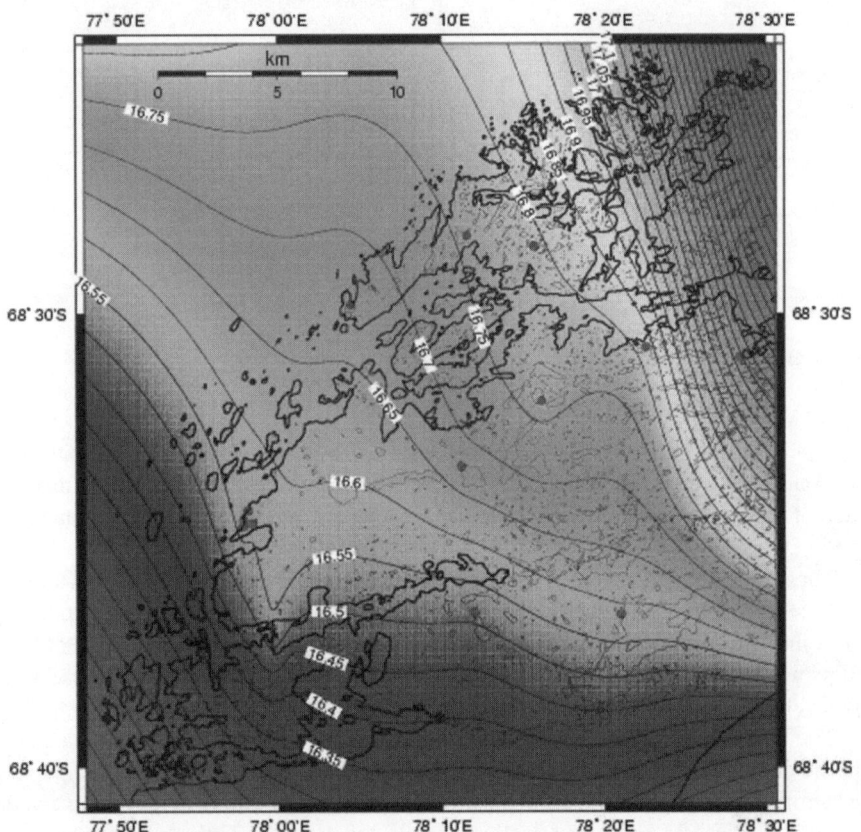

Fig. 7 Vestfold Hills Geometric Geoid

network is diminishing and the need for active Global Navigational Satellite System (GNSS) stations whose data is available to the broader research community is increasing.

Research fields such as neotectonics and post glacial isostatic rebound require accuracies only achievable using CGPS. Effort needs to be placed in developing autonomous remotely operating systems that operate reliably over both summer and winter periods. This will also allow the study of the seasonal effects on tectonic motion not to mention a large variety of glaciological topics. In addition to CGPS, site targeted campaigns of the type undertaken during PCMEGA will continue with the aim of direct geophysical interpretation rather the network enhancement.

VESTFOLD HILLS EGM96 GEOID

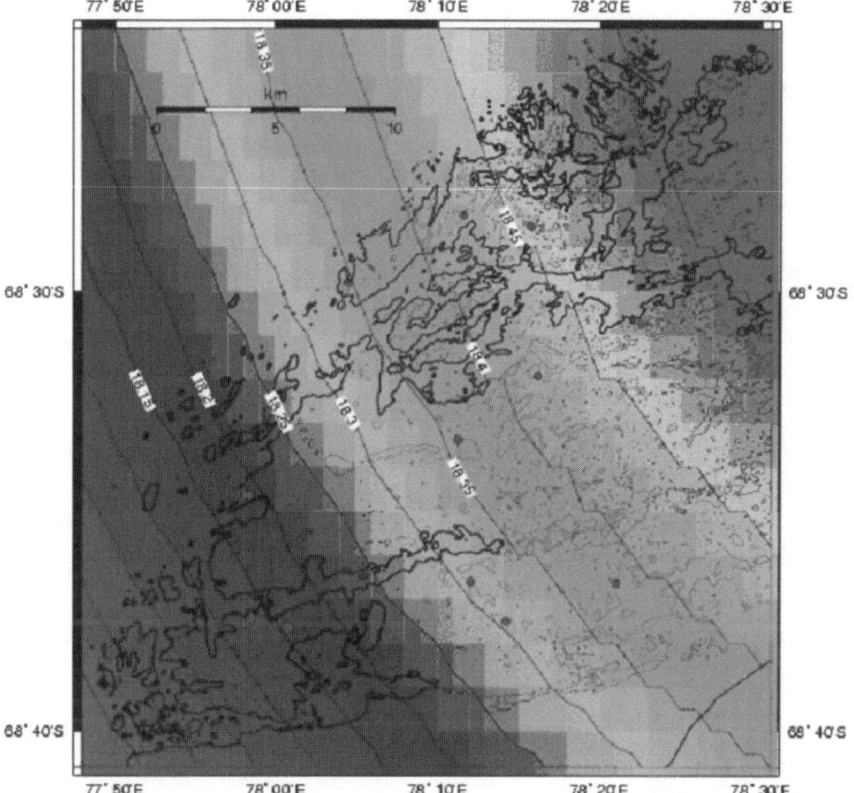

Fig. 8 Vestfold Hills EGM96 Geoid

The integration of GNSS with gravimetry and satellite based observations like GRACE (NASA, 2008) and InSAR will reveal geodetic accuracies not previously imagined.

5 Conclusion

Geodetic networks are a continually evolving infrastructure. As user requirements change and new techniques become available the design and functionality of the network needs to change as well. In the near future, there will be an increasing demand for a more active GNSS network and there is an emerging interest in the use of gravimetry as an independent constraint on height variation. Finally, the integration of all ground based techniques with the satellite techniques will result in a better understanding of the dynamics of Antarctica at local, regional and continental scales.

References

Geoscience Australia, 2001, Australian Antarctic Geodetic Network, viewed 26 March 2008, http://www.ga.gov.au/geodesy/antarc/aagn.txt

Geoscience Australia, 2004, Airborne Gravity 2004, Abstracts from the ASEG-PESA Airborne Gravity 2004 Workshop, viewed 26 March 2008, http://www.ga.gov.au/about/ corporate/ga_authors/ga2004_18/Ga_2004_18_abs12.jsp

Geoscience Australia, 2008, Permanent GPS Tracking Sites in Antarctica, viewed 26 March 2008, http://www.ga.gov.au/geodesy/antarc/antgps.jsp

International GNSS Service (IGS), 2008, International GNSS Service Homepage, viewed 26 March 2008, http://igscb.jpl.nasa.gov/

International Terrestrial Reference Frame (ITRF), 2008, International Reference System Frame Homepage, viewed 26 March 2008, http://itrf.ensg.ign.fr/

NASA, 2004, The NASA GSFC and NIMA Joint Geopotential Model, viewed 26 March 2008, http://cddis.nasa.gov/926/egm96/egm96.html

NASA, 2008, Gravity Recovery and Climate Experiment, viewed 26 March 2008, http://www.csr.utexas.edu/grace/

Geodetic Research on Deception Island and its Environment (South Shetland Islands, Bransfield Sea and Antarctic Peninsula) During Spanish Antarctic Campaigns (1987–2007)

M. Berrocoso, A. Fernández-Ros, M.E. Ramírez, J.M. Salamanca, C. Torrecillas,
A. Pérez-Peña, R. Páez, A. García-García, Y. Jiménez-Teja, F. García-García,
R. Soto, J. Gárate, J. Martín-Davila, A. Sánchez-Alzola, A. de Gil,
J.A. Fernández-Prada and B. Jigena

M. Berrocoso
Laboratorio de Astronomía, Geodesia y Cartografía. Departamento de Matemáticas.
Facultad de Ciencias. Campus de Puerto Real. Universidad de
Cádiz. 11510 Puerto Real (Cádiz-Andalucía). España, e-mail: manuel.berrocoso@uca.es

A. Fernández-Ros
Laboratorio de Astronomía, Geodesia y Cartografía. Departamento de Matemáticas.
Facultad de Ciencias. Campus de Puerto Real. Universidad de Cádiz. 11510 Puerto Real
(Cádiz-Andalucía). España

M.E. Ramírez
Laboratorio de Astronomía, Geodesia y Cartografía. Departamento de Matemáticas.
Facultad de Ciencias. Campus de Puerto Real. Universidad de
Cádiz. 11510 Puerto Real (Cádiz-Andalucía). España

J.M. Salamanca
Laboratorio de Astronomía, Geodesia y Cartografía. Departamento de Matemáticas.
Facultad de Ciencias. Campus de Puerto Real. Universidad de
Cádiz. 11510 Puerto Real (Cádiz-Andalucía). España

C. Torrecillas
Laboratorio de Astronomía, Geodesia y Cartografía. Departamento de Matemáticas.
Facultad de Ciencias. Campus de Puerto Real. Universidad de
Cádiz. 11510 Puerto Real (Cádiz-Andalucía). España.

A. Pérez-Peña
Laboratorio de Astronomía, Geodesia y Cartografía. Departamento de Matemáticas.
Facultad de Ciencias. Campus de Puerto Real. Universidad de Cádiz. 11510 Puerto Real
(Cádiz-Andalucía). España

R. Páez
Laboratorio de Astronomía, Geodesia y Cartografía. Departamento de Matemáticas.
Facultad de Ciencias. Campus de Puerto Real. Universidad de Cádiz. 11510 Puerto Real
(Cádiz-Andalucía). España

A. García-García
Departamento de Volcanología. Museo Nacional de Ciencias Naturales. Consejo Superior
de Investigaciones Científicas. C/ José Gutiérrez Abascal, 2. Madrid

Y. Jiménez-Teja
Laboratorio de Astronomía, Geodesia y Cartografía. Departamento de Matemáticas.
Facultad de Ciencias. Campus de Puerto Real. Universidad de Cádiz. 11510 Puerto Real
(Cádiz-Andalucía). España

Abstract Since 1987, Spain has been continuously developing several scientific projects, mainly based on Earth Sciences, in Geodesy, Geochemistry, Geology or Volcanology. The need of a geodetic reference frame when doing hydrographic and topographic mapping meant the organization of the earlier campaigns with the main goals of updating the existing cartography and of making new maps of the area. During this period of time, new techniques arose in Space Geodesy improving the classical methodology and making possible its applications to other different fields such as tectonic or volcanism. Spanish Antarctic Geodetic activities from the 1987–1988 to 2006–2007 campaigns are described as well as a geodetic and a levelling network are presented. The first network, RGAE, was designed and established to define a reference frame in the region formed by the South Shetlands Islands, the Bransfield Sea and the Antarctic Peninsula whereas the second one, REGID, was planned to control the volcanic activity in Deception Island. Finally, the horizontal and vertical deformation models are described too, as well as the strategy which has been followed when computing an experimental geoid.

F. García-García
Escuela Superior de Ingeniería Cartográfica y Geodésica. Universidad Politécnica de Valencia

R. Soto
Servicio de Satélites. Sección de Geofísica. Real Instituto y Observatorio de la Armada.
San Fernando (Cádiz, España)

J. Gárate
Servicio de Satélites. Sección de Geofísica. Real Instituto y Observatorio de la Armada.
San Fernando (Cádiz, España)

J. Martín-Davila
Servicio de Satélites. Sección de Geofísica. Real Instituto y Observatorio de la Armada.
San Fernando (Cádiz, España)

A. Sánchez-Alzola
Laboratorio de Astronomía, Geodesia y Cartografía. Departamento de Matemáticas.
Facultad de Ciencias. Campus de Puerto Real. Universidad de Cádiz. 11510 Puerto Real
(Cádiz-Andalucía). España

A. de Gil
Laboratorio de Astronomía, Geodesia y Cartografía. Departamento de Matemáticas.
Facultad de Ciencias. Campus de Puerto Real. Universidad de Cádiz. 11510 Puerto Real
(Cádiz-Andalucía). España

J.A. Fernández-Prada
Laboratorio de Astronomía, Geodesia y Cartografía. Departamento de Matemáticas.
Facultad de Ciencias. Campus de Puerto Real. Universidad de Cádiz. 11510 Puerto Real
(Cádiz-Andalucía). España

B. Jigena
Laboratorio de Astronomía, Geodesia y Cartografía. Departamento de Matemáticas.
Facultad de Ciencias. Campus de Puerto Real. Universidad de
Cádiz. 11510 Puerto Real (Cádiz-Andalucía). España

1 Geodetic Reference Frame for South Shetland Island, Bransfield Sea and Antarctic Peninsula

Meteorological, oceanographic, geophysical, geodynamic, biological, glaciology, etc., all require reference frame that establishes a precise time and space of the data. Due to its geographical isolation, realising accurate reference frames in Antarctica has been a large challenge.

The beginning of the Global Positioning System (GPS) and its functionality on April 27, 1985, constituted an important improvement in the productivity and the obtained precision in relation to other spatial positioning systems. It also meant a reduction in the cost of the equipments and field programmes. GPS is based in the interferometric principle of simultaneous observations of NAVSTAR constellation, which provides a reference system called WGS-84 ellipsoid (NIMA 2000), and obtain a relative positioning between the stations with precisions of 1 ppm. To obtain absolute precisions, a first level network formed by geodesic stations situated near VLBI stations is established. Later, a relative positioning of the stations is realized in respect of the stations which belong to that network. This positioning gives absolute coordinates using a later adjustment.

The use of the reference ellipsoid WGS-84, implies the unify of the local reference systems established in different zones of the Earth; In the other hand, the ellipsoid coordinates of the stations are obtained, and later, the values of vertical deviation and geoid undulation are calculated. These values allow obtaining the astronomical coordinates and their respective ortometric altitudes for each station.

In the Antarctic environment (Berrocoso 1997), the lack of a local datum, obtained by direct observation, and therefore, a local ellipsoid, implied that the adoption of global ellipsoid WGS-84 presented some disadvantages. In the Antarctic area we can not obtain transformation parameters between the global and the local ellipsoid for use it to secondary stations.

In 1989–1990 campaign, a fundamental geodetic station was constructed in Juan Carlos I Antarctic Base (Livingston Island), by means of monitoring TRANSIT satellites. The absolute coordinates of this station were calculated by applying Doppler precise point positioning, referred to the WGS-72 Geodetic System (NIMA 2000). These coordinates were transformed to WGS-84 to calculate the coordinates of the other stations by means of relative GPS positioning. During 1989–1990 campaign, a prolongation of the South American Geodetic Network was carried out. A link between Tierra del Fuego and South Shetland Island was made using GPS measurements, where a point constructed in Rio Grande Astronomical Station was elected as the principal point of the link.

The validity of this work was extended and improved during the austral summer of 1991–92, with the accomplishment of the international campaign SCAR'92 (*Scientific Commission Antarctic Researches*). This campaign allowed the establishment of the final Global Geodetic Reference Frame in Antarctic in Geodetic Level A (Seeber 2003) whose station coordinates was provided with accuracy of the order of centimetres in its absolute coordinates.

1.1 The Spanish Antarctic Geodetic Network

The Spanish Antarctic Geodetic Network, RGAE network consists of several stations around the South Shetland Island, the Bransfield Sea and the Antarctic Peninsula. These stations have precise absolute coordinates and they are referred to WGS84 ellipsoid in the ITRF reference frame corresponding. To the initial objective of being the regional geodetic reference frame (due to the accuracies reached in the relative positioning), the objective of establishment the geodynamical frame was added. The GPS surveying of this network will provide the tectonic behaviour of the region.

1.1.1 Design and Evolution of the Geodetic Network RGAE

The design of the geodetic network RGAE was planned taking into account the available equipments and the special conditions of the area. Other important aspects in the location of the stations of the RGAE geodetic network due to the isolated characteristic of the environment, was the logistic, the accessibility of the station, the electric management of the instruments, a sky free from obstacles, the existence of problems in the signal and the safety of the researchers. These considerations mint that this geodetic network, the most extensive in the Antarctic area of Level B, has a specific design with accuracy similar to other networks of classical designs (Seeber 2003). It is important to emphasize that the stations were built on geodetic pillars, with a fix screw to insure the precise placement of the GPS antenna.

Due to the lack of satellites in the GPS constellation during the first campaigns, the observations were planned depending on the geometric configuration of the satellites: the choice of the suitable satellites ensured simultaneous visibility between the points, and the storage capacity of the receiver.

From 1995–1996 campaign, GPS receiver had more storage capacity and lower power requirements. These aspects allowed a greater flexibility in the planning of the GPS surveys (sessions of continue 24 h in all of the campaign). In the latest campaigns, 1 Hz sampling rate and a 0° elevation mask are available and some of the stations and continuously recording data. It is important to emphasize that the surveys were made with dual frequency geodetic receivers in order to remove the ionospheric effect from the observations. Figure 1 shows the site locations.

In the Antarctic campaign 1987–1988 the TRANSIT satellite observations were recorded by JMR-1 receivers, allowing the obtaining of absolute coordinates (ellipsoid WGS-72) using the Doppler Punctual Positioning Method (accuracies of 3–10 m). During the 1988–1989 campaign every station was provided by absolute geocentric coordinates, obtaining an accuracy of 5 m approximately. In the 1989–1990 campaign, RIOG point station was set as the fixed station in the network adjustment, obtaining an accuracy of 1–2 m. During the 1990–1991 campaign, RIOG (Argentina) and PUAR (Chile) were fixed in the processing to obtain the absolute coordinates of the rest of the stations in the RGAE network.

The validity of this approach was confirmed during the austral summer of 1991–1992 with the realization of the SCAR'92 international campaign, whose

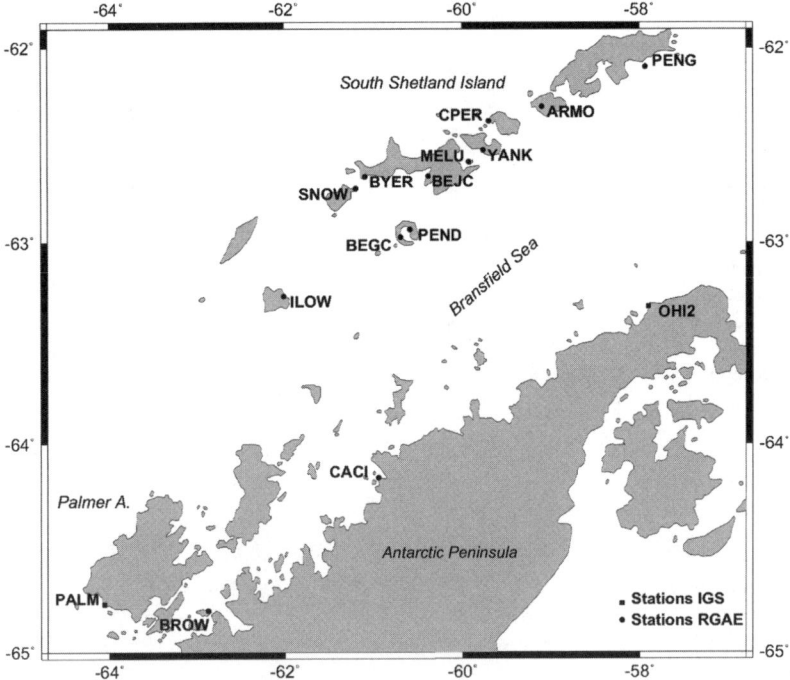

Fig. 1 Distribution of the stations of the REGAE network

main objectives were the establishment of an Antarctic geodetic network covering the whole area, and connected to the IER world network, an extension of the Australian, the New-Zealander, the South-African and the South-American network. During this campaign the final Antarctic polyhedron for a global geodetic reference frame of A Level were established. Its stations covered the Polar Cap and they were provided by absolute coordinates with precisions of the order of one centimetre. Although USHU and BARG were observed, they were not consider in the final adjustment since they were too close compared to the rest of the stations in the network. The calculus and the adjustment of the network were realized by Canberra University and the resulting absolute coordinates obtained were referred to the ITRF 1992.02 frame.

A volcanic event crisis occurring on Deception Island from December 1991 to January 1992 motivated the monumentation of another station in the island, BEGC at the Spanish Base Gabriel de Castilla, which would be included in the RGAE network. In 1999–2000 campaign OHIG mark as the reference point. During 2003–2004 campaign, the observations were realized with a 1 Hz sampling rate and a 0° elevation mask (Table 1).

At present, RGAE geodetic network consists of the following stations: BEJC and BYER (Livingston Island), BEGC (Deception Island), ILOW (Low Island), SNOE (Snow Island), LUNA (Half Moon Island), PRAR and YANK (Greenwich

Table 1 Summary of the occupation for GPS campaigns

ID	Situation	87–88	88–89	89–90	90–91	91–92	99–00	03–04	06–07
BEJC	Livingston I.	X	–	X	X	X	X	X	X
BARG	Deception I.	X	–	X	X	X	X	X	X
PUPO	Livingston I.	–	X	X	–	–	–	–	–
FUMA	Deception I.	–	X	–	X	X	X	X	X
PRAT	Antarctic P.	–	X	–	–	–	–	–	X
BREJ	King George I.	–	–	X	–	–	–	–	–
PALM	Anvers I.	–	–	X	–	–	–	–	–
RIOG	Argentina	–	–	X	X	X	–	–	–
USHU	Argentina	–	–	X	X	X	–	–	–
PEND	Deception I.	–	–	–	X	X	X	X	X
BALL	Deception I.	–	–	–	X	X	X	X	X
PUAR	Chile	–	–	–	X	–	–	–	–
OHIG	Antarctic P.	–	–	–	X	–	–	–	–
BEGC	Deception I.	–	–	–	–	–	X	X	X
CACI	Antarctic P.	–	–	–	–	–	–	X	X
BROW	Antarctic P.	–	–	–	X	–	–	–	X
ILOW	Low I.	–	–	–	–	–	–	–	X
SNOW	Snow I.	–	–	–	–	–	–	–	X
BYER	Livingston I.	–	–	–	–	–	–	–	X
ARMO	Nelson I.	–	–	–	–	–	–	–	X
MELU	Half Moon I.	–	–	–	–	–	–	–	X
YANK	Greenwich I.	–	–	–	–	–	–	–	X
PENG	Penguin I.	–	–	–	–	–	–	–	X
CPER	Robert I.	–	–	–	–	–	–	–	X

Island), CPER (Robert Island), ARMO (Nelson Island), PING (Penguin Island), CACI (Cierva Cove, Antarctic Peninsula), ALBR (Antarctic Peninsula). Other stations previously considered are now administered for other countries, as OHIG and PALM stations, at the O'Higgins Chilean Base and at the American Antarctic Base Palmer on Anvers Island, respectively (Fig. 1).

From 1991–1992 campaign, several episodic campaigns (organized for SCAR) have been realized. With the goal of contributing to the redefinition of the Antarctic reference frame, these campaigns were carried out from January 20th to February 10th each year. The configurations of the observations are: 24 h sessions, 0° elevation mask and 1 s sampling rate. Since 1996, GPS observations from BEJC and BEGC are included in episodic campaigns: EPOCH'96, EPOCH'02, EPOCH'03, EPOCH'04, EPOCH'05, EPOCH'06 and EPOCH'07.

1.1.2 Data Processing and Adjustment of the Network

The data coming from the first campaigns were processed with software TRIMMBP (Trimble 1991). And from campaign 1991–1992 all the data have been reprocess with Bernese v4.2 scientific software (Hugentobler et al. 2001), developed by Astronomical Institute at the University of Berne in Switzerland. It combines specialized surveying knowledge with advanced software techniques. Standard procedures were applied by using IGS products. Final IGS orbits and Earth Rotation Parameters were used according to the proceedings of the Fourth Analysis Centres Workshop in Graz, Austria, September 2003.

Stations coordinate are estimated considering 24 h sessions, 10° elevation masks and a 30 s sampling rate (Hugentobler et al. 2006). Since the software is based on relative positioning, the daily set of baselines were combined in a network adjustment which involves every station for each day. During the parameter estimation process, carrier-phase double-difference data is used in an ionospheric delay free mode.

Tropospheric errors are dealt with by using a combination of the a priori Saastamoinen model and Neill mapping functions. Tropospheric parameters are hourly estimated, ambiguities are dealt for each baseline independently, using the ionosphere free observable. The QIF algorithm is used to solve ambiguities (Mervart 1995). QIF ambiguity resolution requires L1 and L2 to be processed in parallel rather than processing the ionosphere free L3 linear combination. An a priori ionosphere model, ocean tide loading displacement corrections from Onsala Observatory (*www.oso.chalmers.se/~loading/*) and tropospheric parameters are introduced.

Finally, we calculate the station coordinates and normal equations for the daily solutions. ADDNEQ was used to combine normal equations. Two strategies have been applied for the solution adjustment at each epoch: free network adjustment and heavily constrained. The first solution was used to analyse the quality of the coordinates by comparing the IGS fiducial station coordinates in both reference system. Constrained solutions were processed fixing IGS (*International Geodesy Station*) stations to the ITRF2000 reference frame (Altamimi et al. 2002) at the mean epoch of each campaign.

The coordinates for the stations in RGAE network referred to ITRF2000 are listed in Table 2. The stations of network RGAE that are not indicated in this table they have not been processed yet, since it has been surveyed during the last Antarctic campaign for the first time.

Table 2 Coordinates of stations RGAE network (epoch 2006.5)

Station	X (mts.)	σ_X	Y (mts.)	σ_Y	Z (mts.)	σ_Z
BEGC	1423027.717	0.004	−2533143.911	0.004	−5658977.619	0.008
BEJC	1451089.570	0.004	−2553226.315	0.008	−5642854.567	0.017
CACI	1353462.767	0.003	−2437400.371	0.006	−571735.142	0.016
FUMA	1421994.784	0.005	−2535674.066	0.009	−2535674.066	0.018
PEND	1428697.780	0.007	−2534792.711	0.011	−5656765.155	0.023
PRAT	1492479.572	0.085	−2550306.205	0.076	−5633422.474	0.080
BROW	1237367.299	0.085	−2414926.681	0.076	−5752840.021	0.080

1.2 Application of the RGAE Geodetic Network for the Obtention of a Tectonic Model of the Area

The RGAE geodetic network constitutes the geodetic reference frame for the spatial positioning in the South Shetland Islands, the Bransfield Sea and the Antarctic Peninsula. In fact, the maps of the Spanish Antarctic Cartography about Livingston Island and Deception Island were realized according to the stations of the network. Several organizations participated in the cartography elaboration: The Geographic Service of the Spanish Army, the Autonomous University of Madrid in collaboration with the British Antarctic Survey, the Hydrographical Institute of the Spanish Army and the Argentinean Hydrographical Institute.

One in the most interesting application of the RGAE network is the determination of the superficial deformation models for the tectonic activity of the area, which is a very complex tectonic zone where several tectonic plates converge: The South American Plate, the Antarctic Plate and three minor plates, Scotia microplate, Phoenix micro-plate and South Shetland micro-plate (Baraldo 1999).

The Scotia micro-plate is characterized by the presence of vertical faults in its northern and southern boundaries, which separate it from the subduction generated in the West of the South American coast. The East boundary is defined by the presence of a back-arch ridge that separates it from the micro-plate of the South Sandwich Islands. This micro-plate subducts under the South American Plate to the East. The North limit of the Scotia Arch presents a relative movement to the East with respect to the South American.

The South boundary has vertical faults but has some differences to the North boundary. The South boundary is characterized by the presence of an extension basin (pull-apart type) like the Protector, Dove or Scan basins (Galindo-Zaldívar et al. 2006). In the Southwest, the micro-plate borders to the vertical Shackleton Fracture Area, this moves apart from the Phoenix and the South Shetland microplate. In this area there are several tectonic features: The Drake Ridge, the Back-arch basin of the Bransfield Strait and the South Shetland Islands.

The Phoenix micro plate borders the West by the Drake Ridge at North, and in the East on the subduction area of the South Shetland; in the North by the Shackleton Fracture Zone and in the South by the Hero Fracture Zone (Fig. 2). This micro-plate presents a subduction process under the South Shetland basin and the Bransfield Strait. Several studies reveal that the subduction process in the West of the Hero Fracture Zone is no longer active, and thus belongs to the Antarctic Plate.

On the other hand, the Bransfield Central Basin presents a tectonic configuration related to an active subdution. It has an extension area that produces an extension rift with an axis on the NE-SW direction. The central area is the most active of the basin and it borders on the Deception Island at West and Bridgeman Island at East. Two asymmetrical edges limit on the South Shetland Island at North and, on the Antarctic Peninsula at South. It is the 60 km on width, 230 km long and 1950 m deep. The limit of the South Shetland Islands is more abrupt with maximum slope of 25–30°) and it is about 10 km long. (Canals et al. 1997a, González-Ferrán 1985, González-Ferrán 1991).

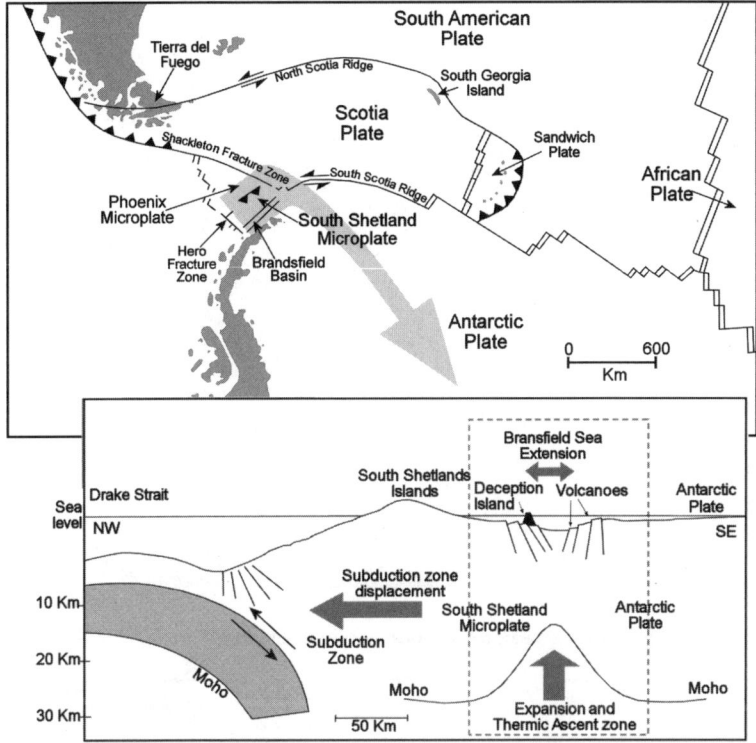

Fig. 2 Tectonic setting of Deception Island, according to the profile of the Hero fracture

To obtain the superficial deformation models related to the tectonic activity of the area, several stations were resurveyed. To determine the limits of the active subduction area, several new stations were built during the last campaign 2006–2007. To evaluate the tectonic displacement for South Shetland Islands, Bransfield Sea and Antarctic Peninsula, GPS data from 2000/2001 to 2005/2006 campaigns were processed according to the methodology previously describe. OHI2 IGS station was set as the fixed site in the processing.

Figure 3 shows the mean displacement rates for the stations included in the processing. The absolute horizontal velocities are estimated to be at the level of 15 mm./year, except in the CACI and BEJC stations. Few data of CACI station are available since it was surveyed during just two campaigns. The estimated BEJC velocity seems to be anomalous and must be studied in depth when data from the present campaign are be processed.

In addition, Fig. 4 illustrate the displacement rates from 1999/2000 to 2001/2002 campaign, 2001–2002 to 2002–2003 campaign, 2002–2003 to 2003–2004 campaign and 2003–2004 to 2004–2005 campaign respectively. This figure shows an attenuation in the module of the deformation on Deception Island after the seismovolcanic crisis happened in 1998–1999 campaign. It can be watched that the displacement direction vector from BEGC station trends to be equal to the Antarctic Plate, which is given by OHI2 and PALM stations.

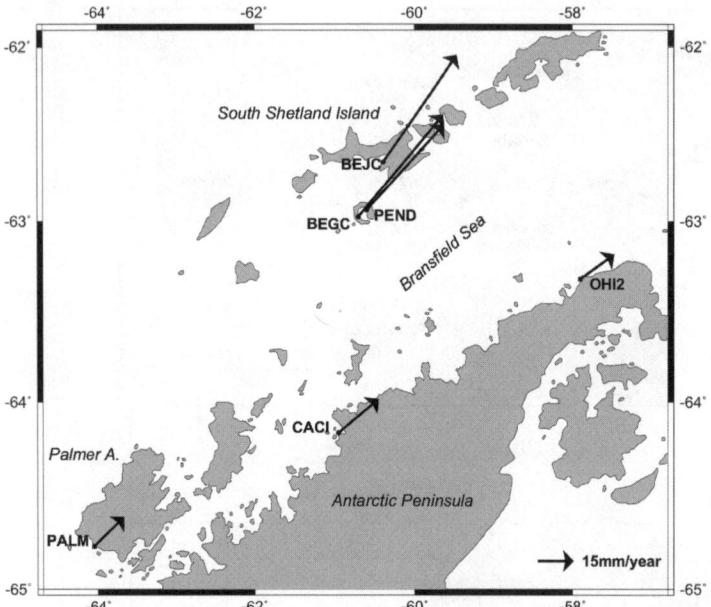

Fig. 3 Mean displacement rate of some of the stations in RGAE network

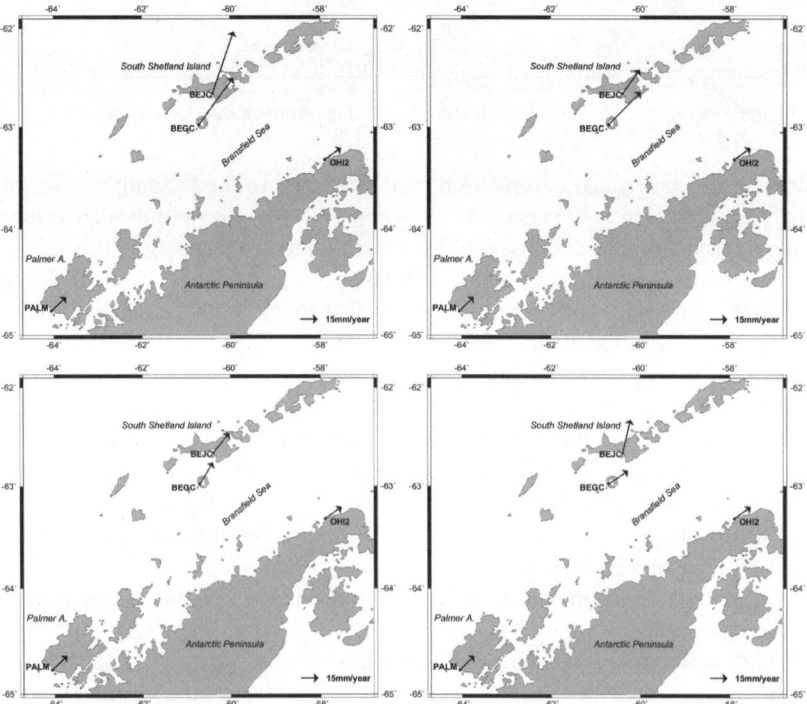

Fig. 4 (from *left* to *right* and up to *bottom*) Tectonic deformation models for the periods 1999–2000 to 2001–2002, 2001–2002 to 2002–2003, 2002–2003 to 2003–2004 and 2003–2004 to 2004–2005

2 Geodetic Activities on Deception Island

The main goals of the geodetic research on Deception Island are the establishment of a local reference frame to get an accurate position of scientific data, the determination of a deformation model for the island and the detection and monitoring of the volcanic activity. In order to achieve these objectives a geodetic network and a levelling network were designed and established along with a local geoid.

The active volcanism in this area is mainly located in the expansion core of the Bransfield Rift and is characterized by the presence of some emergent volcanoes like Deception, Penguin and Bridgeman. In addition there are also numerous submarines volcanic which show different evolution stages, from the volcanic cone to its obliteration, generated by the existing tectonic of divergence (Canals et al. 1997b). These volcanic buildings are 15 km. base diameter and about 450 m. high. In addition, there are around 34 small buildings with a mean diameter of 2.5 km. These buildings are aligned following the main direction of the basin. The continuous volcanic activity in the area is manifested by historical eruptions in Deception and Penguin islands, together with the Bransfield Central Basin rifting and the magmatic processes in the rift during the last two million years. This fact suggests that the volcanic activity in the Bransfield basin goes on in continuous development (González-Ferrán 1991).

Deception Island is located in the beginning of the expansion axis of the Central Bransfield Basin. It is a horseshoe shape stratovolcano of 30 km. diameter in its submerged part and around 15 km of diameter in the emerged zone. The island has a maximum height of 1.5 km over the marine bottom, being its highest point Pond Mount, over 540 m on the sea level (Smellie 2001). It encloses an inner bay, Port Foster, which is a central flooded depression. Throughout its history the island has suffered several periods of different volcanic activity. At present it is the most active volcano of the South Shetlands Islands and the Antarctic Peninsula, with dated eruptions in 1848, 1967, 1969, 1912, 1917 and 1970 as it can be see in Fig. 5 (Martí et al. 1996).

During 1967–1970 eruptive period, the high volcanic activity caused the destruction of the Chilean and British Scientific Bases, located in Pendulum Cove and Whaler's Bay respectively. These eruptions changed the morphology of the island, forming an islet that joined lately to the island around Telephone Bay (Fig. 6). In this period a great amount of ashes was emitted and deposited on the neighbouring islands, as it can be observed in Johnson glacier at Livingston Island. Due to this eruptive event and their destructive consequences the scientific activities by Argentinean, Chilean and British investigators were interrupted.

The activities to control and monitor the volcanic activity by means of geodesic and geophysical techniques were continued in 1986, when the first Spanish Antarctic campaign took place. Nowadays, the superficial evidences of volcanic activity are mainly the presence of fumarolic areas with 100°C and 70°C gaseous emissions in Fumarolas and Whaler's Bay respectively, 100°C hot soils in Cerro Caliente, and 45°C and 65°C thermal springs in Pendulum Cove and Whaler's Bay. In addition, there are numerous areas where a significant seismic activity is detected.

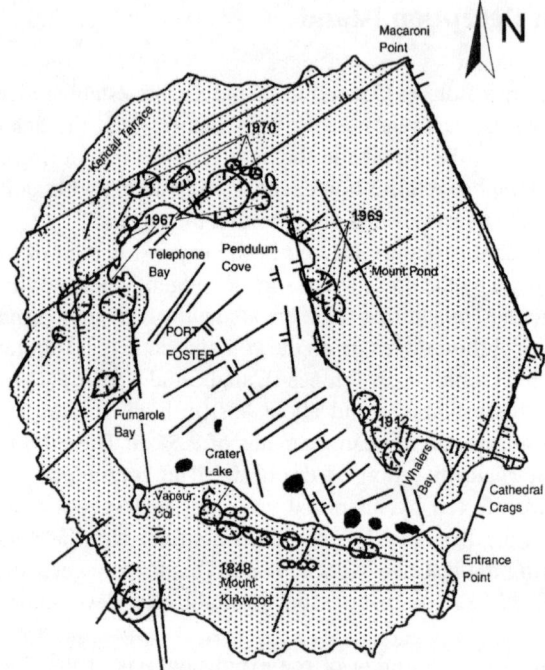

Fig. 5 Deception Island tectonic setting with dated eruptions

This registered seismicity has two different sources: on one hand, the origin of tectonic activity due to the expansion of Bransfield Rift, and on the other hand, the purely volcanic source. In 1991–1992 and 1998–1999 campaigns two seismic-volcanic crises were detected (Ibáñez et al. 2003).

2.1 Regid Geodetic Network

The REGID geodetic network was established from GPS observations in order to set up the basic reference frame of both the observations and the results obtained by other sciences. In this way, the network has been the fundamental reference for the Spanish cartography of the Antarctica, for the geophysical and oceanographic observations and for the hydrographical sounding carried out throughout the Spanish campaigns at the Antarctica. On the other hand, GPS accuracy, especially for horizontal positioning, has made the REGID network very useful for the control of the geodynamic activity of the island.

The construction of the geodetic network began in the 1989–1990 Antarctic campaign. The aim of the studies and the place where they were going to be carried out led to the construction of high stability benchmarks, with no visual impact. They

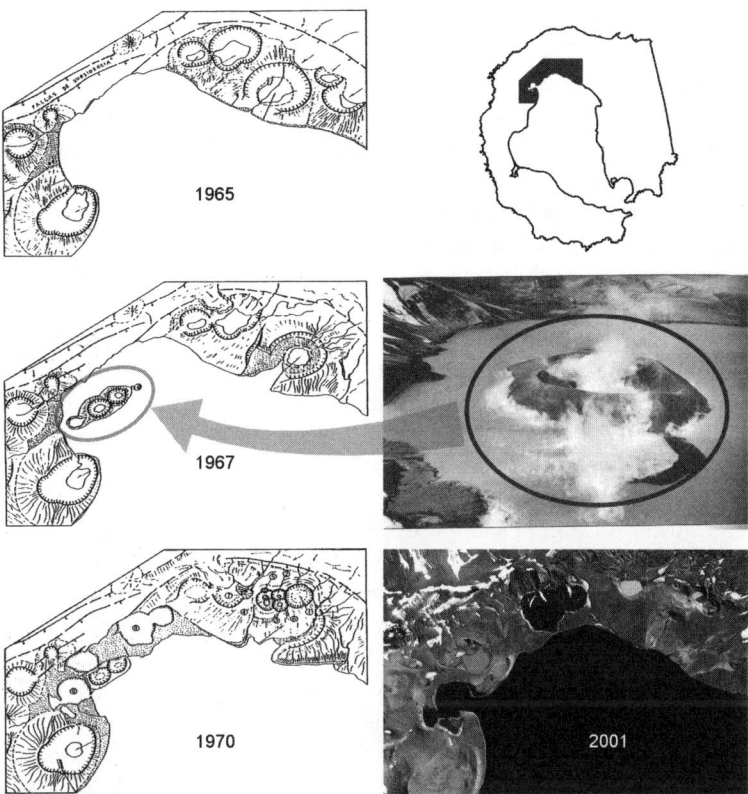

Fig. 6 Morphological evolution in Telephone Bay during the 1967–1970 eruptive process (Brecher 1975)

were built with concrete, well-rooted in the permafrost with steel bars and a low height above the ground. A standard screw was fixed in one of the corners of the post to allow the use of geodetic instrumentation. Furthermore, some of the stations were built using existing structures of the buildings destroyed by the volcanic activity which happened between 1967 and 1970 (Berrocoso 1997).

In the 1990–1991 campaign, the geodetic network consisted of five stations, four of them in Deception Island and another one at the Spanish Base Juan Carlos I in Livingston Island. Their placements was chosen according to several factors, including accessibility, stability of the building, proximity of fumarolic areas, density of founded epicentres, last eruptions, etc. The stations in Deception Island were placed at the Argentinean Base (BARG) next to the Radio hut, at an existing hydrographical station in Fumaroles Bay (FUMA), at Pendulum Cove (PEND) near the remains of the Chilean Base and at the foundations of the old British Base in Whalers' Bay (BALL). BARG station was the main vertex of the network. The station in Livingston Island was built at the Spanish Antarctic Base Juan Carlos I.

During the 1995–1996 campaign a new station was built near the Spanish Antarctic Base Gabriel de Castilla. This station (BEGC) was set as the new main point of the network because it was further from one of the most active areas of the island, Fumaroles Bay. This station has an electricity supply and was built in the same way as the others (Berrocoso 1997). During the 2001–2002 campaign seven more stations were built in the north and south area of Deception Island, in order to fill in its entire inner ring and to make the REGID network more consistent. The designed of the stations is given in Fig. 4. The construction of these new stations was carried out according to the same geodetic directions as the last ones: clear GPS horizon, low multipath effect, etc. The area where they were built was chosen taking into account the volcanic nature of the island, but without considering any prior volcanic activity (Vila et al. 1992; García et al. 1997; Ibáñez et al. 2003).

These new stations are: UCA1, placed in northwest of the island, in the east of Obsidian Hill; TELE, located in the area of flooded craters in Telephone Bay; BOMB, positioned in a volcanic bombs field between 1970 Craters and Telephone Bay; CR70, placed in the area of craters generated by the eruptions in 1970 that destroyed the Chilean Base in Pendulum Cove; GLAN, located near the south of Black Glacier; GEOD, situated next to the Soto crater between Colatinas and the Spanish Antarctic Base Gabriel de Castilla; and COLA, sited in Colatinas, in the southeast part of the island. During the 2006–2007 campaign two more stations were included in the network, in Punta Collins, in the South of the island, and the levelling benchmark LN000, which has become a geodetic mark from now on. The final distribution of the network is shown in Fig. 7.

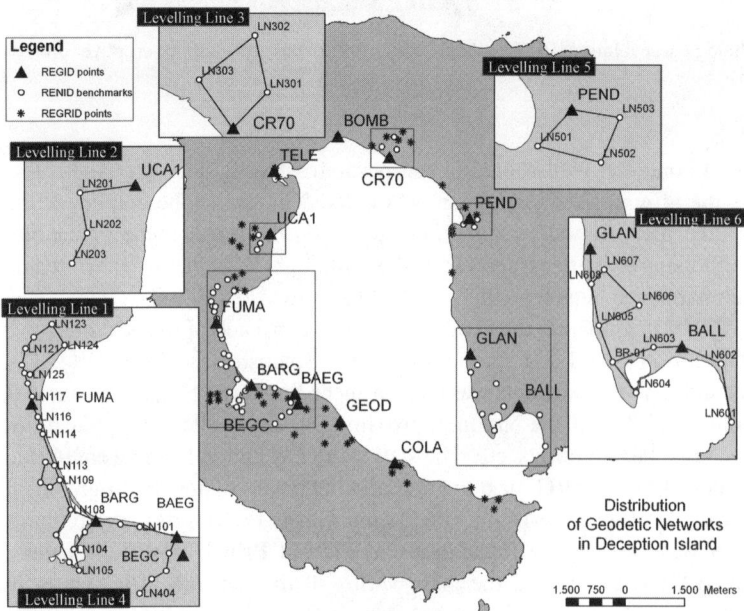

Fig. 7 Distribution of Geodetic Networks in Deception Island

Stations of the initial network -BARG, BEJC, PEND, FUMA and BALL- have been observed during the Spanish Antarctic campaigns from 1991 to 1992. BEGC station was included during the 1995–1996 campaign, when it was observed for the first time. After the monumentation of seven new stations in the 2001–2002 campaign, COLA, GEODEC, UCA1, TELE BOMB, CR70 and GLAN, the whole network was observed during the austral summers from 2001 to 2002 campaign. The network has been surveyed until the last Antarctic campaign. During the 1991–1992 campaign, the fixed station was the one at the Argentinean Base (BARG). Observations from this station and from the one at Livingston Island (BEJC) were sent to the SCAR to be included in the SCAR92 solution.

In the 1999–2000 campaign, the five stations of the whole initial network were observed simultaneously for the first time. From 2002 to 2003 campaign, the joint BEGC-FUMA-PEND was maintained and the rest of the network was observed too.

The processing of the data has been made following the same methodology that previously with network RGAE with Bernese v4.2 software (Hugentobler et al. 2001). The IGS station OHIG was used to the adjustment of the network. Data from OHIG station are referred to the ones corresponding to the first of January of the ITRF96, ITRF97 and ITRF2000 systems, respectively. Every campaign was processed independently, fixing the coordinates of one of the stations of the network. BARG was the fixed station until 2001, and then it was changed to BEGC station.

Once the coordinates of the fixed station were set, the coordinates for the rest of the stations in the network were obtained by means of radial baselines between the fixed stations and the others. For the processing of the data, simultaneous observations were considered. To estimate the tropospheric delays we use the default meteorological data, since the only available data were the medium temperature and pressure and their inclusion could produce more errors than the models provided by the software. In the Table 3 results are referred to the ITRF2000, epoch 2003.1.

2.2 RENID Levelling Network

During the 2001–2002 Antarctic campaign a levelling network was designed to establish a reference frame for the real time monitoring of the vertical deformation taking place in the island. The network consists of six independent lines around the island, which allow a quick action just in the area where the volcanic activity has been detected. The distribution is shown in the Fig. 7 and in Table 4. The first bench mark to be monumented was the LN000 benchmark which to replace the BARG geodetic like origin of altitudes.

Levelling line 1 links the Spanish Base Gabriel de Castilla, the Argentinean Base Deception and Fumaroles Bay. It is supported by the LN000 bench mark and the geodetic stations BEGC; BARG and FUMA; levelling line 2, which extends along the Obsidian Hill, is supported in the geodetic station UCA1; levelling line 3 is supported by the geodetic station CR70 and it is placed in the area where the last eruption in 1970 took place; levelling line 4 is in the surroundings of the River Mekong and it is also supported on the main bench mark LN000; levelling line 5 is supported by PEND, in Pendulum Cove, and finally, levelling line 6 is supported by the geodetic stations GLAN and BALL, in Whalers Bay (Fig. 7).

The realization of the geometric levelling with sub-centimetre precision allows the establishments of a precise reference frame to measure instantaneous vertical deformation. Every bench mark in every line was positioned using a differential GPS. Table 4, shows the vertical data for the RENID levelling network. The LN000 benchmark is the reference point for levelling lines 1 and 4, and the geodetic stations UCA1, CR70, PEND and BALL are the reference points for lines 2, 3, 5 and 6, respectively.

The connections among lines were done during the 2002–2003 campaign, using a Wild T2000 theodolite and a DI5000 distancemeter. Lines 1, 2, 3 and 5 were connected among them, and levelling line 6 was linked to the geodetic station COLA, also connected to the LN000 bench in line 1 along the coastline. The connection between line 4 and line 1 was done during the last stage of that campaign.

2.3 REGRID Gravimetric Network

REGRID gravimetric network was established during the 2002–2003 Antarctic campaign from the geodetic stations of the REGRID network and the levelling benchmarks of the RENID network. The fundamental gravimetric point in the island was set up to be GBEGC, in the surroundings of the Spanish Antarctic Base Gabriel de

Table 3 REGID coordinates referred to ITRF2000 at 2003.1

Station	X (mts.)	σ_X	Y (mts.)	σ_Y	Z (mts.)	σ_Z
BEGC	1423027.714	0.001	−2533143.956	0.001	−5658977.759	0.001
BARG	1422140.963	0.001	−2534034.247	0.003	−5658736.141	0.003
BALL	1427933.006	0.003	−2530514.413	0.003	−5658856.008	0.006
FUMA	1421994.742	0.001	−2535674.049	0.003	−5658043.664	0.003
BEJC	1451089.543	0.003	−2553226.299	0.003	−5642854.597	0.006
PEND	1428697.709	0.001	−2534792.701	0.003	−5656765.170	0.003
COLA	1424578.900	0.003	−2530853.317	0.003	−5659569.913	0.009
GLAN	1427356.936	0.003	−2532112.816	0.006	−5658291.897	0.009
GEOD	1423777.232	0.003	−2532298.366	0.003	−5659121.635	0.009
UCA1	1424097.820	0.003	−2536792.788	0.003	−5657026.513	0.009
CR70	1427518.181	0.003	−2536915.250	0.006	−5656109.896	0.012
TELE	1424808.476	0.003	−2537985.468	0.006	−5656311.865	0.012
BOMB	1426582.129	0.003	−2537924.128	0.006	−5655895.118	0.015

Table 4 Description of the RENID levelling network

Levelling line	Levelling benchmark	Level difference	Levelling line	Levelling benchmark	Level difference
	LN101	+2.132	LINE 2 REF:	LN201	+14.307
	LN102	+8.601		LN202	+2.998
	BARG	-2.868	UCA1;	LN203	+8.968
	LN103	-0.739	LINE 3 REF:	LN301	+7.283
	LN104	-3.464		LN302	+21.333
	LN105	+4.971	CR70	LN303	+1.437
	LN106	+1.177	LINE 4 REF:	LN401	+4.642
	LN107	+2.861	LN000	LN402	+12.430
	LN108	+0.289		LN403	+20.676
	LN109	+3.797		LN404	+32.498
	LN110	+20.390	LINE 5 REF:	LN501	-5.998
	LN111	+24.307	FUMA	LN502	+15.621
LINE 1 REF:	LN112	+10.304		LN503	+22.493
LN000	LN113	-1.745		BR-01	+16.380
	LN114	-4.264		LN601	-5.113
	LN115	-4.917		LN602	+1.039
	LN116	-4.066		LN603	+7.947
	FUMA	-2.468	LINE 6 REF:	LN604	-4.610
	LN117	-4.010	BALL	LN605	+8.684
	LN118	-2.352		LN606	+11.331
	LN119	-2.201		LN607	+17.915
	LN120	-0.642		LN608	+7.061
	LN121	+2.688		GLAN	+7.161
	LN122	+5.515			
	LN123	+11.757	–		
	LN124	-0.220			
	LN125	-4.392			

Castilla. Another gravimetric point at Livingston Island, GBEJC, is included in the network.

Gravimetric measurements were obtained with a Lacoste & Romberg D-203 gravimeter with a priori deviation of $10\,\text{nms}^{-2}$. A gravimetric link among GBEGC (fundamental point) in Deception Island, GBEJC in Livingston Island and APPA gravimetric point in Punta Arenas (Chile) was made, which value is $9813208100\,\text{nms}^{-1}$. Tides, height and drift corrections were applied to the whole set of gravimetric data. The distribution of the secondary points is given in Fig. 7. The values for gravimetric measurements for this network is given in Table 5.

Table 5 Gravimetric measurements for the REGID network geodetic stations

Station	g [nm/s^{-2}]	σ[nm/s^2]
BALL	9822071743	138
BARG	9822106750	174
BEGC	9821962801	145
BOMB	9822075418	259
COLA	9822089209	175
CR70	9822031402	259
FUMA	9822079377	108
GBEGC	9822044682	20
GEOD	9822061136	117
GLAN	9822028138	198
PEND	9822040199	160
TELE	9822077750	198
UCA1	9822035897	198

2.4 Geodetic Network REGID Aplication for Volcanic Deformation Model Determination and Sources Location Estimation

Superficial deformation in volcanic areas is interpreted like reflection of the changes of pressure in the superficial magmatic cameras. Greater deformations happen in limited areas and its interpretation, in terms of schematic models, implies to assume the existence of those superficial or brief coves or reservoirs, intrusions or forts increases of pressure without justification. A possible hypothesis indicates that the observed superficial displacements are the answer of a half elastic space to a pressure increase in magmatic cove. In order to determine the most probable models different geometries and different restrictions are considered (Lavalle et al. 2004; Abidin et al. 1998; Bock et al. 1997).

Deformation models representing Deception Island volcanic activity have been obtained from the GPS observations of the stations conforming the geodetic network REGID. GPS observations have been episodically made from 1991–1992 campaign to present. Taking into account that GPS system provides twice the accuracy in horizontal positioning than in the vertical component, we distinguish between the horizontal and the vertical models, although it is essential to conjugate both results for a proper interpretation. (Donnellan 1993; El-Fiky et al. 1999; Calais et al. 2000, Mantovani et al. 2001; Murray and Wooller 2002).

To obtain the deformation models a topocentric system was established with origin at BARG. In order to obtain the absolute deformation of the island BEJC station on Livingston Island was considered. Data from GPS surveying in Deception Island were processed using the Bernese v4.2 software according the methodology described on previous paragraphs, this is, precise orbits were used downloaded from several sources: SIO (*Scripps Institution of Oceanography*) for 1991–1992 campaign and CODE (*Centre of Orbits Determination*) for the rest of campaigns.

Every campaign was processed independently, fixing the coordinates of one of the stations of the network. BARG was the fixed station until 2001, and then it was changed to BEGC station. In the 1991–1992 campaign, the coordinates for the fixed station BARG were obtained from the international campaign SCAR92. Its precise absolute coordinates for the station were referred to the ITRF91, epoch 2.2.

To get these coordinates for the next campaigns, in 1995–1996 and 1999–2000, the coordinates of the IGS station OHIG corresponding to ITRF96, epoch 96.1 and to ITRF97, epoch 99.9, were fixed. It is situated at the Chilean Base O'Higgins in the Antarctic Peninsula, 150 km. away from Deception Island.

Due to the proximity of the BEGC station to the Spanish Base Gabriel de Castilla, the receiver is easily maintained and it is possible to collect data continuously during the campaign. That is why this station was considered as the main one in the network REGID from the 2001–2002 and 2002–2003 campaigns. Its coordinates were obtained by fixing the ones of OHIG corresponding to the ITRF2000, epoch 2002.1 and 2003.1. Once the coordinates of the fixed station were set, the coordinates for the rest of the stations in the network were obtained by means of radial baselines between the fixed stations and the others. For the processing of the data, simultaneous observations were considered.

In the 1991–1992 campaign, the radial processed baselines were BARG- FUMA, BARG- PEND and BARG- BALL. The ambiguities were calculated by the SIGMA strategy of the BERNESE software, since the observations were made by single-frequency receivers. Finally, the solution is referred to the ITRF91; in 1995–1996, observations of both L1 and L2 frequencies were available, so the ambiguities resolution was solved applying the QIF strategy. The final solution was referred to the ITRF96; in the 1999–2000, the baselines were processed altogether since it was the first time that every station was observed simultaneously. The ambiguities were also resolved with the QIF strategy and the final solution was referred to the ITRF97; from 2001–2002 and 2002–2003 campaigns, solutions for every week were combined with the ADDNEQ program of the BERNESE software. QIF strategy was also used and the final solution was referred to the ITRF2000.

Displacements models are calculated comparing absolute topocentric coordinates, obtained on every observation epoch, between two consecutives campaigns (Fig. 8); determining the velocity fields and its gradient by a finite elements interpolation. Horizontal deformation tensors are calculated to obtain deformation parameters, extension-compression area (Fig 9); finally the components in directions NS and EW are obtained (Dermanis 1985; Eren 1984; Vaníček and Krakiwsky 1986).

Mathematical horizontal deformation models are obtained from these displacements models by dilatation. (Bibby 1982; Drew y Snay 1989; Grafarend and Voosoghi 2003; Dzurisin 2006). From the analysis of the 1991–1992 to 2002–2003 campaign data we obtained a radial extensional process from 1995/1996 to 1999/2000 campaign. Afterwards, the process became compressive; there was an uplift process from 1995/1996 to 2001/2002 campaigns; and subsidence process before and after this period. Extensional process follows the NW-SE direction (Fracture Hero Zone direction), while the compressive process follows the Bransfield Rift extensional direction (Fig. 10).

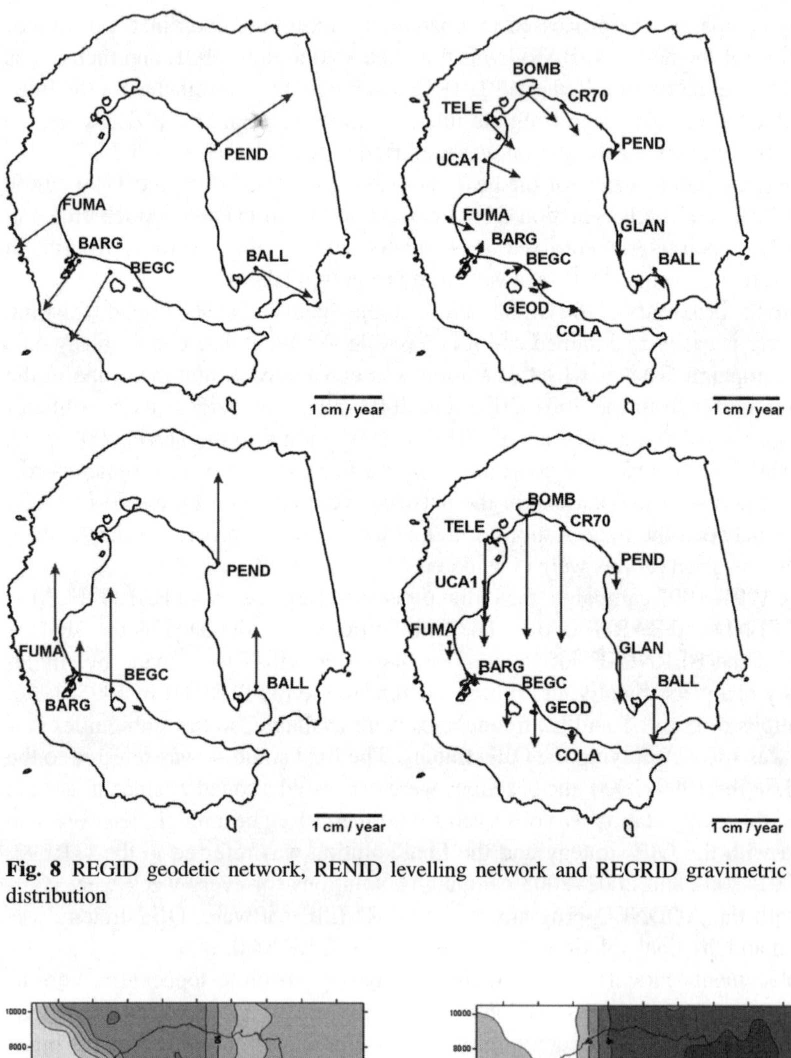

Fig. 8 REGID geodetic network, RENID levelling network and REGRID gravimetric network distribution

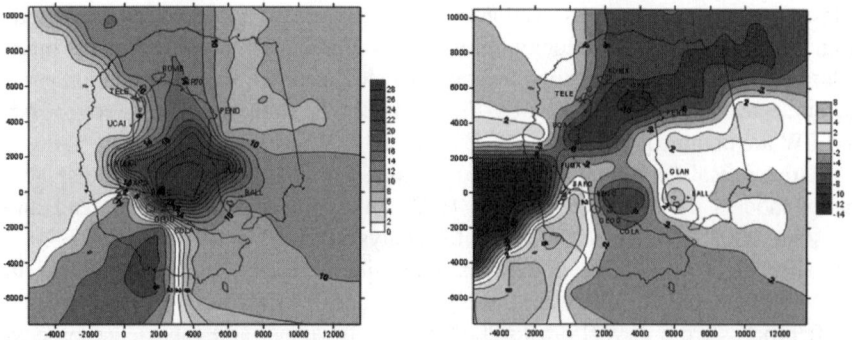

Fig. 9 From *left* to *right*, *top* to *bottom*: horizontal and vertical deformation models for 1996.2/1999.9 and 2002.2/2003.2 epochs (cm./year), respectively

Changes detected from volcanic deformation models in extension-compression as well as subsidence-uplift-subsidence and changes in activity axes direction overlap with 1998/1999 seismovolcanic crisis.

Considering the initial configuration of the geodetic network REGID, Mogi model has been considered for the determination of the source location and variation for the hydrostatic pressure explaining the superficial deformation (Mogi 1958). Experimental data, horizontal and vertical displacements were used, whose Lame elasticity modulus and Poisson coefficient were fixed from theoretical data (Willson 1980; Newhall and Selft 1982; Bonaccorso 1996). Simulated Annealing inversion algorithm has been used for source estimation (Cervilli et al. 2001; Sambridge and Mosegaard 2002).

From 1992 to 1996, displacements can be due to the activity of principal source located in the interior ring, BARG-FUMA-PEND triangle (Fig. 11). From 1992 to 1999, a northern source is estimated at FUMA-PEND direction. The seismovolcanic crisis in 1998 seems to be produced by a chamber activity located at north of Port Foster, between Fumarolas Bay and Pendulum Cove. From 1999, the source is located deeper, indicating volcanic activity decreased, at least, until 2003 (Fernández-Ros 2006).

2.5 Determination of a Local Experimental Geoid

The presence of active volcanism in Deception Island and the existence of both superficial and deep seismicity make the tectonic situation of the area become

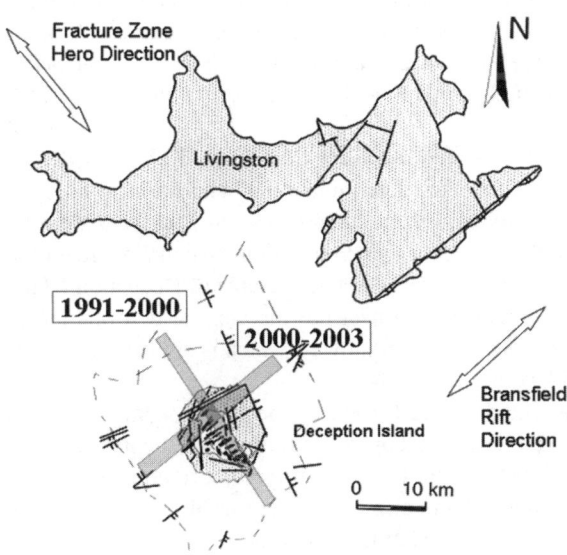

Fig. 10 Superficial dilatation (ppm/year) from 1996.2/1999.9 and 2002.2/2003.2 epochs

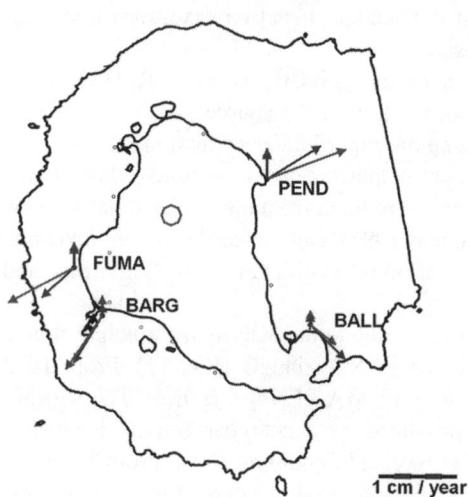

Fig. 11 Volcanic activity responds to two main alignments for Deception Island: Source location estimated for superficial deformation origin zone from volcanic activity in Deception Island

very complex and, therefore, the main goal of the Geosciences studies. Thus, the establishment of a proper geodetic reference frame acquires a high degree of relevance.

The lack of physical meaning of the ellipsoidal height as well as the lack of accuracy in vertical deformation models make not only a mathematical but also a physical reference frame, such us the geoid, necessary in order to calculate the existing deformation. However, global geoid models, such as OSU91A or EGM96, are so inaccurate at small spatial scales in volcanic areas with high geodynamic activity, that determining an experimental geoid model to refine the globally extrapolated models is essential.

This experimental geoid will allow us to detect gravity field changes due to crustal movements and it will also make the determination of a hazard map possible thanks to the combination of gravity equipotential surfaces and data from a digital elevation model. In addition to this, the geoid could permit us to carry out real time levelling measurements by means of GPS receivers. In the area of Deception Island this application would be very interesting to measure the vertical deformation in case of volcanic reactivation in a fast and accurate way, which is almost impossible and dangerous using classical levelling methods.

Therefore, an experimental geoid model for Deception Island, computed from GPS data, geometrical levelling and absolute gravimetric measurements, has been calculated. This mean sea level surface will be a very important reference frame to enhance our knowledge of the area and will allow us to improve the accuracy of the high degree geopotential models in this area of the Antarctic.

Besides the main purpose of each one of the designed networks at Deception Island, other different and complementary measurements were gathered in order to use all the stations and benchmarks for any geodetic application, as the geoid determination. So, the stations of the REGID network were linked to the RENID network by geometric levelling, in such a way that these stations are now provided with levelling measurements with respect to the main benchmark LN000. In the same way, every levelling benchmark was positioned in 'fast-static' mode, keeping one base receiver at BEGC station. Therefore, and taking into account that the gravimetric network was established from REGID and RENID networks, REGRID is provided not only with absolute gravity values, but also with geodetic coordinates and levelling measurements.

Since REGID, RENID and REGRID networks had to be developed around the inner bay and not further than one kilometre offshore, 44 secondary points were also established to spread out the observations to the outer area and to make the measurement set denser. So, GPS observations, levelling and gravimetric measurements were also obtained for these marks; on one hand, every point was positioned by 'fast-static' mode. Regarding GPS data, although the REGID geodetic network has been surveyed from 1988, considered data correspond to the 2002–2003 Antarctic campaign, from November 2002 to February 2003 (Berrocoso et al. 2004a,b).

The adjustment of the network was carried out in two stages: firstly, absolute geocentric coordinates for both BEGC and BEJC stations were obtained by the processing of the network with the IGS station OHI2 respect to ITRF2000 reference frame, epoch 2000; secondly, the coordinates of the remaining stations at Deception Island, calculated respect to ITRF2000 as well, were processed from the coordinates previously obtained.

In relation to levelling measurements, levelling surveys were carried out using a Leica NA2 level, with $\sigma = 0.07$ mm, during the 2001–2002 and the 2002–2003 Antarctic campaigns. Corrections due to refraction effects were applied to the data whereas the sphericity effect was not taken into account because of the size of the area under study. Three different methods were considered to obtain this physical reference frame: remove-restore, collocation and GPS/levelling methods. GPS and gravimetric data are needed when using the two first methods whereas levelling data are also needed with the last one. The problem is that these data are really difficult and expensive to obtain in areas like the Antarctica.

On one hand, the remove-restore method is based on the resolution of the Stokes integral by breaking down the frequency spectrum of the gravity anomaly and undulation into different bands. Then, FFT or wavelets can be used as an integration method. On the other hand, collocation method provides values of the geoid height which best fit the gravity measurements by means of the matrix of covariance. Finally, GPS/levelling technique considers a set of points around the island where absolute gravity, levelling and ellipsoid coordinates are known. By means of the orthometric height in BARG, we are able to obtain the orthometric heights in the remaining points. So, to calculate the geoid, we only have to consider the well-known formula $h = H + N$, where h, H and N are the ellipsoidal, orthometric and geoid heights respectivel

With regards to computational cost, remove-restore method has a higher cost than GPS/ levelling and collocation methods not only due to the use of Global Geopotential Models but also because of the methods of numerical integration. Also, by comparing the acquired values to those geoid heights from global geopotential models, we easily obtain the correction factor for the area of Deception Island. In fact, GPS/levelling and collocation techniques provide more details than remove-restore method in comparison with global geopotential models (Fig. 12).

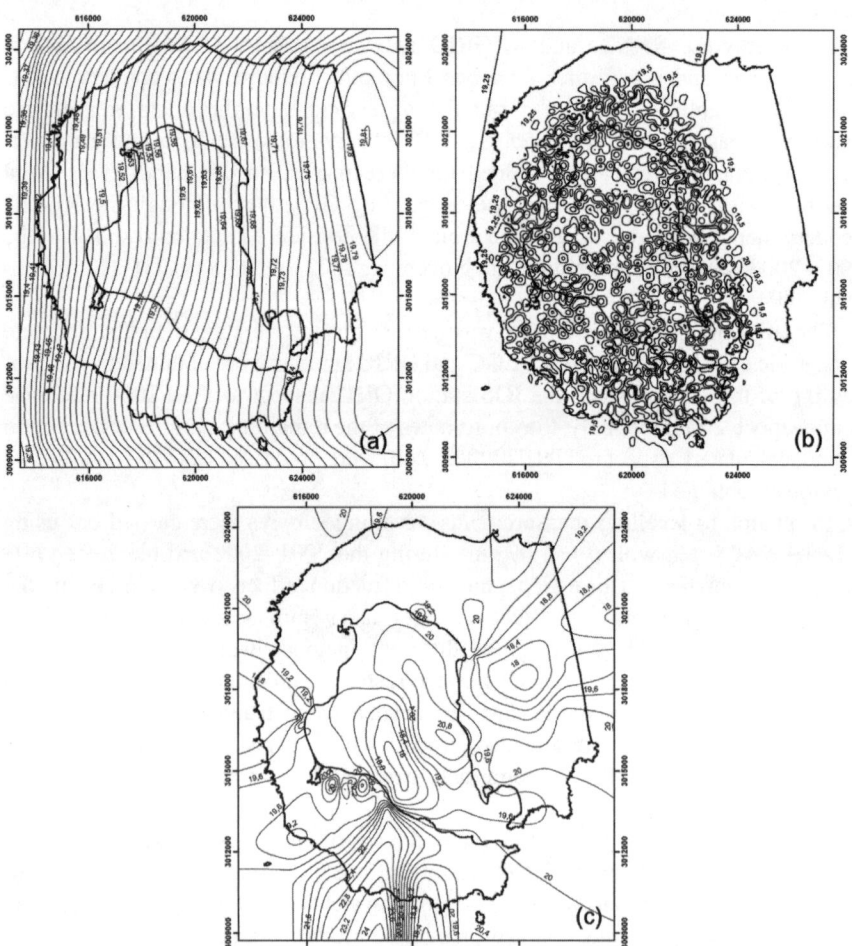

Fig. 12 Geoid height maps: (**a**) remove-restore; (**b**) collocation; (**c**) GPS/levelling

3 Summary and Conclusions

The geodetic networks which have been established in South Shetland Islands, Bransfield Strait and the Antarctic Peninsula have been presented and described in this work. The main aims of RGAE and REGID networks are, on the one hand, to define a proper geodetic reference frame in the area under study and, on the other hand, to make possible the survey of tectonic and volcanic processes taking place in the region. We would like to point out the fact that the OHI2 and PALM stations are not the same as those ones belonging to the current IGS network although they are near the present points. In order to keep the Spanish base located in the island and the deformation of the volcano under surveillance, an accurate levelling network (RENID network) has been also designed and built. These benchmarks will allow us to monitor the real time island deformation as a volcanic precursor.

So as to define a physical reference frame for Deception Island, an experimental geoid model, computed from GPS data, geometrical levelling and absolute gravimetric measurements, has been calculated. This mean sea level surface will be necessary to enhance our knowledge of the area and will allow us to improve the accuracy of the high degree geopotential models in this zone of the Antarctic.

Finally, the need to establish an Antarctic geoid with decimetric accuracy, the need to homogenize the existing cartography, to determine local and regional refraction models in the Antarctica and the need to make the networks in existence denser, they all define the objectives with top priority in the geodetic Antarctic research which must be favoured from the International Polar Year.

Acknowledgments The realization of this geodetic research is supported by the Spanish Ministry of Education and Science as part of the National Antarctic Program. The following research projects have been awarded so far: "Recognition and fast evaluation of the volcanic activity of the island Deception (GEODESY) (ANT1999.1430.E/HESP)"; "Geodetic Studies on Deception Island: deformation models, geoid determination and Information System for Scientific (REN2000.0551.C03.01/ANT)"; "Acquisition of a scientific software for GPS data processing (REN2000.2690.E)"; "Geodetic Control of the volcanic activity of Decepcion Island (CGL2004.21547.E/ANT)"; "Update of the Spanish Cartography for Decepcion Island (CGL2004.20408.E/ANT)"; "Volcanotectonic activity on Deception Island: geodetic, geophysical investigations and Remote Sensing on Deception Island and its surroundings (CGL2005-07589-c03-01/ANT)"; "Geodetic Control of the volcanic activity of the Island Deception". We would like also thank the BIO Hespérides and the Palmas crew for they support in the development of the carried out activities, the logistic crew of both Spanish Antarctic Bases Gabriel de Castilla and Juan Carlos I., the Hydrographic Institute of the Navy, the Geographic Service Army as well as the Royal Naval Observatory of San Fernando.

References

Abidin HZ, Meilano I, Suganda OK, Kusuma MA, Muhardi D, Yolanda O, Setyadji B, Sukhyar R, Kahar J, Tanaka T (1998) Monitoring the deformation of Guntur Volcano using repeated GPS survey method. Proceedings XXI FIG Congress, Brighton.

Altamimi Z, Sillard P, Boucher C (2002) ITRF2000: A new release of the International Terrestrial Reference Frame for earth science applications. Journal of Geophysical Research, 107, B10, pp. 2214–2233.

Baraldo A (1999) Evolución geológica de la Isla Decepción, Islas Shetland del Sur, Antártida. Ph. Tesis, Universidad de Buenos Aires.

Berrocoso M (1997) Modelos y formalismos para el tratamiento de observaciones GPS. Aplicación al establecimiento de redes Geodésicas y Geodinámicas en la Antártida. Boletin ROA vol 1/97. Ministerio de Defensa, San Fernando.

Berrocoso M, Fernández-Ros A, Torrecillas C, Enríquez-Salamanca JM, Ramírez ME, González MJ, Soto R, Pérez-Peña A, Páez R, Tárraga M, García-García A, García-García F (2004a) Investigaciones geodésicas en la isla Decepción. Proceedings IV Asamblea Hispano Portuguesa de Geodesia y Geofísica, Figueira da Foz.

Berrocoso M, García-García F, Fernández-Ros A, Tárraga M, García-García A, Torrecillas C, Ramírez ME, Enríquez-Salamanca JM (2004b) Medidas gravimétricas en la isla Decepción. Proceedings IV Asamblea Hispano Portuguesa de Geodesia y Geofísica, Figueira da Foz.

Bibby HM (1982) Unbiased estimated of strain from triangulation data using the method of simultaneous reduction. Tectonophysics, vol 82, pp. 161–174.

Bock, Y, Wdowinski, S, Fang, P, Zhang J, Williams S, Johnson H, Behr J, Genrich J, Dean J, Van Domselaar M, Agnew D, Wyatt F, Stark K, Oral B, Hudnut K, King, R, Herring T, Dinardo S, Young W, Jackson D, Gurtner W (1997) Southern California Permanent GPS Geodetic Array: continuous measurements of regional crustal deformation between the 1992 Landers and 1994 Northridge earthquakes, Journal Geophysical Research, vol 102, B8, pp. 18013–18033.

Bonaccorso A (1996) Dynamic inversion of ground deformation data for modelling volcanic sources (Etna 1991–1993). Geophysical Research Letters, vol 23 (5), pp. 4261–4268.

Brecher H (1975) Photogrammetric maps of a volcanic eruption area, Deception Island, Antarctica. Institute of Polar Studies, report No. 52, pp. 1–10.

Calais E, Galisson L, Stéphan JF, Delteil J, Decerchère J, Larrouque C, Mercier De Lépinay B, Popoff M, Sosson M (2000) Crustal strain in the Southern Alps, France, 1948–1998. Tectonophysics, vol 319, pp. 1–17.

Canals M, Gracià E, Prieto M. J, y Parson L. M (1997a) The very early stages of seafloor spreading: the Central Basin, NW Antarctic Peninsula. The Antarctic Region: Geological Evolution and Processes, pp. 669–673.

Canals M, Gràcia E, Grupo GEBRA (1997b) Evidence of initial seafloor spreading in the Central Brans_eld Basin, Western Antarctica. Boletín de la Real Sociedad Española de Historia Natural, vol 93(1–4), pp. 53–61.

Cervilli P, Murray MH, Segall P, Auki Y, Fato T (2001) Estimating source parameters from deformation data, with an application to the March 1997 earthquake swarm off the Izu Peninsula, Japan. Journal of Geophysical Research, vol 106 (B6), pp. 11217–11237.

Dermanis A (1985) The Role of Frame Definitions in the Geodetic Determination of Crustal Deformation Parameters. Bulletin Geodesique, vol 59(3), pp. 247–274.

Donnellan A (1993) Geodetic measurement of deformation in the Ventura basin region, Southern California. Journal of Geophysical Research, vol 98, B12, pp. 21727–21739.

Drew AR, Snay RA (1989) DYNAP: software for estimating crustal deformation from geodetic data. Tectonophysics, vol 62, pp. 331–343.

Dzurisin D (2006) Volcano Deformation: New Geodetic Monitoring Techniques. Springer-Verlag, New York.

El-Fiky G, Kato T, Oware EN (1999) Crustal deformation and interpolate coupling in the Shikoku district, Japan, as seen from continuous GPS observation. Tectonophysics, vol 314, pp. 387–399.

Eren K (1984) Strain analysis along the north anatolian fault by using geodetic survey. Bulletin Géodésique, vol 58(2), pp. 137–150.

Fernández-Ros A (2006): Modelización de movimientos y deformaciones de la corteza terrestre mediante observaciones de los satélites GPS (Aplicación al volcán Decepción). Ph. Tesis. Universidad de Cádiz.

Galindo-Zaldívar J, Bohoyo F, Maldonado A, Schreider A, Suriñach E, y Vázquez J T (2006) Propagating rift during the opening of a small oceanic basin: the protector basin (Scotia Arc, Antarctica). Earth Planet Sciences Letters, vol 241, pp. 318–412.

García A, Blanco I, Torta JM, Astiz MM, Ibáñez JM, Ortiz R (1997) A search for the volcanomagnetic signal at deception volcano (South Shetland I., Antarctica). Annali di Geofisica, vol 40(2), pp. 319-327.

Grafarend E, Voosoghi B (2003) Intrinsic deformation analysis of Earth's surface based on displacement fields derived from space geodetic measurements. Case studies: present-day deformation patterns of Europe and the Mediterranean area (ITRF data sets). Journal of Geodesy, vol 77(5–6), pp. 303–326.

González-Ferrán O (1985) Volcanic and tectonic evolution of the northern Antarctic Peninsula- Late Cenozoic to recent. Tectonophysics, 114, pp. 389–409.

González-Ferrán O (1991) The Bransfield rift and its active volcanism. (eds) Thomson RA, Crame JA, Thomson JW. Geological Evolution of Antarctica. Cambridge University Press, Cambridge, pp. 505–509.

Ibáñez JM, Almendros J, Carmona E, Martínez Arévalo C, Abril M (2003) The recent seismo-volcanic activity at Deception Island volcano, Deep Sea Research II: Topical Studies in Oceanography vol 50, pp. 1611–1629.

Hugentobler U, Schaer S, Friez P (2001) *Bernese GPS Software Version 4.2*. Astronomical Institute, University of Berne.

Hugentobler U, Meindl M, Beutler G, Bock H, Dach R, Jäggi A, Urschl C, Mervart L, Rothacher M, Schaer S, Brockmann E, Ineichen D, Wiget A, Wild U, Weber G, Habrichand H, Boucher C (2006) CODE IGS Analysis Center Technical Report 2003/2004. IGS 2003/2004 Technical Reports, Gowey K, Neilan R, Moore A (Eds), IGS Central Bureau, CA, USA.

Lavalle Y, Stix J, Kennedy B, Richer M, Longpre MA (2004) Caldera subsidence in areas of variable topographic relief: results from analogue modeling. Journal of Volcanology and Geothermal Research, vol 129, pp. 219–236.

Mantovani E, Cenni N, Alberello D, Viti M, Babbucci D, Tamburelli C, D'onza F (2001) Numerical simulation of the observed strain field in the central-eastern Mediterranean region. Journal of Geodynamics, vol 31, pp. 519–556.

Martí J, Vila J, Rey J (1996) Deception Island (Bransfield Strait, Antarctica): an example of volcanic caldera developed by extensional tectonics. Geological Society London, vol 110, pp. 253–265.

Mervart L (1995) Ambiguity resolution techniques in geodetic and geodynamic applications of the global positioning system. Philosophisch-natuerwissenschaftlichen Fakultät. University of Bern.

Mogi K (1958) Relations between the eruption of various volcanoes and the deformation of the ground surfaces around them. Bulletin Earthquake Research Institute vol 36, pp. 99–134.

Murray JB, Wooller LK (2002) Persistent summit subsidence at Volcán de Colima, Mexico, 1982–1999: strong evidence against Mogi deflation. Journal of Volcanology and Geothermal Research, vol 117(1–2), pp. 69–78.

NIMA (2000) World Geodetic System 1984: Its Definition and Relationship with Local Geodetic Systems. NIMA TR8350.2, Washington.

Newhall CG, Selft S (1982) The volcanic explosivity index (VEI): an estimate of explosive magnitude for historical volcanism. Journal of Geophysical Research, vol 87, pp. 1231–1238.

Sambridge M, Mosegaard K (2002) Monte Carlo methods in geophysical inverse problems. Reviews of Geophysics, vol 40(3).

Seeber G (2003) Satellite Geodesy. Walter de Gruyter, Berlín.

Smellie JL (2001) Lithostratigraphy and volcanic evolution of Deception Island, South Shetland Islands. Antarctic Science, vol 13(2), pp. 188–490.

Trimble (1991) Trimnet-Plus. Survey network software user's manual. California.

Vanícek P, Krakiwsky E (1986) Geodesy: The Concepts (Second Edition). Elsevier Science. Ámsterdam.

Vila J, Ortiz R, Correig AM, Garcia A (1992) Seismic activity on Deception Island. Yoshida Y, Kaminuma K, y Shiraishi K (Eds), Recent Progress in Antarctic Earth Science, pp. 449–456.

Willson L (1980) Relationships between pressure, volatile content and ejecta velocity in three types of volcanic explosions. Journal of Volcanology and Geothermal Research, vol 8, pp. 297–313.

Validation of the Atmospheric Water Vapour Content from NCEP Using GPS Observations Over Antarctica

Sibylle Vey and Reinhard Dietrich

Abstract An evaluation of the precipitable water (PW) in the reanalysis of the National Center for Environmental Prediction (NCEP) over Antarctica was carried out using observations from eight coastal Antarctic GPS stations. PW time series were derived from tropospheric parameters in conjunction with meteorological observations provided by the British Antarctic Survey for the period from 1994 to 2004 with a 2 h temporal resolution. The tropospheric parameters this study is based on are a product of the common GPS reprocessing from the Universities of Technology in Munic and Dresden.

The validation of NCEP PW reveals an underestimation of the seasonal signal in the PW from NCEP by 25% on the coast of East Antarctica and an overestimation of the PW by about 10% on the Antarctic Peninsula. Subdaily variations in the Antarctic PW are not correctly represented by NCEP due to the coarse spatial and temporal resolution of the global model. The agreement between GPS and NCEP PW is much better for most of the other regions on the earth than it is over Antarctica. The reason for the higher un-certainties in the Antarctic PW from NCEP is the lower availability of water vapour data in other region on the earth than it is over Antarctica. The reason for the higher un-certainties in the Antarctic PW from NCEP is the lower availability of water vapour data in the southern hemisphere especially, over Antarctica. PW values derived from GPS data could help to fill the large gape of water vapour data over Antarctica and could be assimilated in numerical weather prediction models within the near future.

1 Introduction

Atmospheric water vapour is a very important factor for many climate processes. It plays a key role in the heat budget of the Earth. Atmospheric water vapour is a

---

Sibylle Vey
Institut für Planetare Geodäsie, Technische Universität Dresden, Germany

Reinhard Dietrich
Institut für Planetare Geodäsie, Technische Universität Dresden, Germany

major greenhouse gas and it has a large impact on the global redistribution of latent heat (Philipona et al. 2005; Giovinetto et al. 1997). Especially, the polar regions are very important in the global energy cycle as they act as heat sinks and drive the atmospheric circulation. Due to the low water vapour content of the polar regions the heat can easily escape into space over these areas. Additionally, the snow and ice surfaces strongly reflect the solar radiation causing a small heat absorption over the polar regions (King and Turner 1997). Atmospheric water vapour is also the source of all precipitation over Antarctica. It therefore has a significant impact on the ice mass balance of the Antarctic ice sheet (Turner et al. 1999; Huybrechts et al. 2004).

An exact knowledge of the water vapour distribution in the atmosphere and its changes with time is indispensable for the description and understanding of global climate processes. However, compared to other atmospheric parameters the water vapour distribution is known very insufficiently. Especially in Antarctica only few water vapour observations are available. In areas with very cold temperatures and strong winds, measurements of radiosondes are very limited and inaccurate (Gaffen et al. 2000). Satellite observations based on infrared radiometers or solar occultations are restricted by the long periods of cloud coverage and darkness (Rossow 1996). Water vapour can be estimated from satellite based microwave radiometers almost independently of the weather conditions and over the oceans. But, over ice surfaces the correct separation of the emission of water vapour and ice remains still a problem (Wang et al. 2002).

In the last years observations from the Global Positioning System (GPS) provide a large data base for the weather independent estimation of water vapour (Bevis et al. 1992; Johnsen et al. 2004; Wickert et al. 2005). Precipitable water (PW) derived from ground based GPS measurements agree very well with the observations from other systems like radiosondes or water vapour radiometers (Rocken et al. 1993; Tregoning et al. 1998; Hagemann et al. 2003). Besides their good accuracy, PW estimates from GPS show an high temporal resolution. Several studies have been conducted in order to improve the weather forecast in different countries by using PW from GPS measurements (Gendt et al. 2004; Vedel and Huang 2004; Guerova et al. 2006). But, only few emphasis has been placed on the validation of the water vapour in numerical weather prediction (NWP) models over Antarctica. This work concentrates on the validation of PW in the NWP model of the National Center for Environmental Prediction (NCEP) using observations from coastal Antarctic GPS stations.

2 Data Analysis

2.1 PW from GPS

The signal propagation delay due to the constituents of atmosphere including the water vapour along the path from the GPS satellite to the receiver is estimated as additional parameter during the GPS data processing. This study is based on tropospheric parameters estimated in a common GPS reprocessing project of the Universities of Technology in Munic and Dresden (Steigenberger et al. 2006). The reprocessing of a global GPS network covers the period from the beginning of 1994 to the end of 2004 and was carried out with a modified version of the Bernese GPS Software 5.0.

The most important modifications in the reprocessing compared to the processing strategies of the analysis centers of the International GNSS[1] Service are: (1) the use of absolute antenna phase centre variations for GPS receivers and satellites (Schmid et al. 2005), (2) the application of the isobaric hydrostatic mapping function based on data of the NWP model of the European Centre for Medium Range Weather Forecasting (Niell 2000; Vey et al. 2006), (3) the consideration of higher order ionospheric effects (Fritsche et al. 2005) For further details on the strategy of the GPS reprocessing see Steigenberger et al. (2006).

Precipitable water was derived from the tropospheric parameters according to Bevis et al. (1994). For the separation of the hydrostatic and wet delay, surface pressure observations from the British Antarctic Survey (BAS) were used. The transformation of the wet delay in the corresponding amount of PW was performed with data of the mean atmospheric temperature from the ECMWF. PW time series of precipitable water were estimated for the period from 1994 to 2004 with a temporal resolution of 2 h for 8 stations along the Antarctic coast (Fig. 1). The GPS derived PW values were compared with PW estimates from satellite based microwave radiometers. Both estimates agree at the 1 mm level (Vey et al., 2004).

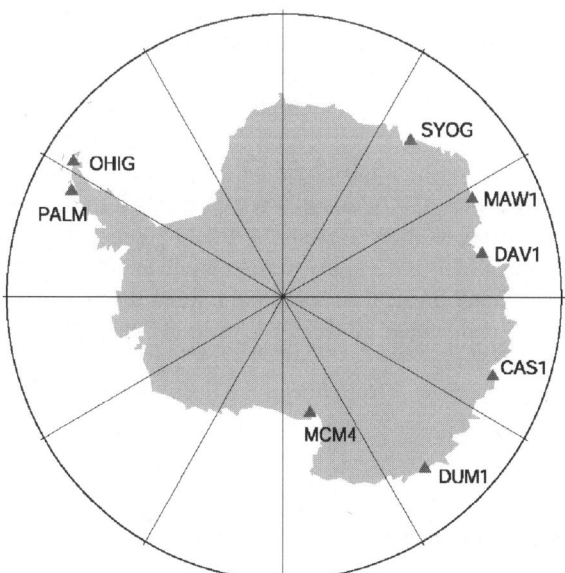

Fig. 1 Permanent GPS sites in Antarctica used in this investigation

---

[1] GNSS: Global Navigation Satellite Service

2.2 PW from NCEP

The two analysis centres NCEP and the National Centre for Atmospheric Research (NCAR) carried out a reanalysis of global atmospheric observations from 1948 to 2002 (Kistler et al. 2001). This reanalysis is based on the NCEP global spectral model with 28 vertical levels and a triangular truncation of 62 waves, equivalent to about 210 km horizontal resolution. The model field analysis is carried out every 6-h in accordance with the radiosonde observation times at 00UTC, 06UTC, 12UTC and 18UTC. The analysis scheme is a three-dimensional variational (3D-VAR) scheme (Kalnay et al. 1996).

The parameters of the reanalysis representing the state of the atmosphere are resolved on a horizontal grid with a resolution of $2.5 \times 2.5°$. These parameters can be classified into three categories (Kalnay et al. 1996). Parameters of the first category, like the temperature, strongly depend on the observations assimilated into the model. They are the most reliable parameters. Parameters of the second category are less reliable, as they depend on both – the model physics and the observations. Water vapour belongs to this category. Parameters of the third category like e.g. precipitation, depend only on the model physics and are less reliable than all other parameters (Kalnay et al. 1996).

A second reanalysis with improved model physics and more consistent observations was performed by NCEP in cooperation with the Department of Energy for the period from 1979 to present (Kanamitsu et al. 2002). The validation of precipitable water in NCEP is based on data of the reanalysis II. The resolution of the reanalysis II is the same as for the first reanalysis. One grid cell of NCEP represents a mean value for an area of $10.000\,km^2$. GPS probes the atmosphere along ray paths. In the 5 km thick atmospheric layer above the Earth's surface the GPS Signal covers a horizontal distance of 57 km to a satellite under an elevation angle of $5°$. The lowest 5 km layer of the atmosphere contain 95% of the total atmospheric water vapour. The GPS analysis, which includes observations to several over the sky distributed satellites, allows for inhomogenities of the atmosphere by 2 azimuthal gradients in North-South and East-West direction. GPS-derived PW represents a mean PW value over an surface with an diameter of about 100 km. In order to compare GPS- and NCEP-derived PW values, the NCEP PW values were interpolated to the GPS station locations by a two-dimensional linear interpolation between the surrounding grid points. As the NCEP PW values are given in 6 h intervals, the GPS PW estimates were combined to 6 h PW values.

3 Results

3.1 Time Series Analysis for Station Davis

Precipitable water time series from GPS and NCEP are presented as an example for station Davis (DAV1) in Fig. 2a. The mean PW content of this station is 3.4 mm. The very low amount of water vapour at station Davis represents the very dry Antarctic climate. The mean atmospheric water vapour of the other Antarctic stations is in the range of 2–7 mm, with more humid climate on the Antarctic Peninsula than on the East Antarctic coast (Table 1). Most regions of the Earth have a mean PW content of more than 10 mm. The mean PW content in the Tropics can even reach up to 40 mm. The seasonal signal in the water vapour shown in Fig. 2c was estimated by data stacking. The range between the minimum PW in the months of July to September and the maximum PW at the beginning of the year is about 4 mm. The interannual variations were estimated by applying a low pass filter with a bandwidth of 1.4 years. The range of the interannual variations at station Davis is 0.4 mm hence, one order of magnitude lower than the seasonal variations.

The differences in PW estimates from GPS and NCEP shown in Fig. 2b have a standard deviation of 0.7 mm, reveal a bias of 1,1 mm and a periodical signal. The periodical signal in the differences is caused by a smaller seasonal signal in PW from NCEP compared to the GPS PW (Fig. 2c). The amplitude spectra in Fig. 2d

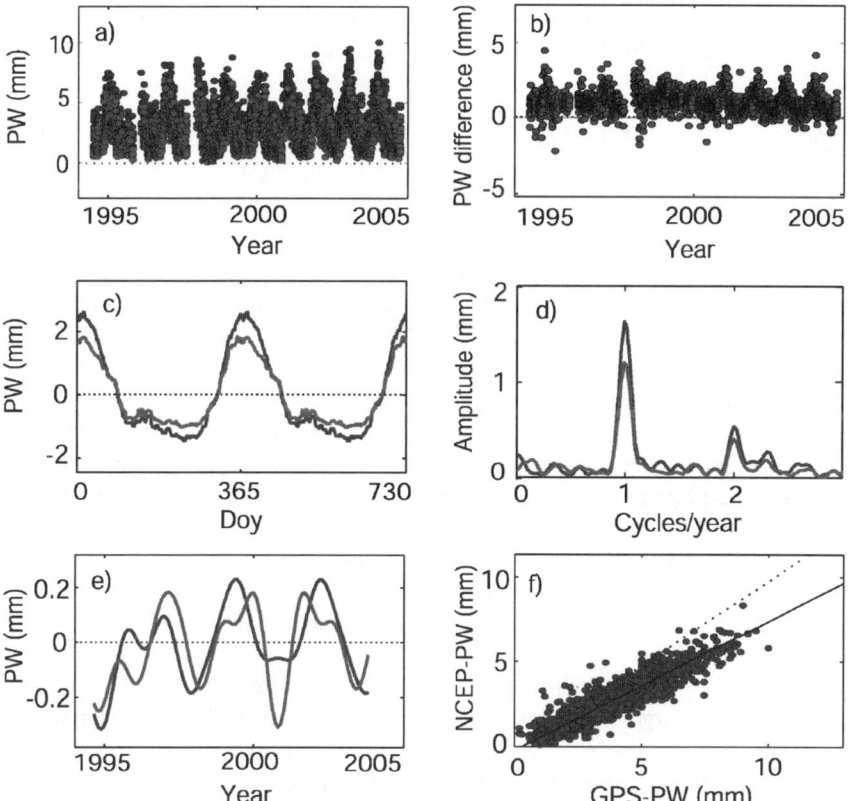

Fig. 2 Comparison of precipitable water (PW) at station Davis (DAV1) from GPS (*dotted grey lines*) and from NCEP (*solid black lines*), (**a**) daily mean values, (**b**) PW-difference (GPS-NCEP), (**c**) seasonal signal estimated by data stacking (monthly mean values), (**d**) spectrum, (**e**) interannual signal (low pass filter with an window size of 1.4 years), (**f**) regression plot of daily values

Table 1 Validation statistics; r: correlation coefficient, bias: difference GPS PW − NCEP PW, std: standard deviations of the PW differences, a: slopes of the regression lines PW(NCEP) = a*PW(GPS) + b

Station	Mean GPS-PW (mm)	r	bias (mm)	std (mm)	a
DAV1	3.4	0.94	1.1	0.7	0.76
CAS1	4.5	0.86	1.2	1.1	0.73
MAW1	2.3	0.94	0.3	0.6	0.77
SYOG	2.1	0.90	−0.5	0.8	0.76
MCM4	1.5	0.85	−0.1	0.7	0.72
PALM	6.8	0.73	−0.2	1.9	1.0
DUM1	3.3	0.89	0.2	0.9	0.91
OHIG	6.0	0.79	−1.6	1.7	1.04

shows that the annual amplitude of GPS PW is 0.5 mm larger than the amplitude of the PW from NCEP. The smaller seasonal signal in NCEP PW is also clearly visible in the regression plot in Fig. 2f. The regression line PW(NCEP) = a*PW(GPS) + b has a slope of a = 0.76. The agreement of the interannual signal at station Davis in Fig. 2e is satisfying.

3.2 Validation Statistics for All Stations

The comparison of the PW from GPS and NCEP reveals very small biases being not significant for most of the stations (Fig. 3a). Significant biases in the order of −1.5–1.3 mm occur for the stations Davis (DAV1), Casey (CAS1) and O'Higgins (OHIG) (Table 1). The bias could be related to both GPS and NCEP PW. In Antarctica mainly humidity measurements from radiosondes are assimilated into the model. The systematic error of relative humidity observations from radiosondes depends on the calibration of the humidity sensor (Soden and Lanzante 1996). Additionally, the accuracy of the humidity measurements from radiosondes strongly degrades in dry and cold conditions (Soden et al. 2004, Miloshevich et al. 2006). Due to the sparse data coverage in the southern Hemisphere, especially over Antarctica, radiosonde observations along the Antarctic coast highly influence the analysis results. Hence, biases in NCEP PW can be introduced by the assimilation of these observations.

Biases in the GPS PW estimates could, to a certain extent, be related to the estimation of the tropospheric parameters. Unmodelled effects of GPS antenna radomes can cause offsets in the ZTD parameters of several millimetres translating into a bias in precipitable water of up to 1 mm (Schupler 2001). In addition, uncertainties in surface pressure directly affect the zenith wet delay. Typical calibration errors of pressure sensors in the order of ±1 mbar causes errors in the PW estimates of up to ±0.5 mm. Furthermore, the elevations of pressure sensors were changed due to snow accumulation at some stations. The height difference between the GPS antenna and the pressure sensor is uncertain in the range of up to 10 m. This uncertainty migrates into the pressure reduction to GPS Antenna height and could result in an additional bias of up to ±0.3 mm in the PW estimates.

Figure 3b shows a smaller standard deviation (std) of the PW differences from GPS and NCEP for the stations on the coast of East Antarctica (std = 0.6 mm) than on the Antarctic Peninsula (std = 1.3 mm). The atmosphere over the peninsula holds about twice as much water than the atmosphere over East Antarctica. In Fig. 3d it is visible, that for all the stations the standard deviation of the PW differences between GPS and NCEP is related to the total water vapour content. The dependence of the standard deviation of the PW differences on the water vapour content indicates uncertainties in the NCEP model (Kistler et al. 2001; Kanamitsu et al. 2002).

Also in the seasonal time scale the differences between GPS and NCEP increase with an increasing amount of water vapour (Fig. 2f). The slope of the regression line between the PW from GPS and NCEP was calculated in order to quantify the relative error in PW. As presented by Fig. 3c the regression line has a slope of 0.75 for most of the stations on the coast of East Antarctica. It reveals an underestimation of the seasonal signal in the PW from NCEP of these stations by 25%. As examples for stations on the coast of East Antarctica the comparison of the seasonal signal in PW is shown for the stations Syowa (SYOG) and Mawson (MAW1) in Fig. 4a,b. For the stations on the Antarctic Peninsula the slope of the regression line is 1.1 which implies an overestimation of the seasonal signal by NCEP of 10%. Fig 4c,d

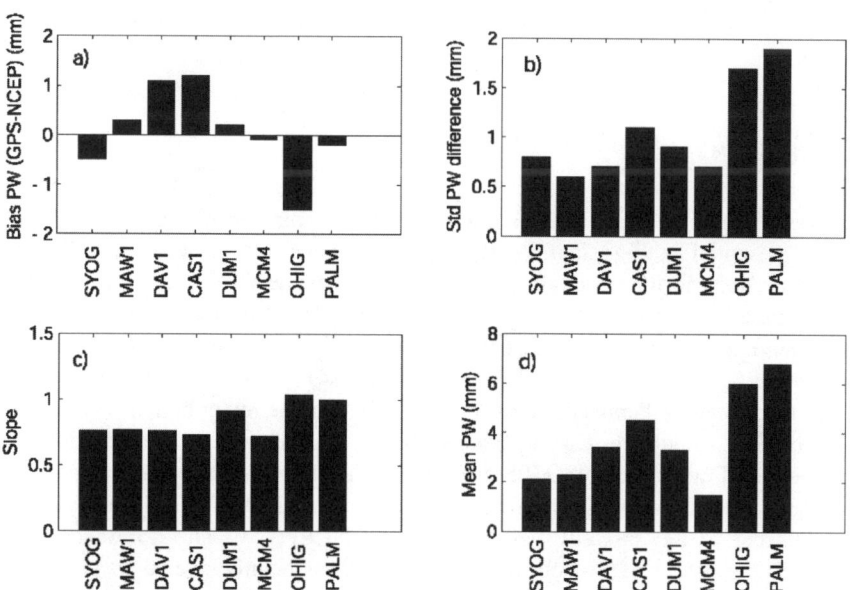

Fig. 3 Validation statistics (**a**) biases between GPS and NCEP PW, (**b**) standard deviations of the PW differences, (**c**) slopes of the regression lines PW(NCEP) = a*PW(GPS) + b, (**d**) means PW from GPS

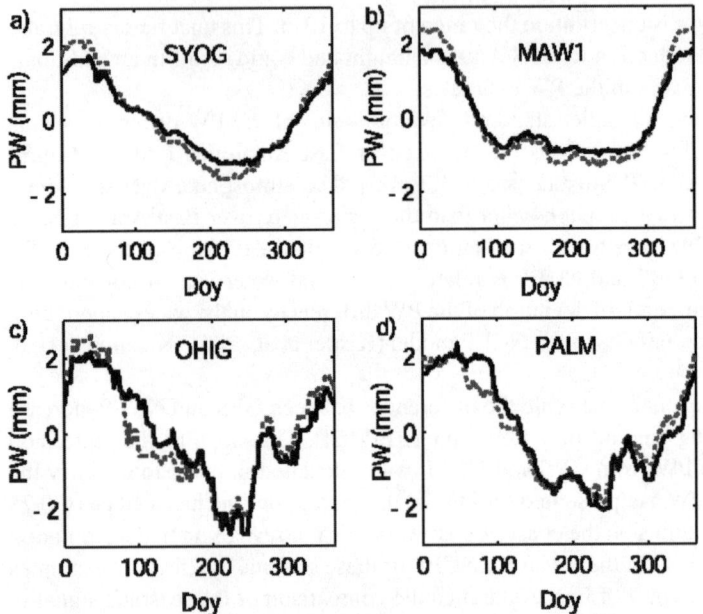

Fig. 4 Comparison of the mean seasonal signal in the PW between GPS (*dotted grey line*) and NCEP (*solid black line*), monthly values (Doy: day of the year)

show for the stations O'Higgens (OHIG) and Palmer (PALM), that this overestimation mainly takes place during the Antarctic autumn in the months of April and May.

3.3 Validation of Subdaily PW Variations

As example, for the coast of East Antarctica the comparison of 6-hourly PW values is presented in Fig. 5a,b for the stations Syowa and Mawson. Small subdaily variations in the order of 1 mm are visible in the GPS PW. However, the small subdaily variations in the PW are not represented by the model. The spatial and temporal resolution of NCEP is too coarse to resolve the small subdaily PW changes.

On the Antarctic Peninsula the subdaily variations are much higher than an the coast of East Antarctica. On the Peninsula, PW can change by up to 5 mm during one day (Fig. 5c,d). Both, GPS and NCEP, PW show subdaily variations in the order of a few millimetres. But, the subdaily variations of the GPS observations are different to the subdaily variations in the NCEP PW. The differences between both datasets can reach up to 5 mm and are mainly due to local climatological effects on the Antarctic Peninsula not considered in the global model. Especially the steep topography of the Peninsula causes such local climate effects.

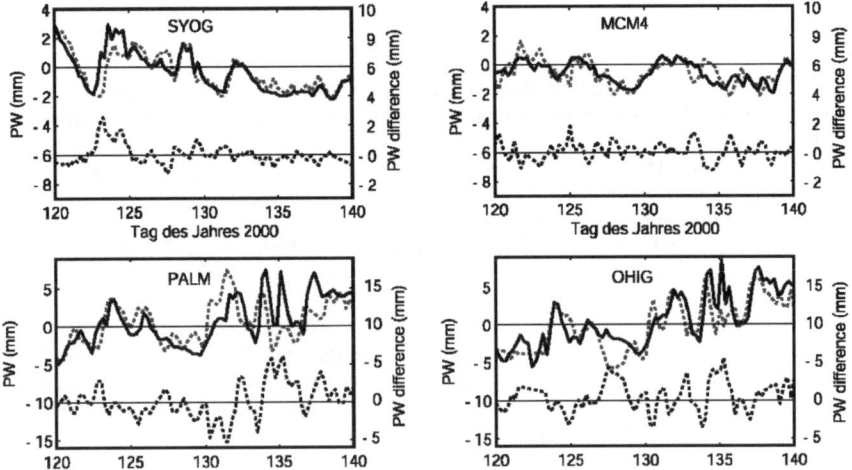

Fig. 5 Validation of the subdaily PW, 6-hourly values, *dotted grey line* GPS, *solid black line*: NCEP, *dotted black line*: PW difference

4 Conclusions

While the seasonal signal in the precipitable water is underestimated from NCEP by 25% at the coast of East Antarctica, it is overestimated by NCEP over the Antarctic Peninsula by about 10%. The overestimation is mainly caused by higher PW values in NCEP during the Antarctic autumn. Subdaily variations in the Antarctic PW are very small on the coast of East Antarctica and can reach up to 5 mm on the Antarctic Peninsula. In both regions NCEP is not able to resolve the subdaily variations in PW correctly due to the coarse spatial and temporal resolution of the model not taking into account local climatological effects.

The reason for the un-certainties in the Antarctic PW from NCEP is probably the low availability of water vapour data in the southern hemisphere. Especially, over Antarctica there are only very few water vapour observations available which can be assimilated in the model. PW values derived from GPS data are of high temporal resolution. They could help to fill the large gape of water vapour data over Antarctica and could be assimilated in numerical weather prediction models within the near future.

Acknowledgments Our thanks go to Peter Steigenberger, Markus Rothacher, Axel Rülke and Mathias Fritsche for their efforts in the reprocessing of the global GPS network. The investigations presented here were supported by the German Research Foundation (DFG). We are grateful to Johannes Böhm, for making the mean atmospheric temperature data of the ECMWF available to us. We also thank NCEP for providing model data of precipitable water on its website. The BAS made Antarctic meteorological data available via its website. One of the authors (S.Vey) was partly supported by the 'Studienstiftung des deutschen Volkes', which is gratefully acknowledged.

References

Bevis M, Businger S, Chiswell S, Herring TA, Anthes RA, Rocken C, Ware RH (1994). GPS meteorology: mapping zenith wet delays onto precipitable water. Journal of Applied Meteorology, 33(3):379–386.
Bevis M, Businger S, Herring TA, Rocken CR, Anthes A, Ware RH (1992). GPS meteorology: remote sensing of the atmospheric water vapor using the global positioning system. Journal of Geophysical Research, 97(D14):15787–15801.
Fritsche M, Dietrich R, Knöfel C, Rülke A, Vey S, Rothacher M, Steigenberger P (2005). Impact of higher-order ionospheric terms on GPS estimates. Geophysical Research Letters, 32. L23311, doi:10.1029/2005GL024342.
Gaffen DJ, Sargent MA, Habermann RE, Lanzante JR (2000). Sensitivity of tropospheric and stratospheric temperature trends to radiosonde data quality. Journal of Climate, 13:1776–1796.
Gendt G, Dick G, Reigber C, Tomassini M, Liu Y, Ramatschi M (2004). Near real time GPS water vapor monitoring for numerical weather prediction in Germany. Journal of the Meteorological Society of Japan, 82(1B):361–380.
Giovinetto MB, Yamazaki K, Wendler G, Bromwich DH (1997). Atmospheric net transport of water vapor and latent heat across 60°S. Journal of Geophysical Research, 102(D10):11171–11179.
Guerova G, Bettems J-M, Brockmann E, Matzler C (2006). Assimilation of COST 716 near-real time GPS data in the nonhydrostatic limited area model used at Meteo-Swiss. Meteorology and Atmospheric Physics, 91:149–164, doi: 10.1007/s00703.005.0110.6.
Hagemann S, Bengtsson L, Gendt G, (2003). On the determination of atmospheric water vapor from GPS measurements. Journal of Geophysical Research, 108(D21):4678, doi:10.1029/2002JD003235.
Huybrechts P, Gregory J, Janssens I, Wild M (2004). Modelling Antarctic and Greenland volume changes during the 20th and 21st centuries forced by GCM time slice integrations. Global and Planetary Change, 42(1–2):83–105.
Johnsen KP, Miao J, Kidder SQ (2004). Comparison of atmospheric water vapor over Antarctica derived from CHAMP/GPS and AMSU-B data. Physics and Chemistry of the Earth, 29(2):251–255.
Kalnay, E. u.a. (1996). The NCEP/NCAR 40-year reanalysis project. Bulletin of the American Meteorological Society, 77(3):437–471.
Kanamitsu M, Ebisuzaki W. Woollen J, Yang S-K, Hnilo JJ, Fiorino M, Potter GL (2002). NCEP-DOE AMIP-II Reanalysis (R2). American Meteorological Society, S. 1631–1643.
King JC, Turner J (1997). Antarctic Meteorology and Climatology. Cambridge University Press.
Kistler, R. u.a. (2001). The NCEP-NCAR 50-year reanalysis: monthly means CD-ROM and documentation. Bulletin of the American Meteorological Society, 82(2):247–267.
Miloshevich ML, Vommel H, Whiteman DN, Lesht BM, Schmidlin FJ, Russo F (2006). Absolute accuracy of water vapor measurements from six operational radiosonde types launched during AWEX-G and implications for AIRS validation. Journal of Geophysical Research, 111:D09S10, doi:10.1029/2005JD006083.
Niell AE (2000). Improved atmospheric mapping functions for VLBI and GPS. Earth Planets Space, 52:699–702.
Philipona R, Dürr B, Ohmura A, Ruckstuhl C. (2005). Anthropogenic greenhouse forcing and strong water vapor feedback increase temperature in Europe. Geophysical Research Letters, 32:L19809, doi:10.1029/2005GL023624.
Rocken C, Ware R, Van HT, Solheim F, Alber C, Johnson J (1993). Sensing atmospheric water vapor with the global positioning system. Geophysical Research Letters, 20(23):2631–2634.
Rossow WR (1996). Radiation and Water in the Climate System: Remote Measurements, Band 45 of NATOASI Series 1: Global Environmental Change, chapter Remote Sensing of Atmospheric Water Vapor, S. 175–191.

Schmid R, Rothacher M, Thaller D, Steigenberger P (2005). Absolute phase center corrections of satellite and receiver antennas. GPS Solutions, 9:283–293, doi 10.1007/s10291-005-0134-x.

Schupler BR (2001) The response of GPS antennas – how design, environment and frequency affect what you see. Physics and Chemistry of the Earth, 26(6–8):605–611.

Soden BJ, Lanzante JR (1996) An assessment of satellite and radiosonde climatologies of upper-tropospheric water vapor. Journal of Climate, 9:1235–1250.

Soden BJ, Turner DD, Lesht BM, Miloshevich LM (2004). An analysis of satellite, radiosonde, and lidar observations of upper tropospheric water vapor from the Atmospheric Radiation Measurement Programm. Journal of Geophysical Research, 109:D04105, doi:10.1029/2003JD003828.

Steigenberger P, Rothacher M, Dietrich R, Fritsche M, Rülke A, Vey S (2006) Reprocessing of a global GPS network. Journal of Geophysical Research, 111:B05402, doi:10.1029/2005JB003747.

Tregoning P, Boers R, O'Brien D, Hendy M (1998) Accuracy of absolute precipitable water vapor estimates from GPS observations. Journal of Geophysical Research, 103(D22):28701–28710.

Turner J, Connolley WM, Leonard S, Marshall GJ, Vaughan DG (1999) Spatial and temporal variability of net snow accumulation over the Antarctic from ECMWF reanalysis project data. International Journal of Climatology, 19(7):697–724.

Vedel H, Huang XY (2004) Impact on ground based GPS data on numerical weather prediction. Journal of the Meteorological Society of Japan, 82(1B):459–472.

Vey S, Dietrich R, Fritsche M, Rülke A, Rothacher M, Steigenberger P (2006) Influence of mapping function parameters on global GPS network analyses: comparisons between NMF and IMF. Geophysical Research Letters, 33:L01814, doi:10.1029/2005GL024361.

Vey S, Dietrich R, Johnsen KP, Miao J, Heygster G (2004). Comparison of tropospheric water vapour over Antarctica derived from AMSU-B data, ground-based GPS data and the NCEP/NCAR reanalysis. Journal of the Meteorological Society of Japan, 82(1B):259–267.

Wang JR, Racette P, Triesky ME, Manning W (2002) Retrievals of column water vapor using millimeter-wave radiometric measurements. IEEE Transactions on Geoscience and Remote Sensing, 40(6):1220–1229.

Wickert J, Beyerle G, König R, Heise S, Grunwaldt L, Michalak G, Reigber C, Schmidt T (2005) GPS radio occultation with CHAMP and GRACE: a first look at a new and promising satellite configuration for global atmospheric sounding. Annales Geophysicae, 23:653–658.

Geodynamics of the Tectonic Detachment in the Penola Strait (Antarctic Peninsula, Archipelago of Argentina Islands)

K.R. Tretyak, Y.I. Golubinka, A.J. Kulchytskyy and L.V. Babiy

The territory of Antarctic peninsula covers a tectonically active zone of both modern and Neogene volcanic activity. Currently, there is great interest in crustal motions of the Antarctic continent and during the last decade methods of satellite geodesy, namely GPS-technologies, are used to study the deformations of the land surface of Antarctic Peninsula. The annual GPS-campaigns implemented under the auspices of SCAR (International Scientific Committee of Antarctic Researches) are a basic step to better understand such active deformations (Deitrich et al. 2001). The locations of these observations are concentrated near polar stations, yet they cover practically whole territory of Antarctic Peninsula. Additionally the GPS-measurements are implemented to form local geodynamic polygons. The purpose of these observations is to study the tectonic activity of terrane separating the polar stations using a collection of various station coordinates, observing epochs and different scientific equipment and experimental procedures. A focused effort has been undertaken by the Ukrainian Antarctic expedition that seeks to investigate the geodynamics and tectonic deformation style of the Argentinean Islands. The targeted terranes include a, here-to-fore poorly understood, but extremely important, tectonic detachment found in the Argentinean Islands archipelago. Centrally located on Galindez Island, and central to the area of tectonic study is the station Academic Vernadskyy. On these territories the precision geodetic network has been created and first cycles of GPS-observations have been collocated with geophysical instruments for determining the regional crustal structure.

Nineteenth Century research in the territory of the Argentinean Islands was undertaken by Belgian scientists under the supervision of Andrian Gerlah. In the

---

K.R. Tretyak
National University "Lvivska Politechnika", Stepana Bandery street, 12, 79013 Lviv, Ukraine, e-mail: kornel@polynet.lviv.ua

Y.I. Golubinka
National University "Lvivska Politechnika", Stepana Bandery street, 12, 79013 Lviv, Ukraine

A.J. Kulchytskyy
National University "Lvivska Politechnika", Stepana Bandery street, 12, 79013 Lviv, Ukraine

L.V. Babiy
National University "Lvivska Politechnika", Stepana Bandery street, 12, 79013 Lviv, Ukraine

results of this early research was the first geological map of these territories. The studies were followed by geological and cartographical mapping conducted by the British Antarctic Service (BAS). A detailed retrospective tectonic-geological overview of the archipelago Argentinean islands was described in (Geological map 1981) and these geological and paleomagnetic maps form the underpinning of much that way can say about the geological environment of this territory today. A number of the main features of what is known from these past tectonic-magnetic studies in the territory of the Argentinean Islands are highlighted in (Bahmutov 1998, 2002). Paleomagnetic lineations form the most solid quantitative record at the site of the "Three little piglets" (Fig. 1) in the Argentinean Islands.

However, the mapped trace of the main detachment fault coincides with waterway of the Penola Strait, with the evidence primarily coming from a regional geomorphological study.

Our results and analysis of this complex structural and tectonic setting now bring the powerful new tools of modern GPS network geodesy to bear on the basic

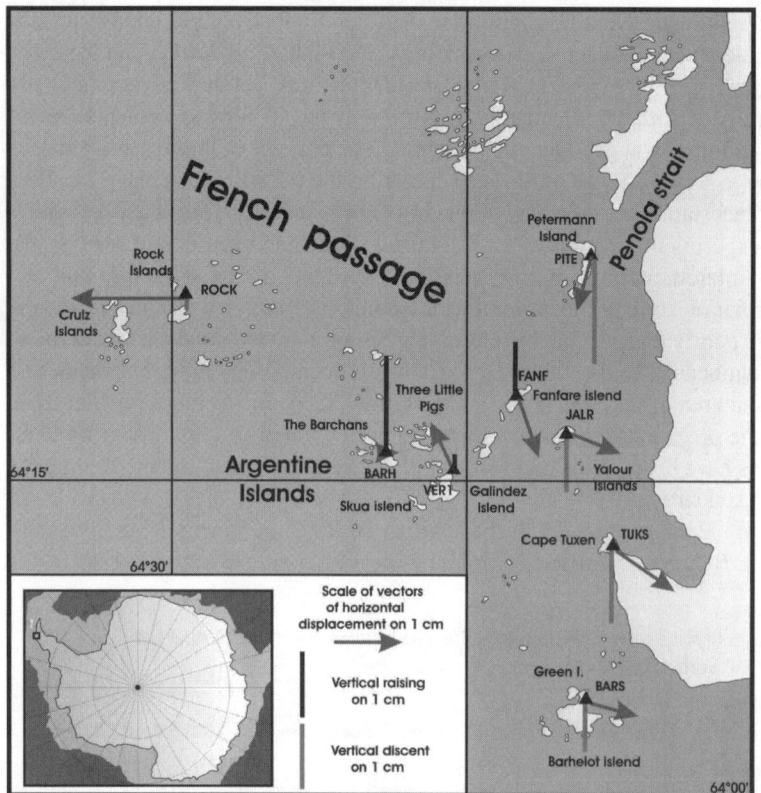

Fig. 1 Scheme of the network and displacements vectors of geodetic points of geodynamic polygon within the Argentinean Islands contiguous to the Ukrainian Antarctic station "Academic Vernadskyy"

geodynamical questions related to the region, for we can now measure the current rates and directions of tectonic block motions.

Due to the relative lack of detailed tectonic structural information in the Antarctic Peninsula, and due to the fundamental differences between geological-geomorphologic and paleomagnetic data mapped traces of the detachment fault through the Argentinean Islands proximal to Academic Vernadskyy station, it was necessary to design a precision geodetic network with repeat measurements in order to isolate and map ongoing deformation.

At the time of the 8-th Ukrainian Antarctic Expedition (February–March, 2003) a regional network had been formed in the surrounding islands by the joint efforts of the scientists of the National University "LvivskaPolitechnika" (K. Tretyak, V. Hlotov) and the Close Corporation "ECOMM" (J. Ladanovskyy, P. Bahmach) and with the support of Ukrainian Antarctic Center (Bahmach et al. 2003). The network was developed not only with purpose of studying the crustal deformations and in the Antarctic Peninsula, but also for providing basic geodetic control points for improving topographic-geodetic mapping at the station "Academic Vernadskyy".

The network requires a premium effort due to the need for special technical methods required to meet the point-to-point accuracy required for measuring tectonic motions.

The geodynamic network spans the northeastern islands in a radius of 15 km, encompassing both the station and a part of the continent. The network utilizes 8 geodetic points, including the point VER1. Monumentation of the sites were designed such that the outside part of the central drill core hole was embedded with a metal rod of length up to 40 cm with a mark placed for the purpose of estimating the stable phase center of the GPS-antenna. At a height 10–15 degrees and more, satellites line-of-site was generally unobstructed. The average baseline length of the measured vectors is 7–8 km, and the longest and shortest are 17 km and 2.5 km, respectively. The heights of points in system WGS-84 are roughly 20–30 m.

Repeated GPS-measurements were implemented during March and April 2005 of the seasonal 10-th Ukrainian Antarctic Expedition campaign.

The characteristics of the repeated cycles are shown in the Table 1. Observations were implemented using dual-frequency GPS receivers Trimble 4800, Leica SR-399 SR-9500, single-frequency receiver Trimble 4600LS (first cycle of observation); and dual-frequency receivers Ashtech Z-12, Leica SR-399 SR-9500, Javad TPSHIPER_GGD, single-frequency receiver Thales and Ashtech Proark2 (second cycle of observations).

In the last column of Table 1 the mean square error (MSE) in the determination of the plane coordinates (upper row) and the altitudes (lower row) is given.

In both repeat observations testing of the antennas of all receivers was conducted, with purpose of calibrating GPS-antenna phase centers, using a rigorous methodology as described in (Tretyak et al. 2002).

Both day-time and night-time measurements were taken and the cut-off elevation for measurements of pseudodistances was 10 degrees. The frequency of pseudodistances measurements was 30 s.

Table 1 Technical characteristics of repeated cycles of GPS observations

No of cycle	Duration of the cycle of measurement	Duration of vectors measurement, hours	Number of repeated measurements of vectors	Number of measured vectors	A priori and a posteriori MSE of unit of weight	MSE, mm Aver.	Max.
1	12.02–28.02 2003	2–12	1–6	37	3	3 5	3 8
2	23.03–01.04 2005	5–24	1–5	68	1.5	1.5 2	1.5 2

Adjustment of these observations was implemented using the software SKI of LEICA. The referencing of the network to the permanent Antarctic stations was accomplished prior to adjustment. The coordinates of points in the ITRF-2000 system were determined and evaluation of the accuracy of point determination was implemented in the result of the network adjusting.

Corrections of phase centers of GPS-antennas were applied and a posteriori optimization of the network according to methodology described in (Tretyak et al. 2002) was implemented with the goal of increasing reliability and accuracy of point's coordinate determination. Owing to the method used for determination of the phase center eccentricities of the antennas and an a posteriori optimization of GPS networks measurements, the accuracy and reliability of point coordinate determinations increased by 15–25%. Figures 2 and 3 detail the scheme used for GPS measurements of the geodetic network, and show the error ellipses and errors in determination of the elevation components of February–March, 2003 and March–April, 2005, respectively. We can see from the two figures that the accuracy of the network of 2005 is considerably more robust than the network of 2003. This is easily explained by the larger number of GPS receivers and the lengthened observational period (Table 1).

The observations determined the spatial displacement vectors of the network benchmarks (Fig. 1). Horizontal and height displacement vectors during two years of observing are as large as 17 mm with unambiguous sensitivity to the sense of direction of the tectonic motions also established.

Displacements of points JALR, TUKS, BARS are directionally oriented to the southeast and their velocity is in the bounded by 5–7 mm/year. Despite this the points are descended relatively to other network points with velocity 5–8 mm/year. Stations PITE and FANF are moving in a southerly direction with velocity 4 mm/year. The elevation of the station PITE decreases with an apparent velocity of 8 mm/year and FANF, in contrast, rises with an apparent velocity of 3 mm/year. The points VER1 and ROCK are moving to the northeastern and eastern directions, respectively, and have essentially no detectable vertical height differences. Point BARH is stable in horizontal position, and the vertical motions have an apparent velocity of 6 mm/year. The geodetically determined kinematics can now be used to decipher the motions of two crustal blocks. These are the eastern and western blocks

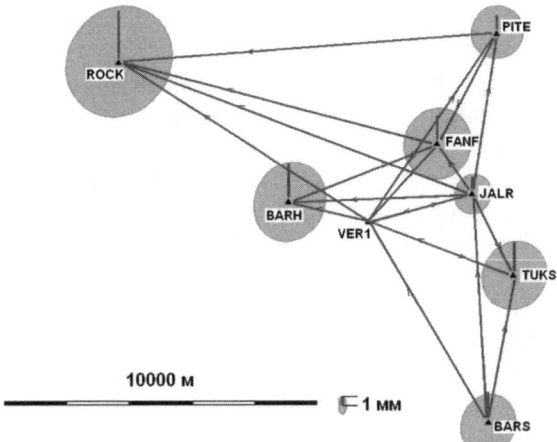

Fig. 2 Scheme of GPS measurements of the geodetic networks implemented in February–March, 2003, and accuracy of coordinates determination

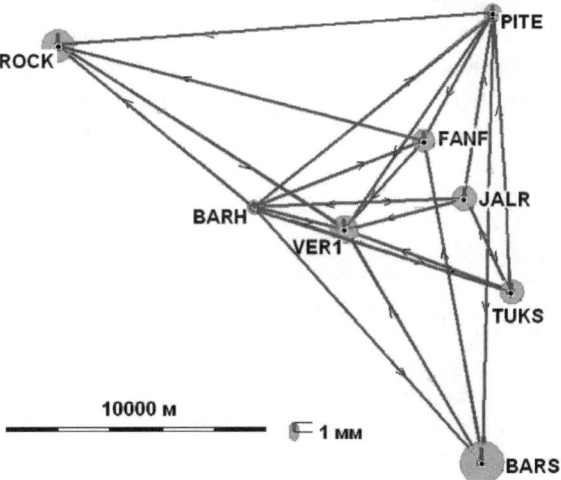

Fig. 3 Scheme of GPS measurements of the geodetic networks implemented in March–April, 2005, and accuracy of coordinates determination

delimited on the conventional line oriented from the north to the south between the station "Academic Vernadskyy" and the island of Fanfare. The opposite directional of movements of the points VER and FANF are consistent with the inference that this line forms the main tectonic detachment.

From a geological point of view the investigated territory consists of two parts: synclinal zones of both the Antarctic Peninsula and of the outboard Argentinean Islands (Bahmutov 1998). Between these two zones British researchers (Geological map 1981) are implementing investigation of detachment violation. Its direction is approximately coincides with direction of waterway part of the strait. Here after we term this structure as the Penola Detachment Fault (PDF). Analysis of vector displacement of geodetic points of a geodynamic polygon (Fig. 1) indicates a trend, in a southwestern direction, of the fault surface and adjacent stepped, sympathetic fault structures (Fig. 4) that are so common in rapidly extending terranes. As in Fig. 4, we infer that this structure has a classic graben structure, quite characteristic of normal faulting structures, wherein the faults have broken the entire continental lithosphere in which they initially formed (Tapponier 1978). Motions are on subhorizontal planes and the two main blocks move away of each other, with the relative vertical displacements as indicated in Fig. 4.

Displacement vectors fixed in the regions of the Peterman and Fanfare Islands are obviously connected to counter-directed motions of stepped blocks 2 and 3 (Fig. 4). The character of these movements can be connected with the wedge structure of these stacked micro-blocks and in a horizontal section these appear to "extrude" in a southernly direction. Additionally we infer that a smaller block 3 inherits the downward motion of a larger block 4, and vice-versa, a smaller block 2 inherits the vertical rise of a larger block 1.

It should be emphasized that the scarcity of geodetic and geological information precludes the disposition of our results to be preliminary in nature. This should be kept in mind concerning the conclusions regarding the tectonic structures we infer for this relatively small portion of the Antarctic Peninsula. For more precise deter-

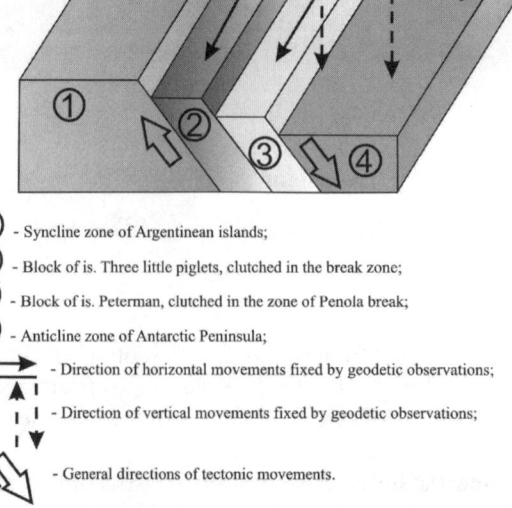

① - Syncline zone of Argentinean islands;
② - Block of is. Three little piglets, clutched in the break zone;
③ - Block of is. Peterman, clutched in the zone of Penola break;
④ - Anticline zone of Antarctic Peninsula;
⟶ - Direction of horizontal movements fixed by geodetic observations;
⇡⇣ - Direction of vertical movements fixed by geodetic observations;
⇗ - General directions of tectonic movements.

Fig. 4 Geological interpretation of vector displacements of geodynamic polygon points of Argentinean islands and contiguous part of Antarctic Peninsula

mination of the location of probable tectonic detachment and more confident rates and directions in the territory of the Argentinean Islands near the station "Academic Vernadskyy", a repeated series of observations with extended network coverage will be necessary.

1 Conclusions

A geodynamic polygon of geodetic observation proximal to the Argentinean Islands clearly and unambiguously differentiate subsets of displacements fields within the frame of the network points. Two opposite block motions are determined and the presence of the main detachment surface currently driving crustal motions has been preliminarily determined.

It is now possible to begin to assemble a geodetically controlled model of the active tectonic structure and kinematics of the PDF.

References

Bahmutov VG (2002) Geological structure and paleomagnetic researches in western Antarctic (region of Argentinean islands) and their importance for paleotectonic reconstructions of Antarctic Peninsula. Bulletin of UAC. – K. – 2002. vol 4. pp 11–24.

Bahmutov VG (1998) Geological overview of archipelago Argentinean islands and contiguous teritorries of Antarctic Peninsula. Bulletin of UAC. – K. – 1998, vol 2, pp 77–84.

Bahmach PG, Hlotov VM, Ladanovskyy JV, Tretyak KR (2003) Geodynamic network of Argentinean islands contiguous to the station "Academic Vernadskyy". Ukrainian Antarctic Journal. No. 1. pp. 149–150.

Tretyak KR, Romanyshyn IB, Golubinka JI (2002) To the question of determination of eccentricity of the phase center of GPS-antenna receiver. Geodesy, Cartography and Aerial Survey. – Lviv. – 2002. – vol 62, pp 87–96.

Deitrich R, Dach R, at all (2001) ITRF coordinates and plate velocities from repeated GPS campaigns in Antarctica – an analysis based on different individual solutions. Journal of Geodesy. – 2001. ВИП. 74. – c. 756–766

Geological map of the Southern Graham Land. – 1:500000 / BAS (500G) Cambridge British Antarctic Survey. 1981

GPS and Radiosonde Derived Precipitable Water Vapour Content and its Relationship with 5 Years of Long-Wave Radiation Measurements at "Mario Zucchelli" Station, Terra Nova Bay, Antarctica

Pierguido Sarti, Monia Negusini, Christian Lanconelli, Angelo Lupi, Claudio Tomasi and Alessandra Cacciari

Abstract The Italian "Mario Zucchelli" Station ($-74° 41' 55''.6997$ N, $164° 06'10''.5887$ E), situated at Terra Nova Bay, Northern Victoria Land, is equipped with a permanent Global Positioning System receiver (TNB1), continuously observing since 1998. "Mario Zucchelli" is an Antarctic experimental facility where a large number of scientific observations are carried out, either permanently or seasonally. In particular, an observatory devoted to atmospheric physics is located at Campo Icaro, 2.5 km from the base: it is a Clean Air Facility where several atmospheric measurements are carried out. Since 2000, long-wave radiation measurements are performed routinely using a Kipp&Zonen CNR-1 net radiometer.

The GPS data set acquired over a six-year period spanning 2000–2005 by the permanent GPS station TNB1 is here analyzed, with the aim of retrieving the Precipitable Water (PW) content.

Water vapour radiative effects on the thermal radiation balance of the atmosphere are of basic importance for the energy budget of the surface-atmosphere system in Antarctica, even though Precipitable Water assumes in general appreciably lower values than in other areas of our planet.

Pierguido Sarti
Istituto di Radioastronomia – INAF, Via P. Gobetti 101, 40129 Bologna, Italy,
e-mail: p.sarti@ira.inaf.it

Monia Negusini
Istituto di Radioastronomia – INAF, Via P. Gobetti 101, 40129 Bologna, Italy

Christian Lanconelli
Istituto di Scienze dell'Atmosfera e del Clima – CNR, Via P. Gobetti 101,
40129 Bologna, Italy

Angelo Lupi
Istituto di Scienze dell'Atmosfera e del Clima – CNR, Via P. Gobetti 101, 40129 Bologna, Italy

Claudio Tomasi
Istituto di Scienze dell'Atmosfera e del Clima – CNR, Via P. Gobetti 101, 40129 Bologna, Italy

Alessandra Cacciari
Carlo Gavazzi Space SpA c/o Istituto di Scienze dell'Atmosfera e del Clima – CNR,
Via P. Gobetti 101, 40129 Bologna, Italy

Therefore, precise calculations and measurements of the mean long-wave radiation flux density reaching the surface at Terra Nova Bay are presented as a function of Precipitable Water, to give evidence of the relationship existing between this radiative balance term and the total atmospheric content of water vapour.

The GPS-derived PW values are compared with radiosonde-derived PW values measured at Terra Nova Bay. The radiosounding data were analyzed by correcting the temperature data for the errors due to radiation and heat exchange processes and lag effects; the air relative humidity data were corrected for the errors and various dry bias following an accurate procedure recently developed by Tomasi et al. (2006) to reduce the errors and bias affecting the moisture measurements.

The analysis strategies that have been applied to GPS and radiosonde data sets for computing PW are presented, including (i) the relation between the measured radiant-flux density and the GPS-derived PW, and (ii) the comparison with the predicted radiant-flux density derived by a model with five different profiles of temperature and humidity, computed from the set of radiosoundings performed at Terra Nova Bay. GPS- and radiosonde-derived water vapour contents at Terra Nova Bay show a good agreement over the whole sample period, with small discrepancies that will be opportunely discussed.

1 Motivations and Introduction

Human-induced variations on aerosols and greenhouse gases appear to have a significant impact on the Earth's climate system. For a variety of different scenarios, numerical models predict worrying climate change in the near future. With little uncertainty, within the end of this century, temperature rise will deeply influence the whole Earth's ecosystem. If climate's response to anthropogenic activities is certain, there are still large uncertainties in quantifying its magnitude. In order to do so, several processes and parameters must be continuously monitored at local, regional and global scale and these observations must be properly combined.

One important role in climate change is played by water vapour: it is the strongest greenhouse gas and its content variations in the low and middle troposphere are expected to originate positive feedback processes. Therefore, water vapour is central to the climate and its variability and change (Houghton et al. 2001, p. 88). It is important to estimate its concentration as continuously and densely as possible. Water vapour observations performed in Antarctica are particularly important: they supply information for depicting the global distribution and circulation of water vapour and its relation with global warming. In the short history of Antarctic science, time series of water vapour records have been obtained using radiosonde observations. These latter are sparse and, often, difficult to perform on a regular basis. Furthermore, an uneven distribution of observations as well as the use of different sensors remarkably complicates the assessments of water vapour trends (Rind 1998). Space geodetic techniques using radio signals, in particular Very Long Baseline Interferometry (VLBI) and Global Positioning System (GPS) have proved to be effective tools for sounding the atmosphere and e.g. inferring about integrated water vapour content (Bevis et al. 1992, Emardson et al. 1998, Niell et al. 2001, Schuh et al. 2005), this quantity being given by the integral of absolute humidity along the vertical path of the atmosphere. Space geodetic networks in Antarctica are far too sparse to attempt the computation of a time varying 3-D tomography of atmospheric moisture: for this purpose, a specific design of the network is necessary. For the moment being, studies have been carried out at mid-latitude regions only (Gradinarsky et al. 2000, Haase et al. 2001). Nevertheless, Antarctic space geodetic permanent sites can be effectively used for computing continuous high-resolution time series of moisture columnar content, as far as observations of local surface pressure and temperature are available (e.g. Vey et al. 2004). With the same requirements, the ever increasing number of episodic Antarctic GPS observations can also be processed for computing site specific PW.

Obviously, GPS observations alone do not allow monitoring the complex impact of water vapour on climate system: its vertical distribution and its radiative forcing effects cannot be determined. They should rather be regarded as an ancillary, although precise, source of information in climatology, as well as in meteorology and glaciology. Nowadays, and with increasing importance as the number of observations increases, they can be processed for densification and validation purposes. Furthermore, GPS-derived time series contain information on high frequency variability of PW, since 24 determinations per day can be easily computed in remote and unmanned sites. This latter characteristic makes GPS very interesting with respect to other sensors.

In order to be useful, PW time series must be properly computed: unbiased absolute values of moisture are what we are looking for. In order to prepare for the analysis of the entire GPS data set collected at Terra Nova Bay, we have focused our attention on fine tuning an efficient processing strategy. We have selected a limited amount of data (December 2002 and January 2003) with the specific purpose of comparing GPS-derived PW with results obtained from radiosonde observations. Since 1987, these latter are performed every 12 h during the opening season of the base, approximately from November to mid-February. No information on water vapour content has ever been collected during the rest of the year. Therefore, at Terra Nova Bay, GPS observations have the capability of providing moisture values filling the temporal gap that is produced by uneven distributed radiosonde observations. In order to do so, it is necessary to test an appropriate analysis strategy and to ensure an unbiased GPS-derived PW.

In Sect. 2, the fundamental physical processes due to water vapour are described, giving evidence of the role of this minor atmospheric constituent in the global climate system. In Sect. 3, the geodetic GPS analysis procedure followed in the present study is described. Section 4 is devoted to atmospheric delays, recalling the different steps that are necessary for the retrieval of the Zenith Wet Delays (ZWD) starting from the adjustment of observations, the estimation of Zenith Total Delay (ZTD) and the modelisation of Zenith Hydrostatic Delay (ZHD). Some tricks applied for fulfilling gaps in surface pressure and temperature readings are also presented, pointing out that their lack makes the computation of ZWD impossible. The radiosoundings and the data processing procedure are described in detail in Sect. 5, along with the procedure used for computing PW from GPS-derived ZWD. Section 5 also contains a comparison between GPS-derived and radiosonde-derived PW values; results are compared and discussed. GPS-derived PW time series are related in Sect. 6 to on site measurements of long-wave radiant-flux density performed during the austral summers from 2000 till 2005, showing the main dependence features of downward radiant flux of the atmosphere on PW.

2 The Role of Water Vapour in the Atmospheric Radiation Budget

Atmospheric water vapour plays a crucial role in the Earth's energy budget, because it absorbs very strongly the incoming (short-wave) solar radiation and contributes efficiently, together with the other greenhouse gases and clouds, to absorb the thermal (long-wave) radiation emitted by the surface-atmosphere system toward the space (Starr and Melfi 1991). The absorption of solar radiation is caused by numerous absorption bands, some of which are in the visible part of the solar spectrum, from 0.57 to 0.70 µm, causing only weak absorption effects, and the other are distributed throughout the near-infrared wavelength range from 0.71 to 3.70 µm, producing considerably stronger absorption effects, due to higher intensities by several orders of magnitude and larger spectral widths (Kondratyev 1969). These bands are commonly identified by means of Greek letters[1], each of those centred at wavelengths $\lambda > 1$ µm being capable of absorbing totally the incoming direct solar irradiance within the central part of its spectral range for values of the vertical atmospheric content of water vapour greater than 1 g cm^{-2} (Leckner 1978; Tomasi and Trombetti 1985). Defining the annual global mean terms of the Earth's radiation balance, Kiehl and Trenberth (1997) evaluated that about 20% of the incoming solar radiation is absorbed by atmospheric water vapour. However, these calculations over the planetary scale are made commonly by means of radiative transfer model, in which parameterisation of water vapour absorption results to be presently inadequate to represent the strong dependence of atmospheric absorption on the columnar content of water vapour (Arking 1996). Calculations made by Tomasi et al. (2000) using the well-known MODTRAN 3.7 code (Kneyzis et al. 1996) for the Midlatitude Summer standard atmosphere model (Anderson et al. 1986) showed that a total columnar content of water vapour equal to 2.9 g cm^{-2} absorbs the incoming direct solar irradiance by 17.1% along the vertical atmospheric path, and by 35.5%, 38.9% and 43.3% along atmospheric slant paths corresponding to relative optical air masses equal to 2, 3, and 5, respectively (i.e. for apparent solar zenith angles equal to 60°, 71° and 79°, respectively). Considering that the most part of water vapour

---

[1] with labels α the band occupying the 0.70–0.74 µm range, 0.8 µm in the 0.79–0.84 µm range, $\rho\sigma\tau$ in the 0.926–0.978 µm range, φ in the 1.095–1.165 µm range, ψ in the 1.319–1.498 µm range, Ω in the 1.762–1.977 µm range, X in the 2.520–2.845 µm range, and X' in the 2.990–3.570 µm range (see Kondratyev 1969).

absorption occurs at tropical and mid latitudes, these calculations indicate that the Kiehl and Trenberth (1997) evaluations are quite realistic.

In addition, atmospheric water vapour absorbs the long-wave (infrared) radiation emitted upward by the surface as well as the thermal radiation emitted downward by the atmosphere. These absorption processes occur through two distinct absorption mechanisms: (i) a selective absorption, mainly due to the vibrorotational band v_2 centred at $\lambda = 6.27$ µm and the rotational bands present in the spectral range $\lambda > 18$ µm (Goody 1964), and (ii) the continuum absorption, which is induced by both foreign and self-broadening mechanisms of the absorption lines that constitute the absorption bands of the water molecule and those of its dimer molecules, distributed throughout the middle and far infrared. The foreign-broadening contribution was evaluated to be proportional to the total dry-air pressure and to increase as a function of air temperature (Bignell 1970), while the self-broadening contribution was determined to be proportional to the water vapour partial pressure and to decrease gradually as the air temperature increases (Bignell 1970, Tomasi et al. 1974, Clough et al. 1989). As a result of the intense selective absorption within both wavelength intervals $\lambda < 8$ µm and $\lambda > 18$ µm, caused by the strong vibrorotational and rotational bands, and of the less intense continuum absorption within the $8 \leq \lambda \leq 13$ µm atmospheric window, water vapour is the most abundant of the greenhouse gases and, hence, causes the dominant contribution to the greenhouse effect occurring in the Earth's atmosphere, giving rise to changes in the energy budget of the terrestrial climate system, which are estimated to be more marked than those produced by clouds and carbon dioxide together (Marsden and Valero 2004).

Water vapour exerts also indirect effects on the Earth's climate, since it is constantly cycling through the atmosphere, evaporating from the surface, condensing to form clouds blown by the winds, and subsequently returning to the surface as precipitation. Heat from the Sun is used to evaporate water, and this latent heat is subsequently released into the air, when water vapour condenses into clouds and precipitates. This evaporation-condensation cycle is an important mechanism for transferring heat energy from the Earth's surface to the atmosphere, and moving heat around the Earth (Chahine 1992). Evaporation, condensation and precipitation give form to the hydrological cycle which regulates in this way the movement of water, in all three phases, within and between the Earth's atmosphere, oceans, and continents. In this picture, the movement of water vapour is strongly coupled to precipitation and soil moisture, thus having important practical implications for the life on our planet. In addition, it is worthwhile mentioning that warm air can sustain a higher concentration of water vapour than cooler air, without becoming saturated. Consequently, as air warms, water vapour can increase in concentration, favouring potentially the occurrence of a positive feedback starting, for instance, with atmospheric warming induced by the increased concentration of carbon dioxide over the planetary scale (Houghton et al. 1990).

In the vapour phase, water moves quickly through the atmosphere and redistributes energy through its evaporation and recondensation. Combined with the variation of temperature with height and geographical coordinates, this rapid turnover causes water vapour to be distributed unevenly in the atmosphere, not only horizontally but vertically as well. Thus, water vapour is not homogeneously distributed over the global scale, and its total columnar content giving the water vapour mass in the vertical atmospheric column of unit cross section (most commonly called Precipitable Water (PW)) assumes usually values higher than 5 g cm^{-2} (measured also in cm) near the Equator and mainly less than 0.8 cm in the polar regions.

In fact, the yearly mean values of PW were found in the polar regions to range mainly between 0.4 and 0.8 cm in the Arctic and to be lower than 0.4 cm in Antarctica. Infrared hygrometry measurements performed at Terra Nova Bay in the austral summer period yielded values of PW varying most frequently between 0.2 and 0.6 cm (Tomasi et al. 1990), while considerably lower values were measured at the high plateau site of Dome C (3250 m a.m.s.l.), where PW was evaluated to vary between 0.04 and 0.12 cm in December and January, and between 0.01 and 0.05 cm in April and May (Tomasi et al. 2006). However, water vapour can produce important effects on the radiation budget of the surface-atmosphere system at the coastal sites of Antarctica, in spite of the relatively low values of PW, playing a significant role in the local energy balance, since it can affect appreciably the solar radiation and attenuate more strongly the outgoing long-wave radiation (Yamanouchi and Charlok 1995). Therefore, determining water vapour accurately over Antarctica is necessary to understand the role of water vapour in the energy balance of the polar regions and to complete the description of the global energy cycle. Conventional radiosonde data and ground-level station measurements of the meteorological parameters are relatively scarce compared to satellite data, because they can provide information on the thermodynamic and moisture characteristics of the atmosphere only a few times per day, and ground data are most commonly used in weather analyses or climate first-guess fields.

In the Antarctic climate studies, it could be very useful to increase the spatial and temporal resolution of PW evaluations by using satellite data together with traditional meteorological data. For instance, the Special Sensor Microwave/ Temperature 2 (SSM/T2) could be employed in order to evaluate PW, as proposed by Miao et al. (2001). However, ice clouds seem to induce considerable underestimation errors, showing that satellite microwave radiometry has problems in measuring PW over ice surfaces. Recently, a substantial progress has been made through special processing of signals received from the GPS to measure PW, demonstrating that this method is suitable for providing long-term measurements of such an atmospheric moisture parameter (Bevis et al. 1992, Rocken et al. 1993), with relatively low bias depending mainly on the assumption of localized atmospheric conditions (Tregoning et al. 1998). A further improvement was achieved through the comparison among the measurements of PW obtained with the GPS method and those derived from measurements taken with radiosondes, water vapour radiometers and VLBI technique, defining the limitations of the GPS method for low elevations and finding discrepancies of less than 5% between radiosonde and GPS evaluations of PW (Niell et al. 2001).

In Antarctica, the combined use of CHAMPS/GPS and Advanced Microwave Sensing Unit-B (AMSU-B) data was found to yield sufficiently accurate measurements of temperature and water vapour, as to compensate for the lack of traditional meteorological observations (Johnsen et al. 2004; Vey et al. 2004).

3 GPS Data Analysis: Estimation of the Zenith Total Delay

We focused our attention on a sub set of GPS data that was continuously acquired at Terra Nova Bay using an Ashtech Z-12 receiver and an Ashtech ASH700936 Dorne Margolin antenna with radome; details on the GPS site installed at "Mario Zucchelli" Station can be found in Capra et al. (2008). The latter paper contains a detailed description of the procedure adopted for optimally estimating coordinates and velocities of TNB1, which will be hereinafter referred to as "positioning strategy". The data set spans six years, from 2000 to 2005, and it was processed with Bernese software version 5.0 (Dach et al. 2007), with automatic processing tool BPE (Bernese Processing Engine). We aimed at estimating a long time series of Zenith Total Delays (ZTD) (see Sect. 4) and subsequently computing values of Integrated Precipitable Water Vapour (see Sect. 5.1). TNB1 positions were estimated using a geodetic network formed by a selection of Antarctic and peri-Antarctic GPS stations, as described in the "positioning strategy". Relevant atmospheric related parameterisations adopted for the present work are worth to be mentioned: an elevation cut-off angle of $10°$ was used; it was considered as a good compromise between quantity and quality of data available at the latitudes of the network. Hourly corrections to a priori ZTD values were computed. The first order term of the ionospheric refraction is eliminated by forming the ionosphere-free linear combination (LC) of the L1 and L2 measurements. Niell (1996) dry and wet mapping functions were used for computing zenith delays.

We did set up three different solutions based on the combination of normal equations obtained through the analysis procedure described in the "positioning strategy". The first solution (S1) was obtained estimating both coordinates and hourly corrections to a priori ZTD values. In the second solution (S2), coordinates and velocities of the geodetic network were fixed to the values estimated following the "processing strategy", while ZTD corrections were estimated every hour. The third solution (S3) used a loosely constrained approach: coordinates and velocities of the geodetic sites were constrained at a 20 cm level to the values determined with the "processing strategy" and ZTD corrections were estimated every hour.

All three solutions are in good agreement: the bias between S2 series and S1 and S3 are well within half mm (0.3 ± 4.1 mm and 0.5 ± 2.8 mm, respectively). Figure 1 presents the results obtained with ZTD series derived by S2 solution; though, the analysis was performed and results were derived for all three solutions.

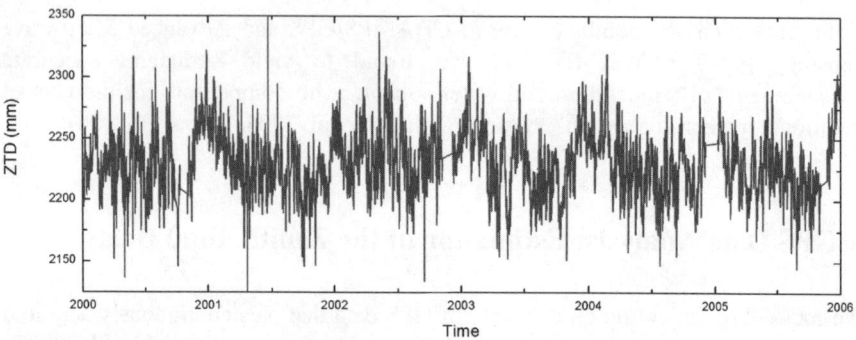

Fig. 1 The long time series of Zenith Total Delay that was estimated according to the procedure described in Sect. 3; the series spans a six-year period. The observations were being performed with the GPS permanent system (TNB1) installed at the "Mario Zucchelli" Station, Antarctica

4 Tropospheric Zenith Delays

The sensitivity of GPS to the composition of the tropospheric plasma is straightforwardly related to the refractive index of the air. The GPS radio signal experiences a delay and bends during its propagation into the neutral atmosphere, which is usually referred to as Tropospheric Delay (TD). It can be expressed as (e.g. Hopfield 1978):

$$TD = \int n(\sigma)d\sigma - \int ds \qquad (1)$$

where n is the index of refraction of air. If signal path bending is small, the first integral of (1) can be evaluated along the geometric path, thus obtaining:

$$TD = \int (n(s) - 1) ds \qquad (2)$$

or, introducing the refractivity N:

$$N = 10^6(n-1) \qquad (3)$$

and subsequently:

$$TD = 10^{-6} \int N(s) ds \qquad (4)$$

The total refractivity N can be conveniently divided into two parts (see e.g. Hofmann-Wellenhof et al. 2001): $N = N_{hyd} + N_{wet}$, where N_{hyd} is related to atoms and molecules in the atmosphere without permanent dipole moment, also referred to as dry refractivity, and N_{wet} is related to atmospheric water vapour, also referred to as wet refractivity. It is therefore possible to express (4) as a sum of two parts, one related to water vapour, and the other to the remaining atmospheric constituents. Thus, the TD will be determined, according to (4), as:

$$TD = 10^{-6} \left(\int N_{hyd}(s)ds + \int N_{wet}(s)ds \right) \qquad (5)$$

i.e. two terms related to the dry and wet refractivity. It will therefore be possible to express the TD as sum of two terms, a Hydrostatic Delay (HD) and a Wet Delay (WD) for which $TD = HD + WD$.

The TD is a function of the distance travelled by the electromagnetic signal through the neutral atmosphere. In GPS, it is therefore related to the elevation of the line of sight between the observing station and the emitting satellite: as the elevation decreases the signal passes through more and more atmosphere, thus increasing the TD. It is convenient to express the TD with respect to a common unique direction, usually taken as the zenith direction. In order to transform the TD expressed by (5) (elevation dependent) into a TD at zenith (ZTD), it is necessary to use proper mapping functions in order to convert the HD and WD into a Zenith Hydrostatic and a Zenith Wet delay, ZHD and ZWD respectively. In this paper we used the Niell mapping functions both for the dry and wet component (Niell 1996).

The final ZTD can therefore be simply expressed as:

$$ZTD = ZHD + ZWD \qquad (6)$$

where ZTD is estimated through GPS data processing, ZHD is computed through proper models and ZWD is obtained by subtracting the latter from the first, according to (6). A computation of ZTD using radiosonde profiles of Sect. 5.2 is presented in Appendix.

4.1 Zenith Hydrostatic Delay

On each ZTD solution S1, S2 and S3, we applied three different Zenith Hydrostatic models: namely the Hopfield model (Hopfield 1969, 1971), the Saastamoinen model (Saastamoinen 1972) and the modified Saastamoinen model (Saastamoinen 1972, 1973; Davis et al. 1985; Elgered et al. 1991). Hopfield's model uses surface temperature and pressure for determining the delay originated by atoms and molecules without permanent dipole moment in the atmosphere:

$$ZHD_{Hop} = \frac{10^{-6}}{5} \frac{77.64 \frac{p}{T}}{\sin\sqrt{90^2 + 6.25}} \cdot [40136 + 148.72(T - 273.16)] \qquad (7)$$

where T and p are surface temperature and pressure (in Kelvin and hPa) respectively and the argument 90 of the sine function is the elevation angle E of the satellite which is observed (E is equal to 90 deg in Zenith HD computation) (see e.g. Hofmann-Wellenhof et al. 2001).

The other two hydrostatic models that were extensively used for comparison in the present paper determine the Zenith Hydrostatic Delays as:

$$ZHD_{Saas} = 0.002277 \cdot p \qquad (8)$$

and as:

$$ZHD_{Saas\_lat} = 0.002277 \cdot p/(1 - 0.00266 \cdot \cos 2\phi - 0.00000028 \cdot h_t) \quad (9)$$

where p is the surface pressure expressed in hPa, ϕ is the latitude and h_t is the geoidal height of the tracking station. According to Hopfield (1969), a computation of ZHD is performed using radiosounding data; Appendix contains a comparison between ZHD values derived using (7) and radiosounding profiles determined in Sect. 5.2.

The values of ZHD computed with the three different hydrostatic models differ significantly: biases between the series obtained using Eq. (7) with respect to (8) and (9) are 2.8 ± 1.3 mm and 8.2 ± 1.1 mm, respectively, being the ZHD computed by (7) larger then the corresponding values derived using (8) and (9).

The evaluation of (7), (8) and (9) was possible because pressure and temperature are continuously recorded at Eneide, an Automatic Weather Station (AWS) installed at the "Mario Zucchelli" station. Eneide and TNB1 can be regarded as co-located (see Fig. 2), although their heights differ and such a height difference sensibly impacts on pressure values to be used for computing ZHD.

In particular, the height difference between TNB1 and Eneide is 36 m, being TNB1 located at a higher altitude; this leads to a correction of -5 hPa to the pressure series recorded at Eneide.

Finally, a remark on the discontinuities found on surface pressure readings performed at Eneide. Parameter p is used in all (7), (8) and (9); it is therefore essential for computing ZHD. Any lack of data or discontinuity encountered in the time series prevents a computation of ZHD which, according to (6), prevents an evaluation of ZWD from GPS observations. In order to fill in the gaps encountered in Eneide's time series, data of the nearby station Rita, situated a few km apart, were

Fig. 2 Overview of the area nearby "Mario Zucchelli" Station. The location of the permanent GPS station "TNB1" is marked with a red triangle on the upper left part of the picture. On the centre-right part of the picture, the locations of the Automatic Weather Station "Eneide" and the radiosoundings station are identified by a black square and a blue circle, respectively. The distances between the three locations do not exceed a few hundred meters; the three observing sites and their sensors can therefore be regarded as co-located

used. Missing data in temperature and pressure series were obtained computing a mean difference between Eneide and Rita over the corresponding month, using data of other years (e.g. a lack of data in Eneide's March time series would be filled with Rita's data, accounting for the mean differences computed over the months of March belonging to different years). This was the case in December 2002, when a comparison between GPS-derived and RS-derived PW has been tempted: the mean difference between Rita's and Eneide's recordings was computed in $-1.2°C$ and -21.9 hPa. Adding these values to Rita's data and filling the gaps into Eneide's series permitted to compute the ZHD as shown in Fig. 3, where the ZHD corresponding to the gap has been drawn in blue and light blue, respectively for Hopfield and modified Saastamoinen models.

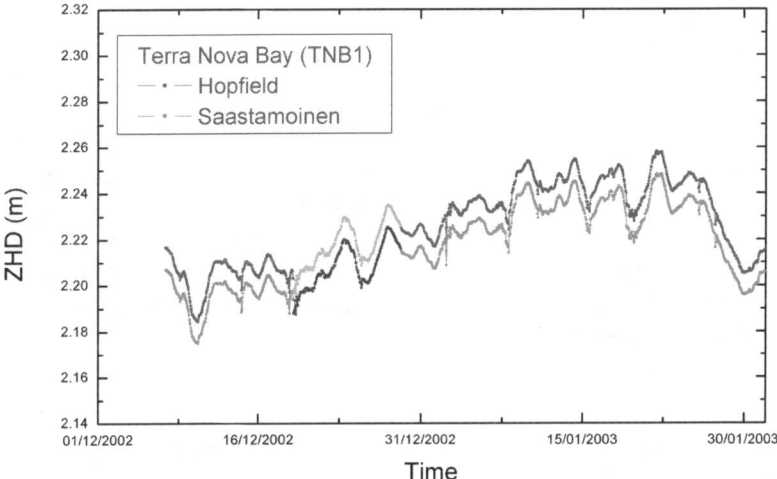

Fig. 3 Two months of Zenith Hydrostatic Delay computed with Hopfield (Eq. 7) and Saastamoinen (Eq. 9) models. The blue and light blue values were derived using the recordings performed at the Automated Weather Station "Rita", corrected as described in Sect. 4.1; this was necessary to fill in the gaps found in surface temperature T and pressure p series recorded at "Eneide"

4.2 Zenith Wet Delay

The ZWD has been computed according to (6), subtracting each of the three ZHD, given by (7), (8) and (9), to the ZTD estimated through S1, S2 and S3 solutions. We thus obtained nine ZWDs: the ZWD corresponding to ZTD estimated with solution S2 and ZHD computed according to (9) over the six 2000–2005 years is shown in Fig. 4.

An accurate evaluation of ZHD is crucial for computing unbiased values of ZWD (and therefore PW). In particular, in those areas where the atmosphere is dry and water vapour content very low, ZHD values computed according to (7), (8) and (9) can occasionally be higher than ZTD estimated values. This obviously leads to a

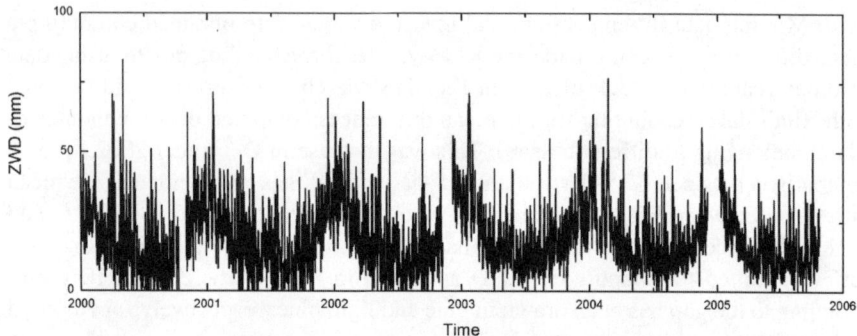

Fig. 4 GPS-derived long time series of Zenith Wet Delay: at least three breaks are evidently present. They are associated with GPS observing system failures causing temporary loss of data. Nevertheless, the annual signal is clearly visible over the entire series, with minimum ZWD values computed during the Antarctic winter

computation of negative PW values (see Sect. 5.1); in order to reduce the occurrences of negative ZWD values, the smaller values of ZHD (derived using Eq. 9) were used and implemented. A deeper investigation about Hydrostatic Delay models and their calibration for the peculiar moisture conditions of the Antarctic atmosphere at coastal sites is necessary, especially during dry winter seasons. These aspects are important and are going to be investigated in detail in the research that we are planning for the near future.

5 Radiosounding Measurements: Mean Temperature and Precipitable Water

Radiosounding measurements were taken at Terra Nova Bay from December 1, 2002, to January 30, 2003, by the meteorologist group of the Antarctic Project (ENEA, C. R. Casaccia, Rome) using the Vaisala (Helsinki, Finland) radiosondes, model RS80-A. The sensors employed in the RS80-A model for measuring air pressure p, temperature T and relative humidity f, are the following:

(i) BAROCAP, a capacitive aneroid with a measurement range from 1060 to 3 hPa, resolution of 0.1 hPa and accuracy of ± 0.5 hPa;
(ii) THERMOCAP, a small capacitive bead in glass encapsulation, with a measurement range from 333 to 183 K, resolution of 0.1 K, total uncertainty in sounding (accuracy) of ± 0.2 K, repeatability in calibration of 0.2 K, and reproducibility in sounding equal to 0.2 K from 1060 to 30 hPa levels, 0.3 K from 50 to 15 hPa, and 0.4 K from 15 to 3 hPa; and
(iii) HUMICAP (model A), a capacitive thin film humidity sensor, with a sensitivity range from 2% to 100%, resolution of 1%, total uncertainty in sounding (accuracy) of less than ± 3%, repeatability in calibration of 2%, and reproducibility in sounding lower than 3%.

During the ascent, the three sensor signals were sent by the transmitter to the ground station every 10 s, turning out into significant levels distributed along the vertical path in steps of $50 \div 60$ m from the ground-altitude ($\cong 65$ m a.m.s.l.). For each triplet of parameters p, T and f given at a certain significant level, the corresponding height z was calculated by determining the virtual temperature and then integrating, step by step (from one level to the subsequent one), the differential term of height given by the well-known hydrostatic equation, starting from the ground values of pressure and height (Tomasi et al. 2004).

5.1 Computation of Precipitable Water From GPS-derived ZWD

In order to compute PW from GPS-derived ZWD, it is necessary to calculate the mean temperature T_m according to the procedure described by Bevis et al. (1992, 1994). It is essential to fine tune the value of T_m according to the area where and the season when GPS observations are performed. This implies the availability of radiosounding data, in particular partial pressure of water vapour p_v and temperature T along the ascent of the sonde, from which the T_m can be computed as (Davis et al. 1985):

$$T_m = \frac{\int (p_v/T) dz}{\int (p_v/T^2) dz} \quad (10)$$

For each T_m computed over the December 2002 – January 2003 period, we computed a linear function which allowed to relate the surface temperature T to the mean temperature T_m for any T recorded by the AWS Eneide over the six years of GPS observations. Figure 5 shows the linear relationship between the two parameters, which results in the function $T_m = 0.78 \cdot T + 46.15$, where T is expressed in Kelvin.

A better computation would be derived by using a considerably larger amount of radiosounding profiles; as it was stated in the previous Section, we only had a limited amount of radiosounding data. From the mean temperature T_m the derivation

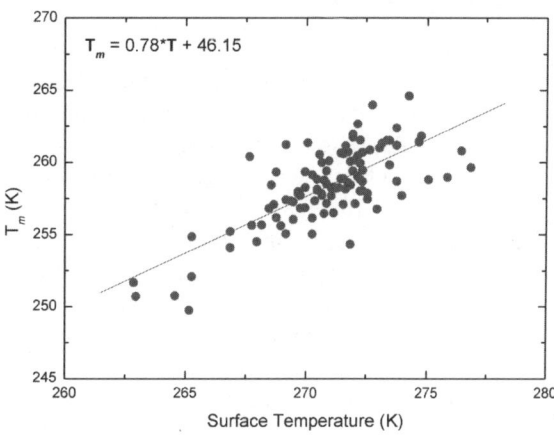

Fig. 5 Scatter plot of mean temperatures T_m, computed using 108 radiosounding profiles (Eq. 10), and the corresponding surface temperatures T, as recorded at Eneide. The corresponding regression line is shown and used for computing T_m from T

of PW follows through a transformation of ZWD (Bevis et al. 1994). In particular, $PW = \Pi \cdot ZWD$, whit Π factor equal to:

$$\Pi = \frac{10^6}{\rho R_v [(k_1/T_m) + k'_2]} \tag{11}$$

where ρ is the density of liquid water, R_v is the specific gas constant for water vapour and k_1 and $k'_2 = k_2 - mk_1$ and k_2 are physical constant contained in the formula for refractivity N (Smith and Weintraub 1953) and $m = M_w/M_d$ is the ratio of the molar masses of water vapour and dry air.

The PW computed over six years of GPS observations at the "Mario Zucchelli" station is shown in Fig. 6.

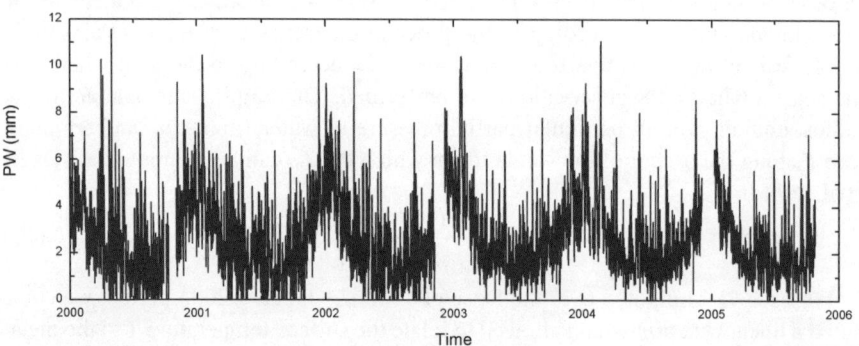

Fig. 6 Long time series of Precipitable Water values derived by GPS. The series has been obtained multiplying the values of ZWD shown in Fig. 4 by the factor Π computed according to (11). The annual water vapour content variation is evident: dryer conditions are determined during the Antarctic winters

5.2 Calculation of Precipitable Water from Radiosounding Measurements

In order to correct (i) the temperature errors, due to contamination by heating from sources other than the air itself (i.e. solar and infrared irradiation of the sensor, heat conduction to the sensor from its attachment points, and radiation emitted by the sensor (Luers and Eskridge 1995), (ii) temperature and relative humidity lag errors (Huovila and Tuominen 1991, Miloshevich et al. 2001), a careful method was adopted by Tomasi et al. (2004) (hereinafter referred to as T04) to analyze a 12-year radiosounding data-set collected at Terra Nova Bay. The values of p, T and f measured for each radiosounding were determined at each significant level by following the T04 method, taking into account the above-mentioned errors in temperature and humidity, which are commonly made using the radiosonde RS80-A.

Then, for each level and, hence, for each triplet of parameters p, T and f, the water vapour partial pressure e was calculated by multiplying f by the saturation vapour pressure $E(T)$ in the pure phase over a plane surface of pure water, the pressure $E(T)$ being evaluated in terms of the general formula proposed by Bolton (1980),

$$E(T) = 6.112 \cdot \exp\left(\frac{17.67(T-273.15)}{T-29.65}\right) \tag{12}$$

with T measured in K. For each value of e, the corresponding value of dew-point T_d was then determined through the inverse formula of (12). Thereupon, for each set of values of p, T, e and T_d, the absolute humidity ρ was calculated in terms of the well-known equation of state for water vapour:

$$\rho = 216.685 \cdot \left(\frac{e}{T}\right) \tag{13}$$

where ρ is measured in g m^{-3}, e in hPa and T in K.

Depending mainly on the vertical ascent velocity of the radiosonde balloon, each vertical profile of temperature was found to include at least 130 significant levels in the troposphere, a comparable number of levels from the tropopause level (varying on average from 9.7 km in early December to 8.1 km in late January) to the 15 km altitude, and many other levels within the upper stratosphere, up to the highest altitude reached by the radiosonde, which varied in general between 15 and 25 km and in many cases exceeded 30 km. When the radiosonde passes from the troposphere to the stratosphere, the air relative humidity decreases very sharply to values of only a few percent and, then, very often lower than the repeatability level of 2% established by the manufacturers.

The above evaluations of relative humidity f obtained from the HUMICAP-A measurements at all the tropospheric levels are generally affected by important dry bias (Wang et al. 2002) due to temperature dependence (T-DEP), chemical contamination (CC), basic calibration model (BCM), sensor aging (SA), ground check (GC), and sensor arm heating (SAH). Among these errors, the first four dry biases were evaluated by Wang et al. (2002) to cause the predominant underestimation errors of f, which are expected to be particularly large for the very cold air temperature conditions usually observed in the Antarctic atmosphere. Moreover, Miloshevich et al. (2004) pointed out that the measurements of f are also affected by lag errors and need to be appropriately corrected. Thus, a new correction procedure was defined by Tomasi et al. (2006) (hereinafter referred to as T06) for analyzing the radiosounding data taken at the high-altitude site of Dome C, on the Antarctic Plateau, by using various Vaisala radiosonde models (RS80-A, RS80-H, RS90 and RS92). Following the T06 method, the RS80-A data recorded at Terra Nova Bay were examined by adopting a procedure consisting of the following sequence of correction subroutines:

(1) the temperature lag errors were corrected using the lag coefficient algorithm proposed by Huovila and Tuominen (1991);
(2) the lag errors affecting the relative humidity data were corrected following the procedure described by Miloshevich et al. (2004);

(3) the values of $f < 3\%$ were then discarded, taking into account that the total uncertainty in sounding (accuracy) was evaluated to vary between $\pm 3\%$ and $\pm 5\%$ throughout the whole troposphere;
(4) the remaining set of relative humidity data were then analyzed following the preliminary procedure suggested by Miloshevich et al. (2004) to define a more schematic "skeleton" of the measured vertical profiles, within each altitude interval where f presented constant values;
(5) the T-DEP, CC, BCM, SA, GC and SAH dry bias of the HUMICAP-A sensor were subsequently corrected using the set of appropriate algorithms suggested by Wang et al. (2002) for the various dry bias; and
(6) the resulting relative humidity data were finally re-examined through the complex procedure proposed by Miloshevich et al. (2004), consisting of a first smoothing step, a further lag correction procedure, and a final smoothing procedure, to remove other possible lag errors often related to the presence of clouds.

Applied to the radiosounding data taken at Dome C, the above procedure was found to correct the absolute values of relative humidity data by $+(5 \div 10\%)$ throughout the tropospheric region from 3.2 to 10 km, suggesting that the use of the T04 method can lead to underestimate the relative humidity and, hence, to obtain lower values of PW than the real ones, by not negligible percentages. In fact, considering that relative humidity assumes relatively low values in the Antarctic atmosphere, with median values usually lower than 50% at all the tropospheric levels, a difference of $5 \div 10\%$ implies variations of more than 20% between the absolute values of PW. In order to evaluate the weight of such underestimation effects, the T06 procedure was used to examine the whole 12-year set of radiosoundings taken at Terra Nova Bay from 1986/87 to 1997/98 and already examined following the T04 method. In this way, the monthly mean vertical profiles of pressure, temperature, relative humidity and the other moisture parameters were determined up to altitudes of around 10 km. In addition, new monthly average vertical profiles of mixing ratio and moisture parameters were determined from 10 to 30 km, following a more rigorous procedure than that defined in the T04 method and utilizing the water vapour mixing ratio evaluations derived at stratospheric levels from various limb satellite observations (HALOE, MLS/UARS, SAGE II/ERBS) by Harries et al. (1996), Lahoz et al. (1996), Chiou et al. (1997) and Randel et al. (2001).

The monthly mean vertical profiles of air pressure, temperature and relative humidity were estimated for the monthly sets relative to November, December, January and February, by analyzing the radiosounding 12-year data-set, consisting of an overall number of 1331 radiosounding measurements taken at Terra Nova Bay, by using first the T04 method and then the more recent T06 procedure. No appreciable differences were found between the vertical profiles of pressure p and temperature T determined using the two procedures. On the contrary, appreciable differences were found between the monthly mean vertical profiles of relative humidity f obtained following the two correction procedures, as shown in Fig. 7.

The vertical profiles of f result into new values that are largely higher than the old ones throughout the whole troposphere in the four months, by (i) about 7 % at all levels in November, (ii) about 7 % from the ground level to 7 km in December,

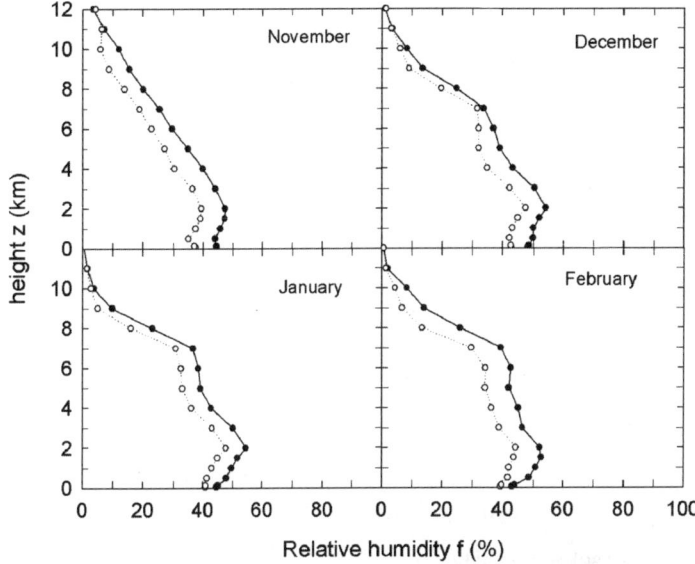

Fig. 7 Comparison between the monthly average vertical profiles of relative humidity f determined at the tropospheric levels applying the T04 method (*open circles*) and the advanced T06 correction procedure for both dry bias and lag errors (*solid circles*). These methods were used to analyze the monthly HUMICAP-A data sets relative to the four months from November to February, which have been selected from the 12-year radiosounding measurement set performed at Terra Nova Bay during the period from the 1986/87 field campaign to that of 1997/98

(iii) $4 \div 7\%$ in January, and (iv) $7 \div 12\%$ in February, confirming the above evaluations on the possibility that the T04 method could provide appreciably underestimated values of PW.

Correspondingly, the monthly mean vertical profiles of absolute humidity were defined for the monthly sets of the four months from November to February, by analyzing the same radiosounding 12-year data set, first following the T04 method and, then, the more advanced T06 procedure. The comparison between the vertical profiles of absolute humidity ρ obtained in the four months is shown in Fig. 8, giving evidence that the values of ρ determined following the new procedure are appreciably higher at tropospheric levels than those evaluated with the less sophisticated T04 method, especially in the upper part of the troposphere, where the temperature is lower than 240 K and, hence, the T-DEP dry bias are expected to be particularly marked. The lower stratospheric levels of absolute humidity have been obtained through the use of a more accurate choice of the satellite data at the various stratospheric levels and a more correct methodology followed to calculate the mixing ratio values, avoiding some systematic errors affecting the T04 method when evaluating the water vapour mixing ratio at stratospheric levels.

Integrating the vertical profiles of ρ from the ground level up to 30 km, the mean monthly values of PW were obtained for the four months, with considerable differences between the two procedures in all the four months:

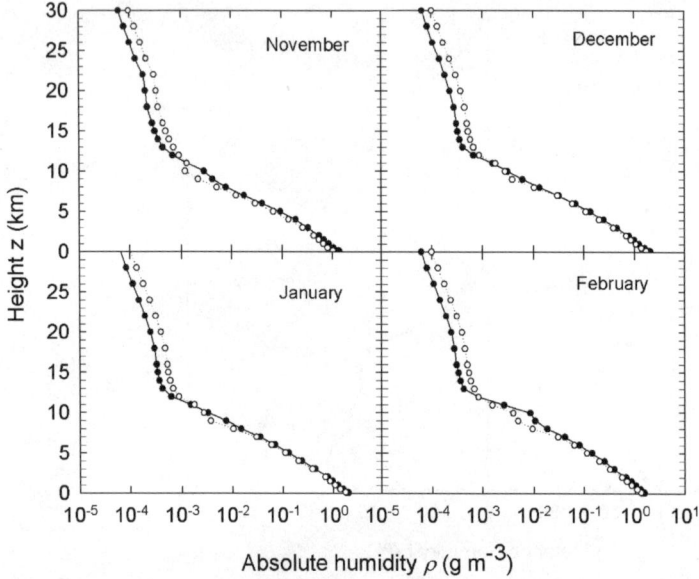

Fig. 8 Comparison between the monthly average vertical profiles of absolute humidity ρ determined throughout the height range from 0.065 to 30 km by using the monthly mean vertical profiles of air temperature T and relative humidity f found for the four months from November to February and calculated following the T04 method (*open circles*) and the T06 procedure (*solid circles*) to analyze the 12-year radiosounding data set collected at Terra Nova Bay from 1986/87 to 1997/98

(i) in November, PW was found to assume an average value of 0.26 cm using the T06 procedure, while a value of 0.20 cm was obtained adopting the T04 method, showing that the average value of PW turns out to be higher than that estimated neglecting the dry bias errors, by 29% on average;

(ii) in December, a value of PW equal to 0.38 cm was obtained following the T06 procedure, against a value of 0.33 cm achieved with the T04 method, showing that the disposability of more realistic measurements of the moisture parameters enable to obtain a monthly average value of PW higher by 17%, on the average, than that found using the first more approximate procedure;

(iii) in January, PW was estimated on average to be equal to 0.37 cm with the T06 procedure and 0.33 cm with the T04 method, presenting an average difference of about 14%; and

(iv) in February, a monthly average value of PW equal to 0.33 cm was obtained with the T06 procedure and 0.27 cm with the T04 method, defining an average difference of 21%.

Assuming these average percentage differences in PW to be indicative of the average underestimation effects made in examining the Terra Nova Bay radiosounding measurements with the T04 method and referring such monthly average correction terms to the mid-November, mid-December, mid-January and mid-February dates, respectively, the appropriate underestimation correction factors were determined for each measurement day from December 1, 2002, to January 30, 2003.

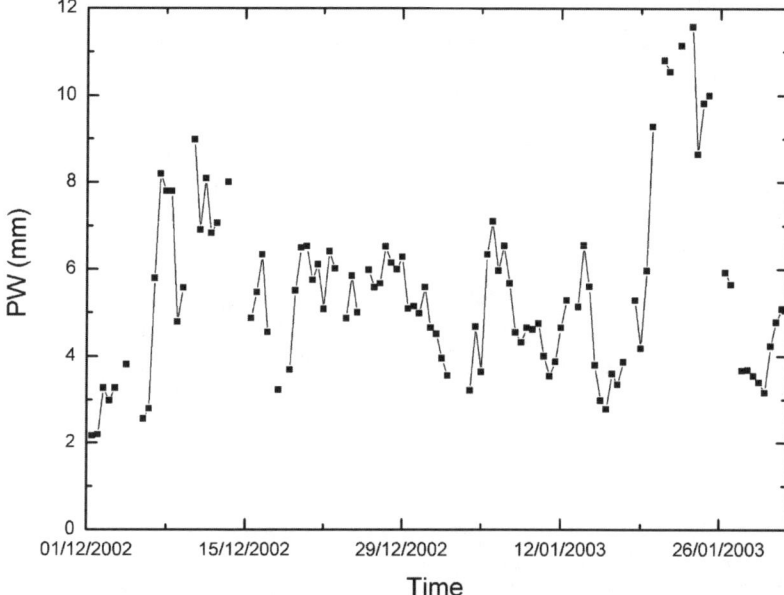

Fig. 9 Time patterns of Precipitable Water determined at Terra Nova Bay during the period from December 1, 2002, to January 31, 2003, by correcting the daily values of PW derived analyzing the Vaisala RS80-A radiosonde data with the T04 procedure using the underestimation correction factors determined in the present paper by means of the T06 procedure

Using these daily correction factors, the time-patterns of PW determined following the T04 method were corrected over the entire period, obtaining the time patterns shown in Fig. 9.

5.3 Comparison of PW Computed by Radiosoundings and GPS

GPS capability of retrieving atmospheric PW can be evaluated through a comparison with PW content computed with radiosounding data. This comparison can be done over the two months period spanning December 2002 and January 2003; the results shown in Fig. 6 can be selected and superimposed to the PW determined using radiosonde data and the procedure described in Sect. 5.2, so as to obtain the graph of Fig. 10, showing the time series of PW computed with both methods.

The agreement of the two series is very good, with small discrepancies greater than 1 mm at a few points of the series only. The hourly GPS-derived PW time series has a higher resolution than the radiosounding series (12 vs. 1 value, respectively) and represents an efficient method for evaluating the total content of water vapour in the atmosphere, its variability as well as the presence of water vapour when no radiosoundings are performed (this is always the case for the closing season of the "Mario Zucchelli" station).

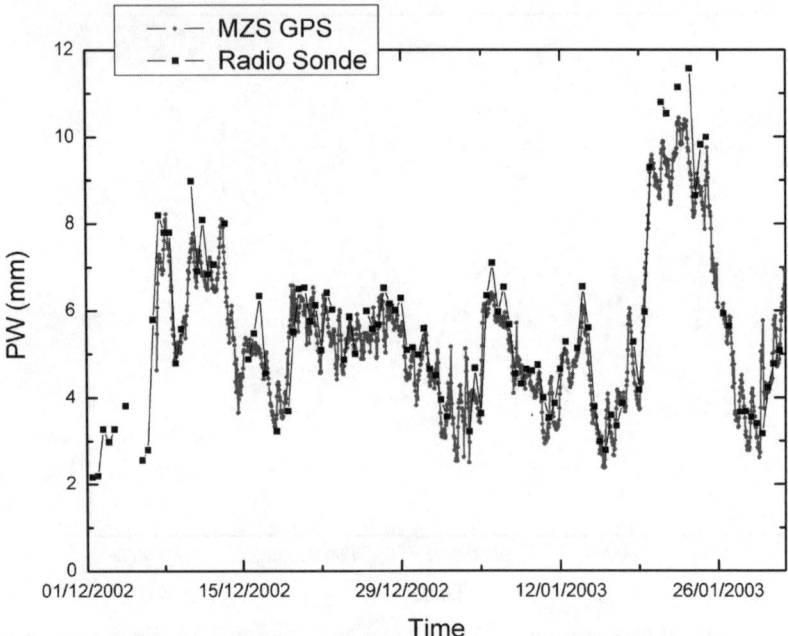

Fig. 10 Comparison between the time patterns of PW obtained from the GPS observations (see Fig. 6) and those derived from the Vaisala RS80-A radiosonde data (see Fig. 9).

The correlation between radiosonde and GPS PW is shown in Fig. 11, with a correlation coefficient very close to 1 ($R = 0.96$).

TNB1 GPS permanent station performs continuous observations over the whole year and, as far as surface pressure (and temperature) are available or can be

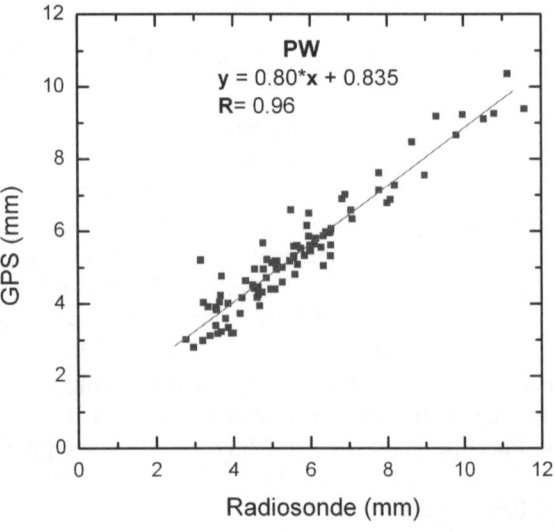

Fig. 11 Scatter plot of the PW values shown in Figure 10 determined from the GPS observations over the test period December 2002 – January 2003 versus the PW values obtained from 108 radiosonde profiles. The corresponding regression line is also shown, along with its analytical form, and the regression coefficient R which is close to unity

recovered at the observing site, it is possible to precisely compute and monitor atmospheric PW.

6 Measurements of Downwelling Infrared Radiant-Flux Density at Terra Nova Bay

Short-wave and long-wave radiation flux measurements were regularly being performed during austral summer periods from 1999 to 2004, at the Clean Air Facility of Campo Icaro to define the upwelling and downwelling terms of the radiation balance. These measurements aim at determining the surface radiative effects of aerosols and clouds, these latter being dependent mainly on cloud coverage and cloud type characteristics. The four radiation balance terms were obtained by using a Kipp&Zonen CNR-1 net radiometer, while simultaneous measurements of the diffuse and direct components of the incoming solar radiation flux were carried out by using a MFR-7 shadow-band radiometer (YES Inc.). Both instruments were located at 2.5 m above the ground, in order to achieve an overall view of the sky with a 2π sr field-of-view. Observations performed with the four sensors of CNR-1 were separately sampled in steps of 3 s and their average, computed over 60 s, stored along with their standard deviations.

The long-wave downwelling $(L\downarrow)$ radiant-flux density measurements taken from November 22, 2003 to February 5, 2004 are shown in Fig. 12 together with the

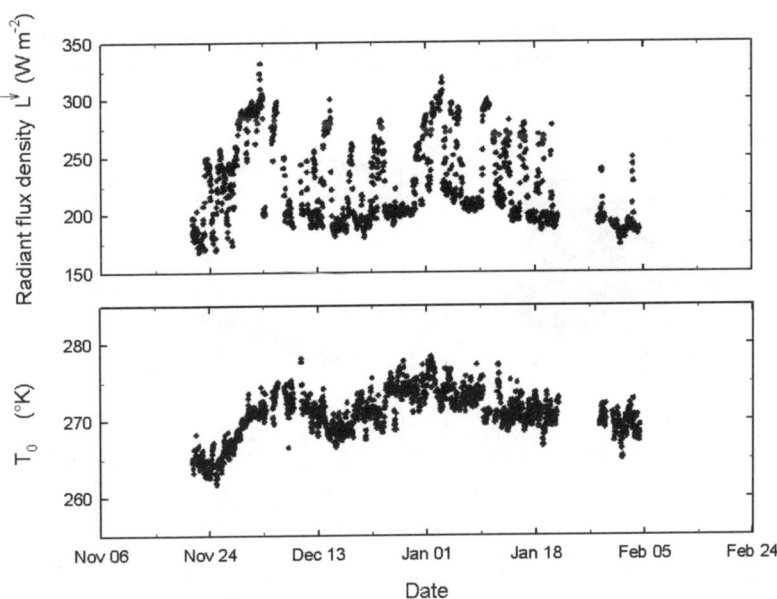

Fig. 12 Time patterns of the radiant flux density $L\downarrow$ (*upper part*) and surface temperature T_0 (*lower part*) measured at Campo Icaro during the period from November 22, 2003 to February 5, 2004

simultaneous measurements of surface temperature T_0. The time patterns of radiant-flux density turn out to range between 170 and more than 300 W m^{-2}, being the lower values associated with clear sky conditions, whereas the rest of such data were taken for cloudy sky conditions. In particular, the higher values are presumably related to the presence of thicker and lower clouds having higher emission temperature. On the other hand, the presence of thin cirrus clouds is expected to cause lower radiant-flux density values not too different from those measured for clear sky conditions. The comparison between the radiant-flux density and the surface temperature time patterns gives evidence of the fact that the clear sky radiant-flux density $L\downarrow$ varies in time in a similar manner to surface temperature.

The overall 5-year set of $L\downarrow$ measurements were plotted vs. T_0 in Fig. 13a, where the scatter of data is evident since. $L\downarrow$ is mainly related to the emission temperature of the cloud bottom and not to the surface temperature T_0, the latter giving only a general measure of the thermal conditions of the lower atmospheric layer. In fact, the density flux of the radiant flux emitted downward by clouds follows the well-known Planck equation:

$$L\downarrow(T) = \varepsilon\sigma T^4 \qquad (14)$$

where $L\downarrow(T)$ represents the radiant flux density emitted by a real grey body at temperature T, $\sigma = 5.6697 \ 10^{-8}$ W m^{-2} K^{-4} is the Stefan-Boltzmann constant, and ε is the cloud emittance, as given by the ratio between the observed value of the radiant flux density of the cloud and the flux density of the radiation emitted by a black body at the same temperature. Thus, the downward radiance depends on (i) the temperature of the lower part of the cloud (and, hence, on its altitude), and (ii) its emittance ε, which is more or less close to unity, to an extent varying with the absorption (and, hence, emission) characteristics of the cloud in the infrared range.

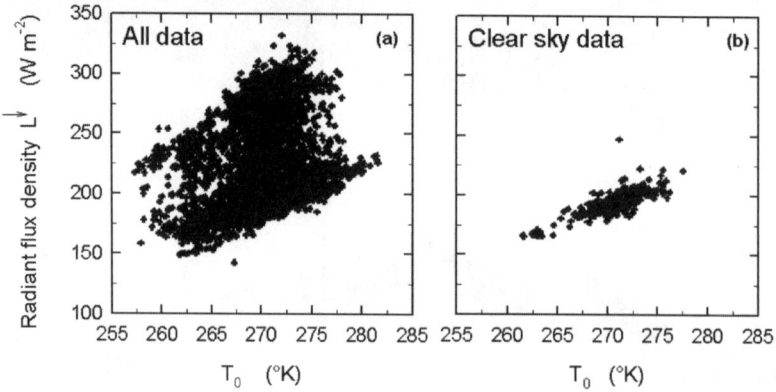

Fig. 13 (a) Scatter plot of the incoming radiant flux density at the surface and the corresponding surface temperature T_0 over the whole five-year period from 2000 to 2004; (b) Clear sky data scatter plot obtained applying the Long and Ackerman (2000) selection method to the five-year data set shown in (a)

In order to relate the GPS-derived PW (Sect. 5.1) with the measurements of $L\downarrow$ performed at Terra Nova Bay, only the data associated with clear sky conditions were selected. This was achieved using the Long and Ackerman (2000) method based on four tests made examining the simultaneous measurements of global and diffuse components of incoming short-wave radiation at the ground. The combined selection procedure allows an automatic cloud screening on the entire data set. The selected clear sky data are presented in Fig. 13b, showing a gradual increase of $L\downarrow$ with temperature T_0. This behaviour is due to the fact that the clear sky atmosphere emits long-wave radiation according to (14) for emission temperatures T which are expected to increase gradually as T_0 increases, and for infrared absorption and emission properties defined by values of ε which are in general largely lower than unity and tend to increase as PW assumes gradually higher values, since the absorption and emission properties of the atmosphere tend to increase with the air humidity conditions.

The measurements shown in Fig. 13b are certainly affected by experimental errors due to the trend of the calibration constant and other instrumental responsivity variations, from one year to another. Therefore, despite the rigorous application of the Long and Ackerman (2000) criteria, it cannot be excluded that the data set shown in Fig. 13b includes not only clear sky data but also data pertaining to thin cirrus clouds, as it is suggested by the presence of evident outliers.

In order to (i) ascertain the reliability of the measurements and of the selection criteria adopted by the Long and Ackerman (2000) procedure and (ii) evaluate the dependence of the clear-sky radiant flux density on PW, calculations of $L\downarrow$ were performed for monthly mean models of the Antarctic atmosphere above Terra Nova Bay, using the SBDART model (Ricchiazzi et al. 1998) over the 4.5–40 µm wavelength range, for the monthly mean atmospheric models defined in Sect. 5.2, which presents the vertical profiles of moisture parameters given in Figs. 7 and 8, and those of air temperature and pressure shown in Fig. 14.

The vertical profiles of $L\downarrow$ are presented in Fig. 15 for the five monthly models (October–February) showing that about 2/3 of the total flux reaching the ground is given by the first 5 km of the atmosphere, with values of the radiant flux density that are gradually higher as the low part of the troposphere tends to become warmer.

Calculations of $L\downarrow$ were made not only for the 5 monthly atmospheric models described in Figs. 8 and 14, providing the results shown in Table 1a, but also for a set of atmospheric models derived from the December model, in which the profiles of relative humidity f were multiplied by factors increasing from 0.3 to 2, so as to obtain a wider range of PW and $L\downarrow$ values (Table 1b).

The results show that the simulated values of $L\downarrow$ can vary from about 140 to 200 W m^{-2} as PW increases, fitting the lowest values of $L\downarrow$ measured for cloudless conditions at Terra Nova Bay. Considering that the values of $L\downarrow$ should be closely related to PW, a comparison between measured and simulated values of $L\downarrow$ is presented in Fig. 16 as a function of PW. It can be noticed that the simulated values of $L\downarrow$ given in Table 1b agree very well with the lower values of $L\downarrow$, presenting discrepancies of a few percents, while the higher values of $L\downarrow$ were found to differ considerably from the modelled values, up to 20% in the worse cases. These

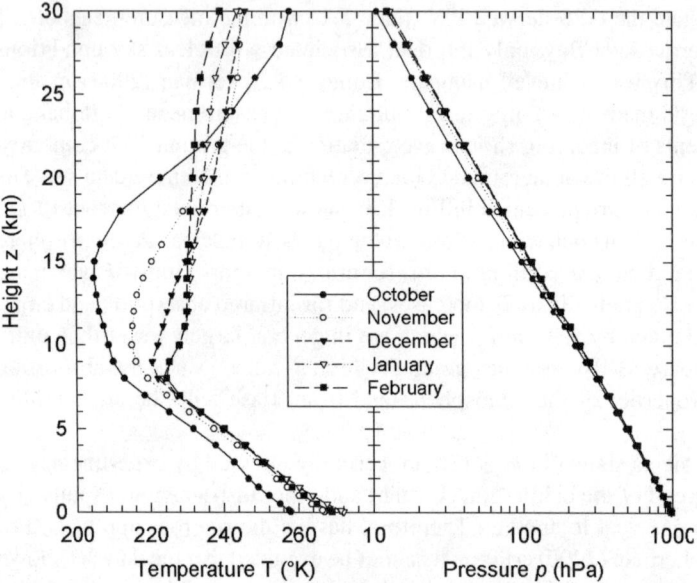

Fig. 14 Monthly mean vertical profiles of temperature (*left*) and pressure (*right*) obtained from the 12-year radiosounding data set recorded at Terra Nova Bay from the 1986/87 to the 1997/98 campaigns (Tomasi et al. 2004)

Fig. 15 Incoming long-wave radiant flux density profiles determined with SBDART model (Ricchiazzi et al. 1998) applied to the monthly atmospheric models of Fig. 14

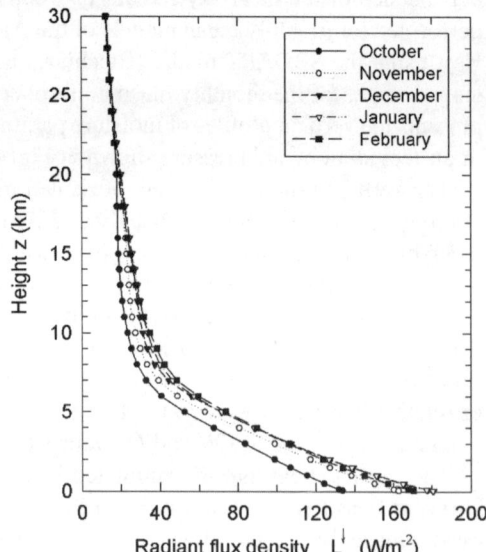

differences are plausibly explained as due partially to (i) the presence of thin cirrus clouds not identified with the selection criteria of the Long and Ackerman (2000) selection method, and (ii) the values of T_0 defined by the monthly models, which are lower by 2–3 K than the monthly average values derived from the ground-level

Table 1 Values of Precipitable Water radiant flux density $L\downarrow$ computed with the SBDART model (Ricchiazzi et al. 1998) and water vapour partial pressure e_0 at the surface, relative to the five monthly atmospheric models of Sect. 5.2 (Part (a)) and to the December models modified according to different multiplying values relative humidity (f) (Part (b))

Part (a)				Part (b)			
Monthly models	PW (mm)	Radiant flux density $L\downarrow$ (W m^{-2})	e_0 (hPa)	Dec. f multiplying factor	PW (mm)	Radiant flux density $L\downarrow$ (W m^{-2})	e_0 (hPa)
Oct	1.4	133.86	0.65	0.3	1.2	162.10	0.79
Nov	2.6	162.62	1.33	0.7	2.8	176.88	1.85
Dec	3.9	180.73	2.28	1.3	5.3	188.45	3.44
Jan	3.8	180.29	2.28	1.7	6.9	194.17	4.50
Feb	3.3	170.53	1.70	2	8.1	198.10	5.30

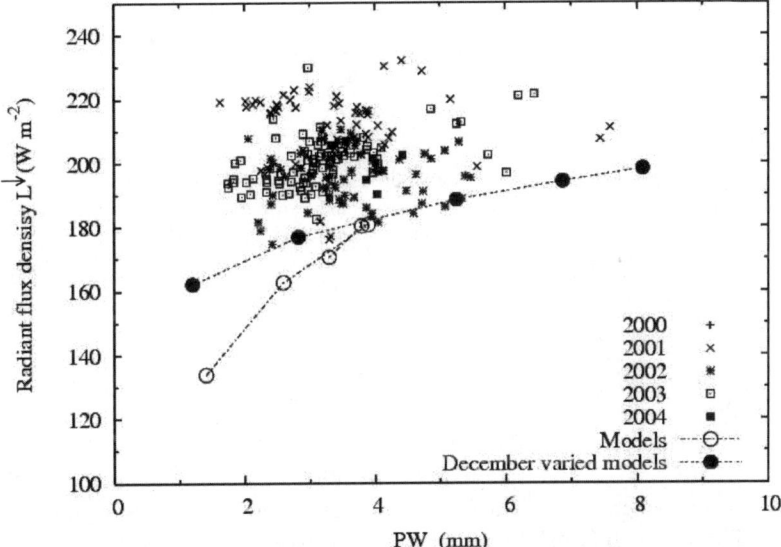

Fig. 16 Plot of the 15-min average values of the downwelling radiant flux density $L\downarrow$ measured by the Kipp&Zonen CNR-1 net radiometer during five years (2000–2004) versus GPS-derived values of PW compared with the values of $L\downarrow$ computed applying the SBDART code (Ricchiazzi et al. 1998) to (i) the 5 monthly mean atmospheric models listed in Table 1a (*open circles*), and (ii) the five variation of December atmospheric models listed in Table 1b (*solid circles*)

measurements of meteorological data taken over the whole daily period and not only at the radiosounding time.

The above discrepancies can also be partially ascribed to the variability of the atmospheric emittance characteristics as a function of the vertical distribution of the moisture conditions of the atmosphere and to the radiance contributions provided by the clouds escaped to the above selection criteria. This is better shown in Fig. 17 where the apparent emittance ε^* defined in terms of the ratio,

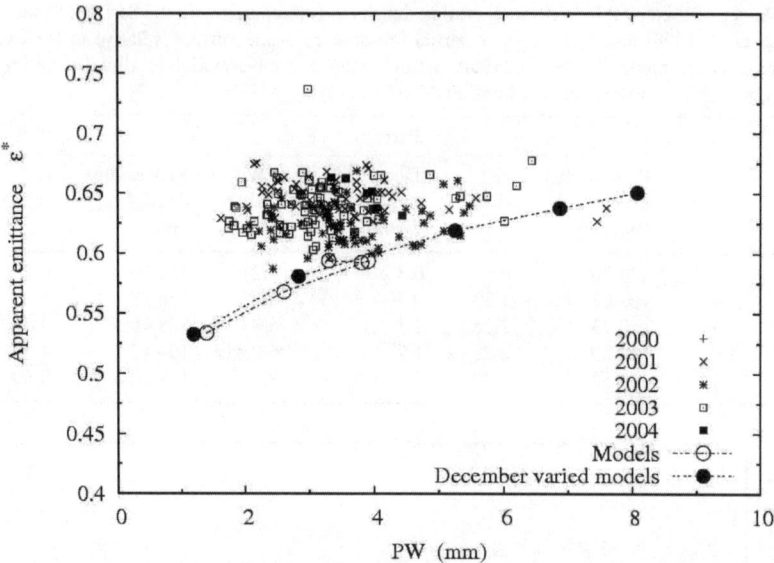

Fig. 17 Values of apparent emittance ε^* (calculated in terms of Eq. (15) for the data set shown in Fig. 16) plotted versus the corresponding GPS-derived values of PW and compared with the results obtained for (i) the 5 monthly mean atmospheric models listed in Table 1a (*open circles*), and (ii) the five December atmospheric models listed in Table 1b (*solid circles*)

$$\varepsilon^* = \frac{L\downarrow}{\sigma \cdot T_0^4} \tag{15}$$

is plotted as a function of the GPS-derived values of PW. Parameter ε^* gives a measure of the apparent emittance of the atmosphere with respect to the radiant flux density emitted upward by the surface in terms of the Planck Eq. (14) for a perfect black-body. This normalisation procedure allows the considerable reduction of data dispersion, since the radiant flux measurements are divided by a radiation term depending on the thermal conditions of the atmospheric ground-layer. Thus, Fig. 17 shows that ε^* assumes field values varying mainly between 0.58 and 0.67. Among them, the lower ones agree closely with those calculated for the two sets of atmospheric models presented in Table 1, in which the surface-level water vapour partial pressure e_0 varies from 0.65 to 2.28 hPa for the monthly models in Table 1a, and from 0.79 to 5.30 hPa for the December models in Table 1b. In fact, for these values of e_0, the formulas of Brutsaert (1975) and Idso (1981) provide values of ε^* ranging between 0.54 and 0.68, which are in good agreement with those shown in Fig. 17, obtained for the atmospheric conditions described in Table 1.

7 Conclusions

The ever increasing number of scientific observations in Antarctica allows a wider and deeper understanding of local as well as global phenomena. As time goes by, observational records form longer time series of increasing accuracy; these data sets can be used either for specific as well as multidisciplinary scientific investigations. Antarctic scientific stations, such as "Mario Zucchelli", offer the opportunity to develop multidisciplinary research, merging the data acquired through co-located sensors and enhancing the cooperation between different scientific groups.

The analysis of a six-year set of continuous Global Positioning System observations was performed aiming at sensing the water vapour content and its variations with time in the area of Terra Nova Bay. GPS capability of retrieving Precipitable Water vapour was confirmed, despite the very low humidity that characterize the Antarctic atmosphere, especially during the winter period. A validation of PW content was achieved through a comparison with PW computed using 108 radiosonde profiles. The test was performed over a two-month period (December 2002–January 2003), paying a careful attention to both the processing of GPS and radiosonde data. In particular, for this latter, the original relative humidity data have been analyzed following a careful procedure based on rigorous selection criteria and using the algorithms of Wang et al. (2002) and those of Miloshevich et al. (2004) to reduce the dry bias and other errors (Tomasi et al. 2006). The results fully confirm that GPS observations of PW at Terra Nova Bay can be considered as a useful ancillary information for atmospheric studies. To this respect, with the purpose of relating the PW measured with GPS to the radiative properties of the atmosphere at Terra Nova Bay, more accurate studies are planned in analyzing more extended sets of GPS and radiosounding observations through more advanced procedures taking into account the day-to-day variations in the thermal and moisture conditions of the atmosphere that can be observed at Terra Nova Bay in different seasonal periods. In addition, the analysis of the measurements of downwelling radiant flux density performed routinely at Campo Icaro with a Kipp&Zonen CNR-1 net radiometer showed that an important influence is exerted by atmospheric water vapour on the thermal energy balance of the atmosphere also at Antarctic sites, suggesting that a significant contribution to the knowledge of the radiative transfer of thermal atmospheric radiation can be provided by relating these long-wave radiation measurements to simultaneous measurements of Precipitable Water.

This scientific paper and its results are a strong motivation for continuing the multidisciplinary approach that was hereby being presented and applied on Antarctic scientific observations. A deeper understanding of how further GPS observations can benefit atmospheric science investigations in Antarctica will be pursued in the future.

Acknowledgments The authors would like to acknowledge that the research activities were supported by the Programma Nazionale di Ricerche in Antartide (PNRA) and developed as parts of the Sub-project 2006/6.01 "POLAR-AOD: a network to characterize the means, variability and trends of the climate-forcing properties of aerosols in polar regions", Sub-project 2004/2.4 "Accurate surface-based radiation measurements at Mario Zucchelli and Concordia stations" and the Sub-project 2004/2.3 "Geodetic monitoring of Northern Victoria Land". Authors would like to thank Dr. A. Pellegrini and his group for providing the radiosounding.

Appendix

The computation of the ZHD may be performed with (7), according to Hopfield (1971); the height parameters are obtained with a least square adjustment procedure over observed data. The refractivity at a given height is expressed as (Smith and Weintraub 1953):

$$N = \frac{77.6}{T}p + \frac{77.6}{T} \cdot \frac{4810 e}{T} = N_{dry} + N_{wet} \tag{16}$$

where T is the temperature in Kelvin, p is the pressure in hPa and e is the partial pressure of water vapour, all these parameters measured at a given height. According to (5) and (16) the refractivity is the sum of two parts, one related to the hydrostatic or "dry" part of the atmosphere, the other related to the water vapour or "wet" part of the atmosphere. Hopfield (1971) extensively used data recorded from radiosonde launched in various locations of the world in order to obtain (7), which uses surface atmospheric parameters for the prediction of tropospheric effects on electromagnetic measured range.

A computation of dry and wet delays along the vertical can be done, according to (5), numerically integrating the dry and wet refractivity (16). In the area of Terra Nova Bay, this can also be achieved using the corrected relative humidity profiles of Fig. 7, as well as the vertical profiles of temperature and pressure shown in Fig. 14.

These profiles also allow a computation of parameter PW, which can be derived integrating the absolute humidity ρ along the vertical path, according to:

$$PW = \int_0^\infty \rho(z) dz \tag{17}$$

where ρ is computed from the relative humidity f as:

$$\rho = \frac{f \cdot E(T)}{4.615 \cdot 10^{-3} \cdot T} \tag{18}$$

where $E(T)$ is the saturation pressure of water vapour expressed as a function of the temperature (Bolton 1980), as given in (12). Inserting (12) into (17), (18) can be computed.

According to (16) and (17), numerical integration based on the values of the profiles of Figs. 8 and 14 allows the computation of monthly values of ZHD and ZWD as well as PW. Table 2 summarizes these values along with ZHD predicted using (7); the lapse rate, the surface temperature T_0, the total pressure at 60 m above ground-level and PW are also shown.

A comparison between the ZHD computed with (7) and with numerical integration of the hydrostatic refractivity N_{hyd} of Fig. 18 shows that this latter method yields lower values.

The profiles shown in Fig. 14 are limited to 30 km. Nevertheless, observing Fig. 18 it is quite clear that the value of N_{hyd} at 30 km is not negligible; this might

Table 2 Lapse rate α, surface temperature T_0, surface pressure p_0, Precipitable Water PW, integrated Hydrostatic and Wet Zenith Delays ZHD and ZWD and ZHD computed according to Eq. (7)

Month	α(°C/km)	T_0 (K)	P_0 (hPa)	PW (mm)	ZHD (mm)	ZWD (mm)	ZHD (eq. 7) Hop '69
Oct	−5.43	257.9	978.6	1.4	2204.5	–	2231.1
Nov	−5.97	266.6	982.1	2.6	2210.3	33.9	2240.0
Dec	−6.25	270.7	979.8	3.9	2203.8	50.0	2235.2
Jan	−6.29	270.7	982.3	3.8	2208.9	48.5	2240.9
Feb	−5.69	267.9	982.1	3.3	2207.4	41.8	2240.2
Mean values	−5.93	266.8	981.0	3.0	2207.0	43.5	2237.5
St. dev.	0.37	5.3	1.7	1.0	2.8	7.3	4.0

be the cause of an underestimation of ZHD. In order to take into account the contribution of refractivity due to higher atmospheric layers, the value of $N_{hyd}(z > 30\,km)$ was extrapolated fitting an exponential function $N_{hyd}(z) = N_{hyd,0}\exp(-h \cdot z)$ to the N_{hyd} values from 20 to 30 km, being the parameter $N_{hyd,0} = 313.74$ and $h = 0.1443$. The integration performed from 30 to 100 km can be analytically obtained. The

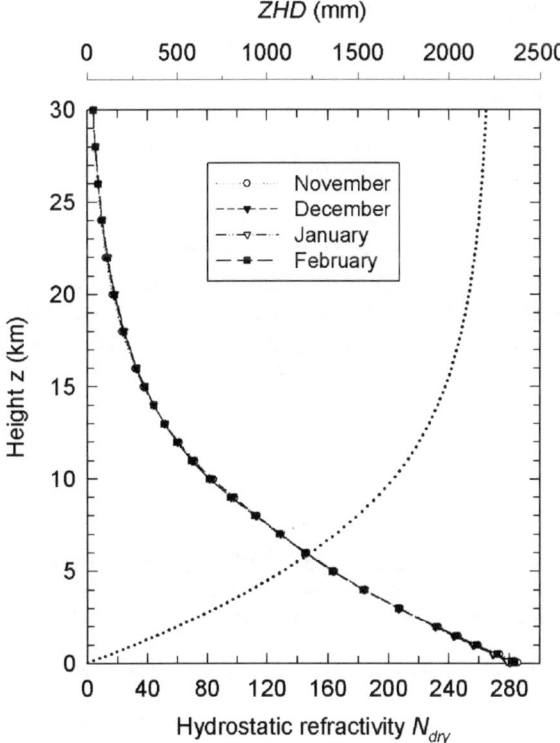

Fig. 18 Hydrostatic refractivity computed for four monthly atmospheric models. The dotted curve represents the integral of the December profile of the hydrostatic refractivity

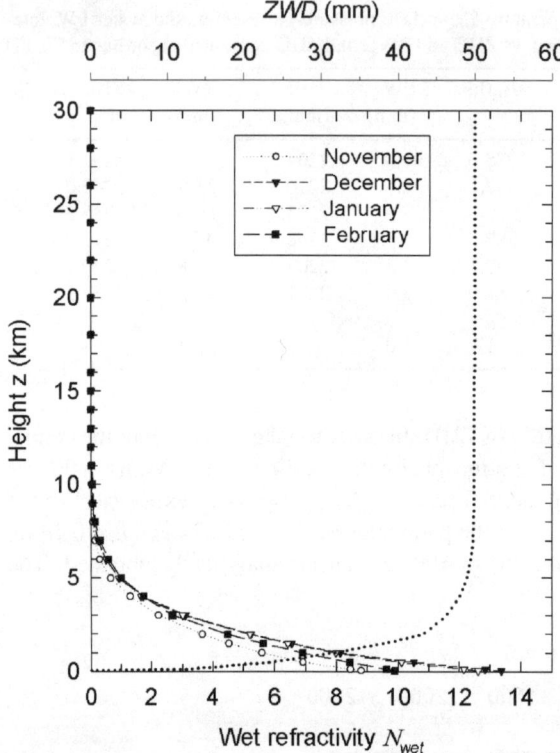

Fig. 19 Wet refractivity computed for four monthly atmospheric models. The dotted curve represents the integral of the December profile of the wet refractivity

value computed e.g. using the December monthly profiles of temperature and pressure is approximately 30 mm, which covers completely the gap between the modelled and integrated values of ZHD (Table 2). On the other hand, being the water vapour distribution concentrated in the lower part of the troposphere, the wet refractivity N_{wet} (16) vertical profiles tend very quickly to zero, as it is shown in Fig. 19.

References

Anderson G. P., S. A. Clough, F. X. Kneizys, J. H. Chetwind, and E. P. Shettle (1986), AFGL Atmospheric Constituent Profiles (0–120 km), Environmental Research Papers, No. 954, AFGL-TR-86-0110, Optical Physics Division, Air Force Geophysics Laboratory (AFGL), L. G. Hanscom Field, Mass. 01731, 43 pp.

Arking A. (1996), Absorption of solar energy in the atmosphere: discrepancy between model and observations, *Science*, **273**, 779–782.

Bevis M., S. Businger, T. A. Herring, C. Rocken, R. A. Anthes, and R. H. Hare (1992), GPS meteorology: Remote sensing of atmospheric water vapour using the Global Positioning System, *J. Geophys. Res.*, **97(D14)**, 15787–15801.

Bevis, M., S. Businger, S. Chiswell, T. Herring, R. Anthes, C. Rocken, and R. Ware (1994), GPS meteorology: Mapping zenith wet delays onto precipitable water, *J. Appl. Meteorol.*, **33**, 379–386.
Bignell K. J. (1970), The water-vapour infrared continuum, *Q. J. R. Met. Soc.*, **96**, 390–403.
Bolton D. (1980), The computation of equivalent potential temperature, *Monthly Weather Rev.*, **108**, 1046–1053.
Brutsaert W. (1975), On a derivable formula for long-wave radiation from clear skies, *Water Resour. Res.*, **11**, 742–744.
Capra A., M. Dubbini, A. Galeandro, L. Gusella, A. Zanutta, G. Casula, M. Negusini, L. Vittuari, P. Sarti, F. Mancini, S. Gandolfi, M. Montaguti, G. Bitelli (2008), VLNDEF project for geodetic infrastructure definition of Northern Victoria Land, Antarctica, *this volume*.
Chahine M. T. (1992), The hydrological cycle and its influence on climate, *Nature*, **359**, 373–380.
Chiou E.-W., M. P. McCormick, W. P. Chu (1997), Global water vapor distributions in the stratosphere and upper troposphere derived from 5.5 years of SAGE II observations (1986–1991), *J. Geophys. Res.*, **102**, 19105–19118.
Clough S. A., F. X. Kneizys, R. W. Davies (1989), Line shape and the water vapor continuum, *Atmos. Res.*, **23**, 229–241.
Dach R., U. Hugentobler, P. Fridez, M. Meindl (2007), *Bernese GPS Software Version 5.0*, Astronomical Institute of University of Berne, 612 pp.
Davis J. L., T. A. Herring, I. I. Shapiro, A. E. E. Rogers, G. Elgered (1985), Geodesy by radio interferometry: Effects of atmospheric modeling errors on estimates of baseline length, *Radio Sci.*, **20**, 1593–1607.
Elgered G., J. L. Davis, T. A. Herring, I. I. Shapiro (1991), Geodesy by radio interferometry: Water vapor radiometry for estimation of the wet delay, *J. Geophys. Res.*, **96(B4)**, 6541–6555.
Emardson T. R., G. Elgered, J. M. Johansson (1998), Three months of continuous monitoring of atmospheric water vapour with the network of Global Positioning System receivers, *J. Geophys. Res.*, **103**, 1807–1820.
Goody R.M. (1964), *Atmospheric Radiation, I.– Theoretical Basis*, Oxford, Clarendon Press, 436 pp. (see 67232 pp).
Gradinarsky L. P., R. Haas, G. Elgered, and J. M. Johansson (2000), Wet path delay and delay gradients inferred from microwave radiometer, GPS and VLBI observations. *Earth Planets Space*, **52(10)**, 695–698.
Haase J., E. Calais, J. Talaya, A. Rius, F. Vespe, R. Santangelo, X.-Y. Huang, J. M. Davila, M. Ge, L. Cucurull (2001), The contributions of the MAGIC project to the COST 716 objectives of assessing the operational potential of ground-based GPS meteorology on an international scale, *Phys. Chem. Earth*, **26**, 433–437.
Harries J. E., J. M. Russell III, A. F. Tuck, L. L. Gordley, P. Purcell, K. Stone, R. M. Bevilacqua, M. Gunson, G. Nedoluha, W. A. Traub (1996), Validation of measurements of water vapor from the Halogen Occultation Experiment (HALOE), *J. Geophys. Res.*, **101**, 10205–10216.
Hofmann-Wellenhof B., H. Lichtenegger, J. Collins (2001), Global Positioning System, Theory and practice. 5th ed., Springer Wien NewYork, 382 pp., ISBN 3-211-83534-2.
Hopfield H. S. (1969), Two-quartic troposheric refractivity profile for correcting satellite data. *J. Geophys. Res.*, **74(188)**, 4487–4499.
Hopfield H. S. (1971), Troposheric effect on electromagnetically measured range: Prediction from surface weather data, *Radio Sci.*, **6(3)**, 357–367.
Hopfield H. S. (1978), Tropospheric correction of electromagnetic ranging signals to a satellite: A study of parameters. In: *Proceedings of International Symposium on Electromagnetic Distance Measurement and the Influence of Atmospheric Refraction* (P. Richardus, ed.), Wageningen, The Netherlands, 23–28 May 1977, Netherlands Geodetic Commission, 205–215.
Houghton J. T., G. J. Jenkins, and J. J. Ephraums (1990), *Intergovernmental Panel on Climate Change (IPCC) Report, Climate Change, The IPCC Scientific Assessment*, U.K. Meteorological Office, Bracknell, England, 364 pp.
Houghton J. T., Y. Ding, D. J. Griggs, M. Noguer, P. J. van der Linden, X. Dai, K. Maskell, and C. A. Johnson (Eds.), (2001), IPCC, 2001: Climate Change 200:, The Scientific Basis.

Contribution of Working Group I to the Third Assessment, Report of the Intergovernmental Panel on Climate Change. Cambridge University Press, 881 pp.

Huovila S., and A. Tuominen (1991), Influence of radiosonde lag errors on upper-air climatological data. In *Seventh Symposium on Meteorological Observations and Instrumentation, Special Sessions on Laser Atmospheric Studies*, Boston, American Meteorological Society, New Orleans, January 14–18, 1991, 237–242.

Idso S. B. (1981), A set of equations for full spectrum and 8 to 14 μm and 10-5 to 12.5 μm thermal radiation from cloudless skies, *Water Resour. Res.*, **17(2)**, 295–304.

Johnsen K. P., J. Miao, and S. Q. Kidder (2004), Comparison of atmospheric water vapor over Antarctica derived from CHAMP/GPS and AMSU-B data, *Phys. Chem. of the Earth*, **29**, 251–255.

Kiehl J. T., and K. E. Trenberth (1997), Earth's annual global mean energy budget, *Bull. Am. Met. Soc.*, **78**, 197–208.

Kneizys F. X., L. Abreu, G. P. Anderson, J. H. Chetwind, E. P. Shettle, A. Berk, L. S. Bernstein, D. C. Robertson, P. Acharya, L. S. Rothman, J. E. A. Selby, W. O. Gallery, and S. A. Clough, (1996), *The MODTRAN 2/3 Report and LOWTRAN 7 MODEL* (Abreu L. W., and G. P. Anderson, eds.). Contract F19628-91-C.0132, Phillips Laboratory, Geophysics Directorate, PL/GPOS, Hanscom AFB, Mass., 261 pp.

Kondratyev K. Ya. (1969), *Radiation in the Atmosphere*, New York, Academic Press, 912pp. (see 85–159pp).

Lahoz W. A., A. O'Neill, A. Heaps, V. D. Pope, R. Swinbank, R. S. Harwood, L. Froidevaux, W. G. Read, J. W. Waters, and G. E. Peckham (1996), Vortex dynamics and the evolution of water vapour in the stratosphere of the southern hemisphere, *Q. J. R. Met. Soc.*, **122**, 423–450.

Leckner B. (1978), The spectral distribution of solar radiation at the Earth's surface – Elements of a model, *Solar Energy*, **20**, 143–150.

Long C. N., and T. P. Ackerman (2000), Identification of clear skies from broadband pyranometer measurements and calculation of downwelling shortwave cloud effects, *J. Geophys. Res.*, **105(D12)**, 15609–15626.

Luers, J. K., and R. E Eskridge (1995), Temperature corrections for the VIZ and Vaisala radiosondes, *J. Appl. Met.*, **34**, 1241–1253.

Marsden D., and F. P. J. Valero (2004), Observation of water vapour greenhouse absorption over the Gulf of Mexico using aircraft and satellite data, *J. Atmos. Sci.*, **61**, 745–753.

Miao J., K. Kunzi, G. Heygster, T. A. Lachlan-Cope, and J. Turner (2001), Atmospheric water vapor over Antarctica derived from Special Sensor Microwave/Temperature 2 data, *J. Geophys. Res.*, **106(D10)**, 10187–10204.

Miloshevich L. M., H. Vömel, A. Paukkunen, A. J. Heymsfield, and S. J. Oltmans (2001), Characterization and correction of relative humidity measurements from Vaisala RS80-A radiosondes at cold temperatures, *J. Atmos. Oc. Techn.*, **18**, 135–156.

Miloshevich L. M., A. Paukkunen, H. Vömel, and S. J. Oltmans (2004), Development and validation of a time-lag correction for Vaisala radiosonde humidity measurements. *J. Atmos. Oc. Techn.*, **21**, 1305–1327.

Niell A. E. (1996), Global mapping functions for the atmosphere delay at radio wavelengths, *J. Geophys. Res.*, **101**, 3227–3246.

Niell A. E., A. J. Coster, F. S. Solheim, V. B. Mendes, P. C. Toor, R. B. Langley, and A. Upham, (2001), Comparison of measurements of atmospheric wet delay by radiosonde, water vapor radiometer, GPS, and VLBI, *J. Atmos. Oc. Techn.*, **18**, 830–850.

Randel W. J., F. Wu, A. Gettelman, J. M. Russell III, J. M. Zawodny, and S. J. Oltmans (2001), Seasonal variation of water vapor in the lower stratosphere observed in Halogen Occultation Experiment data, *J. Geophys. Res.*, **106**, 14313–14325.

Ricchiazzi P., S. Yang, C. Gautier, and D Sowle (1998), SBDART: A research and teaching software tool for plane-parallel radiative transfer in the Earth's atmosphere, *Bull. Am. Meteor. Soc.*, **79(10)**, 2101–2114.

Rind D. (1998), Climate change: Just add water vapor, *Science*, **281(5380)**, 1152–1153.

Rocken C., R. Ware, T. Van Hove, F. Solheim, C. Alber, J. Johnson, M. Bevis, and S. Businger (1993), Sensing atmospheric water vapor with the Global Positioning System, *Geophys. Res. Lett.*, **20**, 2631–2634.

Saastamoinen J. (1972), Contributions to the theory of atmospheric refraction, *Bull. Geod.*, **105**, 279–298; **106**, 383–397.

Saastamoinen J. (1973), Contributions to the theory of atmospheric refraction, *Bull. Geod.*, **107**, 13–34.

Schuh H., J. Boehm, G. Engelhardt, D. MacMillan, R. Lanotte, P. Tomasi, M. Negusini, I. Vereshchagina, V. Gubanov, and R. Haas (2005), Determination of tropospheric parameters within the new IVS Pilot Project. In: *A Window on the Future of Geodesy, Proceedings of the IAG General Assembly, Sapporo, Japan, June 30–July 11, 2003* (F. Sanso, ed.), Springer, IAG, **128**, pp. 125–130.

Smith E. K., and S. Weintraub (1953), The constants in the equation for atmospheric refractive index at radio frequencies, *Proc. IRE*, **41**, 1035–1037.

Starr D., and S. H. Melfi (1991), The Role of Water Vapour in Climate, A Strategic Research Plan for the Proposed GEWEX Water Vapour Project (GVaP), *NASA Conference Publ.* **3210**, 60pp.

Tomasi C., R. Guzzi, and O. Vittori (1974), A search for the e-effect in the atmospheric water vapor continuum, *J. Atmos. Sci.*, **31**, 255–260.

Tomasi C., and F. Trombetti (1985), Absorption and emission by Minor Atmospheric Gases in the Radiation Balance of the Earth, *La Rivista del Nuovo Cimento*, **8(No. 2)**, Monograph, pp. 89.

Tomasi C., V. Vitale, M. Tagliazucca, and L. Gasperoni (1990), Infrared hygrometry measurements at Terra Nova Bay, *SIF Conference Proceedings*, **27**, 187–200.

Tomasi C., S. Marani, V. Vitale, F. Wagner, A. Cacciari, and A. Lupi (2000), Precipitable water evaluations from infrared sun-photometric measurements analyzed using the atmospheric hygrometry technique, *Tellus*, **52B**, 734–749.

Tomasi C., A. Cacciari, V. Vitale, A. Lupi, C. Lanconelli, A. Pellegrini, and P. Grigioni (2004), Mean vertical profiles of temperature and absolute humidity from a twelve-year radiosounding data-set at Terra Nova Bay (Antarctica), *Atmos. Res.*, **71**, 139–169.

Tomasi C., B. Petkov, E. Benedetti, V. Vitale, A. Pellegrini, G. Dargaud, L. De Silvestri, P. Grigioni, E. Fossat, W. L. Roth, and L. Valenziano (2006), Characterization of the atmospheric temperature and moisture conditions above Dome C (Antarctica) during austral summer and fall months, *J. Geophys. Res.*, **111**, D20305, doi:10.1029/2005JD006976.

Tregoning P., R. Boers, D. O'Brien, and M. Hendy (1998), Accuracy of absolute precipitable water vapor estimates from GPS observations, *J. Geophys. Res.*, **103**, 28701–28710.

Vey S., R. Dietrich, K.-P. Johnsen, J. Miao, and G. Heygster (2004), Comparison of tropospheric water vapour over Antarctica derived from AMSU-B data, ground-based GPS data and the NCEP/NCAR reanalysis, *Jour. Met. Soc. Japan*, **82**, 259–267.

Wang J., H. L. Cole, D. J. Carlson, E. R. Miller, K. Beierle, A. Paukkunen, and T. K. Laine (2002), Corrections of humidity measurement errors from the Vaisala RS80 radiosonde – Application to TOGA COARE data, *J. Atmos. Oc. Techn.*, **19**, 981–1002.

Yamanouchi T., and T. O. Charlok (1995), Comparison of radiation budget at TOA and surface in the Antarctica from ERBE and ground surface measurements, *J. Climate*, **8**, 3109–3120.

Results of the Investigations of the GNSS Antennae in the Framework of SCAR GIANT Project "In Situ GNSS Antenna Tests and Validation of Phase Centre Calibration Data"

Jan Cisak and Yevgen M. Zanimonskiy

Abstract The calibration of geodetic GPS antennae has been a topic of intensive research over the last 10 years. Different methods have been tested to determine the phase centre position and its variations. Not all of these methods have been used "in situ", where the antenna is receiving real measurements.

The paper describes the influence of the surrounding ground conditions on multipath effects of GNSS antennas, as well as the impact of snow cover on antenna. This is particularly important in polar regions where the antennae are located to the nearest rocky outcrop.

The influence of changes in different local conditions on the phase centre position of the antenna located at different heights above the ground is also discussed in this paper.

A project on the development of the method of GNSS antennae calibrations taking into account the parameters of local conditions was established in the Institute of Geodesy and Cartography, Warsaw, Poland. A simple and portable prototype of the device for antenna calibration was constructed and it is currently undergoing intensive investigation.

1 Introduction

For precise geodetic and geophysical applications the site positions should be known to the nearest millimetre. Obstacles encountered on the way to such accuracy are systematic errors appearing due to the presence of an uncertain shift of phase of the registered satellite signal in the antenna system. This means, that the system consists of the antenna itself and the bodies surrounding the antenna are modifying its electrodynamics' properties. The configuration of antenna systems is thus unique for a

---

Jan Cisak
Institute of Geodesy and Cartography, Warsaw, Poland, e-mail: jcisak@igik.edu.pl

Yevgen M. Zanimonskiy
Institute of Radio Astronomy National Academy of Sciences of Ukraine, Kharkiv, Ukraine,
e-mail: yzan@poczta.onet.pl

site. At permanent GNSS stations, the electrodynamics' properties of the antenna may become modified depending on weather conditions, dampness of the surrounding surface, or the amount of snow covering the antenna and snow in its vicinity (Niell, 1997; Ray, 2006). To eliminate the influence of these factors by modelling is impossible because of their complexities.

Experimental work with GNSS antennae has been carried out for solving several tasks. They concerned a definition of parameters of the antenna in "ideal" conditions; the problem of defining how much the real conditions of the antenna arrangement differ from the "ideal" conditions of calibration and what is the minimal set of requirements to the antenna installation site; and finally mostly relevant nowadays problem, i.e., how to calibrate the antenna in real conditions on an installation site. That last one is a major task of the SCAR project "In situ GNSS Antenna Tests and Validation of Phase Centre Calibration Data" which was established in Bremen at the SCAR Conference in 2004 and was extended for further 2 years at the XXIX SCAR meeting in Hobart in 2006.

2 GNSS Antenna Phase Centre Offset and Phase Centre Variations

To fulfil the requirements of millimetre level accuracy it is important to know the exact position of the phase centre of the receiving GNSS antenna. Antennae can be characterized by mean phase centre offset and by phase centre variations (PCV) (Fig. 1).

The antenna phase centre offset is defined as the average of phase centre locations relative to a physical reference point (antenna reference point – ARP). Unfortunately, the theoretical concept of a mathematical point representing the intersection of all ranges measured from the satellites cannot be realized because the environment of the antenna distorts the properties of the incoming signals (Meertens et al., 1997; Niell, 1997; Ray, 2006). The errors of these ranges are at the level of single millimetres.

The calibration of geodetic GNSS antennae has become the subject of intensive research over the last 10 years. Different methods have been elaborated to determine

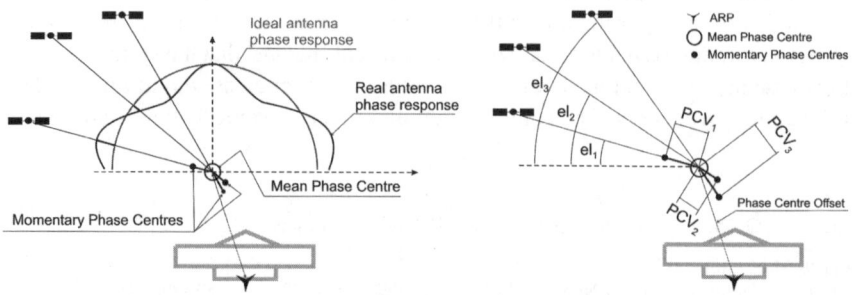

Fig. 1 GNSS antenna phase centre offset and phase centre variations

Fig. 2 Examination of vertical component differences in phase centres of 6 different antennae

the phase centre position and its variations in the "ideal" situation (Rothacher and Schmid 2002). Those methods are laboratory ones: anechoic chamber (Görres et al., 2006); NGS test field; (Mader, 1999) or used in special field conditions like robot calibration system (Wübbena et al., 1997). Test field measurements represent the relative methods while anechoic chamber and robot systems give the absolute data. The necessity of absolute antenna calibration was reflected in the Resolution 1 of the EUREF 2005 Symposium in Vienna and in the Resolution 3 of the EUREF 2006 Symposium in Riga. http://www.euref-iag.net/html/resolutions.html

The Institute of Geodesy and Cartography, Warsaw, has been involved in GPS antenna phase centre investigations since 1999. In the framework of the project on the Polish Geodynamic Network the results of the test GPS measurements on the pillars of Bemowo baseline (Fig. 2), along with well-known heights and distances, were analysed in terms of antennae phase centres determinations. Six types of antennae: Leica SR399, Trimble geod. L1/L2 22020, Dorne Margolin Ashtech, Dorne Margolin Trimble, Ashtech 700718 and Ashtech 700228 were used (Dobrzycka and Cisak, 2001) (Fig. 2).

3 The Snow Effect

At the beginning of February 2006 (GPS week 1360) some local effects on coordinate changes of IGS stations were observed. This effect is well seen on the graph in Fig. 3. The very snowy winter of 2005/2006, particularly in the region of Lamkowko, could cause the observed disturbances of co-ordinates.

Using Ashtech Z12 receivers and the Ashtech Choke Ring antenna, the investigation of the effect of snow was performed at the end of winter 2006 at Lamkowko (Fig. 4). The antenna was placed about 70 m from the LAMA station, on a concrete pillar (1 m diameter). Different layers of snow covering the antenna were put on the pillar. The 24-hour observations of each level of snow were performed. Data collected from the antenna covered by snow and from LAMA station (without snow) were divided for one hour sessions. The GPPS software was used for postprocessing of the observations data of each session. Results for height components of calculated

Fig. 3 The effect of snow on vertical component of LAMA (Lamkowko UWM Observatory) co-ordinates changes is seen on the graph published on IGS web page

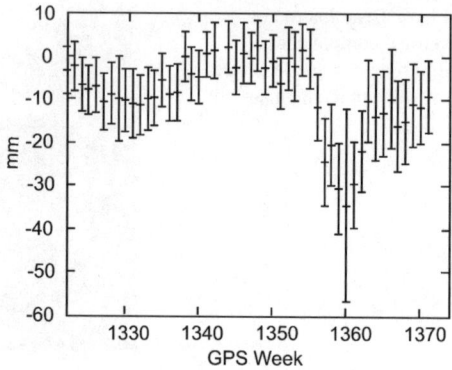

co-ordinates of the mean phase centre of the tested antenna are shown in Fig. 4. Each point on the graph (Fig. 4) represents the result of one hour session. The experiment was performed twice.

The dependence of errors in vertical solutions vs height of snow layer covering the antenna consists of two parts. Snow cover up to mean phase centre position located 25 cm over the pillar's top faintly modifies the scattering by concrete pillar's top. It is the reason of small variations of mean phase centre elevations and consequently the variations of "apparent height". Snow cover above the mean phase centre acts as the refractive medium delaying the signal pro rata to the cover thickness. In this case the "apparent height" varies noticeably. The greater variations shown in Fig. 4 relatively to those shown in Fig. 3 are the consequence of a greater density of hand made snow cover compared to natural snow cover.

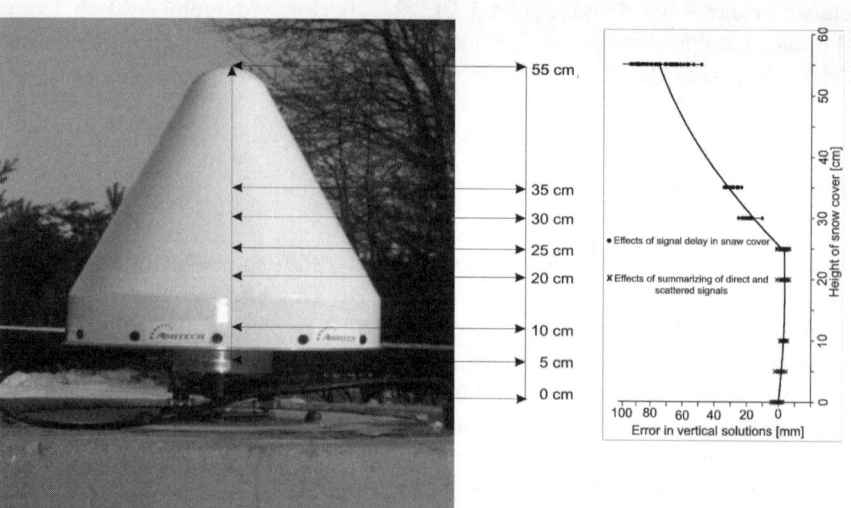

Fig. 4 Antenna used in "snow effect experiment". Different levels of the snow layers are shown

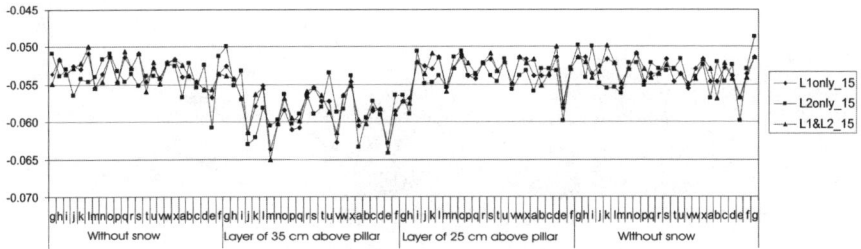

Fig. 5 Changes in the vertical component with the effect of snow cover on BOGI station

Similar experiment was performed on BOGI station at Borowa Gora. The highest layer of snow cover was 35 cm above the pillar. The heights calculated with reference to BOGO station using the Pinnacle software are presented in Fig. 5. Similar effects of snow covering antennae are described in (Meertens et al., 1997; Ray, 2006).

4 Phase Centre Changes due to the Different Height and Types of Antenna Mount

High-accuracy GNSS observations can be affected not only by the antenna type but also by the height and types of the mount (Ray, 2006). The multipath far-field effects, i.e. reception of the reflected signal with a propagation difference larger than a few meters, may be reduced due to averaging over time. Near-field (scattering) effects in general cannot be reduced (Wübbena et al., 2006; Niell, 1997). Moreover, the multi-path effects are practically absent at the high elevation angle of a satellite (Fig. 6 left). Scattering produces errors that vary insignificantly with elevation angle of a satellite (Fig. 6), so it is difficult to detect them under standard antenna calibration procedures.

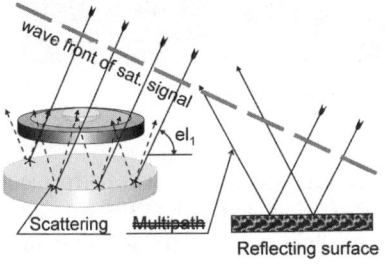
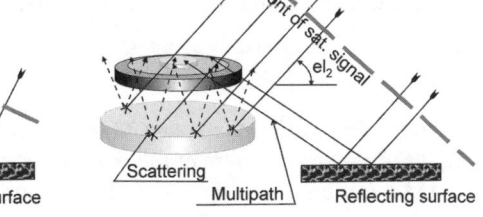

Fig. 6 The effect of multipath and scattering

Fig. 7 Different types of antenna mounted on Antarctic GNSS stations

Some years ago, when the majority of permanent tracking GNSS stations were established, the problem of type of antenna mount was not taken into account. Some examples of Antarctic stations' antenna installation are presented in Fig. 7. It seems that scattering effects from the top of any monument or rocky outcrop causes the changes in the measured height of sites of some millimetres.

To detect elevation-dependent height errors of the antenna, test measurements were performed at Lamkowko Observatory. The elevation of tested antenna was consecutively changed from 0 to 5, 10, 30 and 50 cm above the soil (Fig. 8).

Fig. 8 Antenna was placed on the long screw

At each height of the antenna one 23-hour measurement session was performed. The heights of the phase centre of the antenna were calculated with GPPS software, using LAMA station as the reference. The results obtained from the test measurements are illustrated in Fig. 9. Each point on the graph (Fig. 9) represents the result of one hour session.

Date of observ.	Elevation (mm)	Mean (mm)	Std. dev. (mm)
2006-04-23	0	10.2	0.3
2006-04-25	100	−0.7	0.4
2006-04-27	300	−0.5	0.3
2006-04-29	500	0.5	0.6
2006-05-01	500	0.1	0.5
2006-05-03	400	0.3	0.4
2006-05-05	300	−0.2	0.3
2006-05-07	100	11.7	0.7
2006-05-09	50	16.9	0.6
2006-05-11	0	24.6	0.7

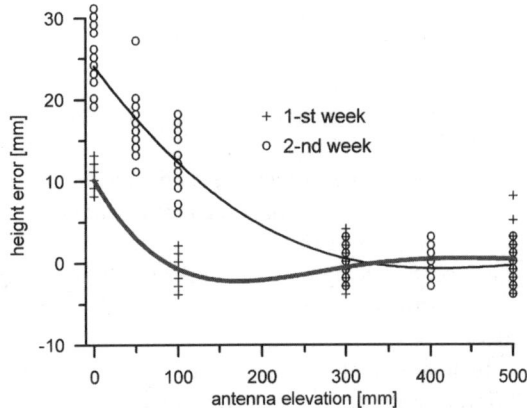

Fig. 9 The phase centre changes vs antenna elevation above the soil in Lamkowko. Observations in two consecutive weeks in different weather conditions

Variations in phase centre height, of up to 25 mm were found, due to different elevation of antenna mounts above the ground. These results were surprising. Our preliminary conclusions are that soil water, as the effective scattering medium, had influenced the measurements of phase centre height in Lamkowko. Additional experiments in different weather conditions are planned at the same location.

Two experiments were performed in Borowa Gora Observatory. In summer 2006 the same choke ring antenna was placed on a tribrach above the large concrete pillar. The elevation of the antenna reference point was changed from 0 to 8, 23, 30, 48, 55, 70 and 78 cm with the use of a set of spacer rods with fixed length with an accuracy of a tenth of millimetre. Mean values of antenna heights obtained from eight 6-hour sessions at each elevation together with their experimentally determined RMS are shown in Fig. 10 (left). Torrid summer provided us with the chance to perform measurements at a dry as well as an artificially moistened top of the concrete pillar. On the other hand in winter 2006/2007 a changeable weather during two consecutive weeks of observations randomised the influence of scattering signals. The variations of height errors vs antenna elevation were difficult to separate from random ones (Fig. 10 right).

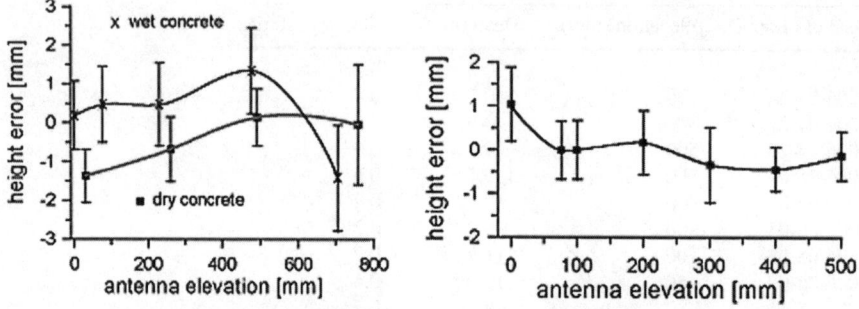

Fig. 10 Phase centre changes vs antenna elevation above concrete pillars in Borowa Gora. Summer observations were based upon different pillar dampness criteria (*left*). Winter observations were based upon combinations of different weather conditions (*right*)

Small variations (≤2 mm) of antenna phase centre occurred; they were faintly dependent on the concrete pillar top characteristics.

The next experiments were performed with the wooden tripod (without any scattering or reflecting surfaces) and with an aluminium plate (0.5 × 0.5 m) on the top of the pillar, like in cases of some permanent GNSS stations. There was no significant or systematic trend of "apparent" height vs antenna elevation above the tripod. The phase centre changes vs antenna elevation over metal plate are shown in Fig. 11. Variation of the antenna elevation in this case gives a distinguishable and stable effect.

It should be emphasized that near-field effects have different phases and slowly differing amplitude for carrier signals of both frequencies as well as multipath effects (Niell, 1997; Wübbena et al., 2006). This is the one of the reasons for the different results of measurement with different carrier combinations. In principle, the use of a few carriers in some antennae configurations may serve for the dampening of near-field effects. The application of a third frequency in the GPS system and design of special antennae for near rock placement are the ways to solve the "near-field problem", particularly for geodesy and geodynamics in Antarctic.

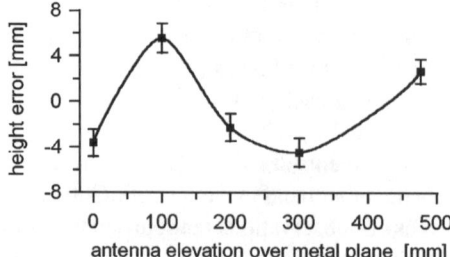

Fig. 11 Phase centre changes vs antenna elevation over metal plate

5 The Kinematic Device for Calibration of GNSS Antennae

A project on the development of GNSS antennae calibration, taking into account the parameters of local conditions, was established as a continuation of the previous investigations by the Institute of Geodesy and Cartography, Warsaw (Cisak and Zanimonskiy, 2005). A simple and portable prototype of the device for antenna calibration was constructed. In this kinematic instrument the test antenna is mounted on the rocking lever (Fig. 12). First test measurements were performed in Borowa Gora and Lamkowko Observatories in Poland as well as in the Ukrainian Antarctic "Akademik Vernadsky" station.

Using the antenna-tilting device it is possible to determine the mean phase centre offset as well as the PCV in both the real multipath and scattering situations. Algorithms and data processing schemes are very similar to those used for absolute antennae calibration with robot device (Wübbena et al., 1997). Differences in data processing arise due to continual angular movement (tilting) of the antenna in this instrument contrary to the discrete tilting of robot carrying antenna. In this situation one may use the simple filtering procedures for extraction of sinusoidal variations of antenna position and carrier phases of received signals. Parameters of mechanical movement of ARP are being measured by means of a special sensor. The swing period of the pendulum-like device in the same plane lasts 30 seconds. Complete investigation of antenna parameters needs two sets of measurements with an axial rotation of the antenna through 90 degrees.

Phase centre offset is being measured with the use of a reference site, whose antenna does not need a precise or absolute calibration. On the other hand, PCV may be measured in single site mode with usage of satellite-to-satellite carrier differences.

In Fig. 13 the example of variations of phase centre offset (PCO) averaged over observed satellites in one minute intervals are shown.

Variations in consecutive days have apparently a similar character as a consequence of recurrence of a configuration of GNSS satellites. Differences on the graphs are the result of noise. Variations of PCO averaged over one hour are characterized by dispersion at the level of a few millimetres. Observed variations of PCO

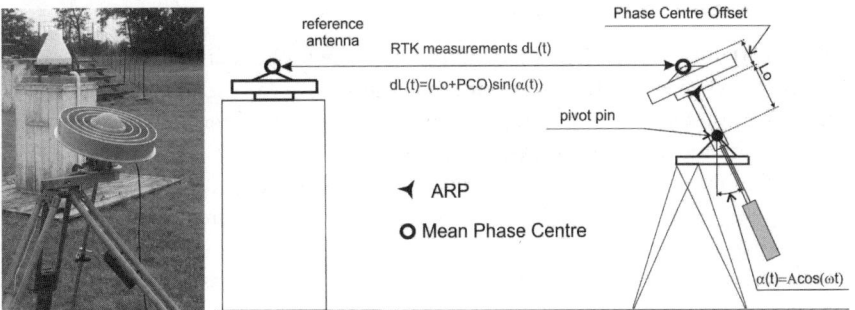

Fig. 12 Prototype of the tilting device with counterbalance for antenna calibration

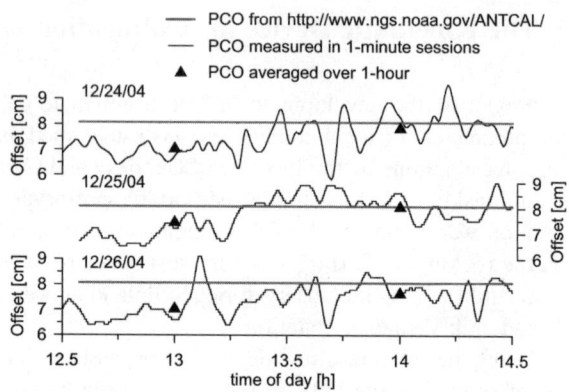

Fig. 13 Comparison of the results of the mean phase centre offset obtained from the observations with the results of NGS calibration

of antennae appreciably explain the dispersion of the results of GNSS solutions in hourly sessions shown in (Krynski and Zanimonskiy, 2001; Grejner-Brzezinska et al., 2006). Constant value of PCO shown in Fig. 13 was taken from web page http://www.ngs.noaa.gov/ANTCAL/

The results of numerous tests of two prototypes of kinematic devices placed on different elevations over the ground show that:

- mean phase centre offset of the antenna placed near the ground depends on soil dampness properties;
- parameters of antenna measured "in situ", under certain weather conditions, may not be coincide with those obtained in different weather conditions. This is the main contradiction for "in situ" antenna calibration and validation investigations.
- in the case of a metal plate under the antenna, the stable scattering and near-field effects may be corrected. Similar conclusions have been found in (Wübbena et al. 2006).

6 Conclusions and Future Work

On the basis of the results from the experiments presented in the paper it is possible to draw the following conclusions:

- the uncertainty of the measured height of site due to the scattering of an electromagnetic wave on the top of the pillar may reach one to several millimetres;
- the effect of scattering on a top of the concrete pillar is unstable and strongly depends on the humidity level found at the pillar basement;
- scattering on the metal surface found under an antenna at some permanent stations gives regular and stable in-time changes of the measured height of the site of up to 5 millimetres;
- the most stable results were obtained in the absence of disseminating surfaces near the antenna, namely the installation of antenna on a standard wooden geodetic tripod at height approximately 1.5 m above a grass-covered ground;

- the measurements executed with the antenna located on elevation from several centimetres up to half a meter above a soil, have yielded the most unstable results with variations of the measured height up to 2–3 cm;
- changes of the mean phase centre elevation due to snow cover between the top of the pillar and the antenna do not exceed one millimetre;
- changes of the "apparent height" of the antenna due to snow cover above the mean phase centre are at the centimetre level;
- investigation of antennae in real conditions of installation, i.e. "in situ", should be carried out. This would allow for testing the stability of scattering of satellite signals and multipath and, in case of stable conditions, there could be a calibration conducted "in situ".

It is planned, that further work in the Institute of Geodesy and Cartography, Warsaw, Poland, will be carried out to:

- repeat measurements with antennae at small elevations above the ground;
- measure antennae PCV on a kinematic device; and
- measure various types of metal plate antennae in order to determine the antennae configuration required for the installation nea-ground (rock), particularly in Antarctic conditions.

It is suggested to continue the collaborative efforts of the international scientific community for solving these problems connected with GNSS measurements at the millimetre level of accuracy.

Acknowledgments A part of the investigations were performed in the frame of the research grant 4T12E 048 30 of the Polish Ministry of Science. The authors express their thanks to Larry Hothem and other collaborators of SCAR GIANT project In situ GNSS Antenna Tests and Validation of Phase Centre Calibration Data for the discussions.

References

Cisak J., Zanimonskiy Y. (2005): Wyznaczanie centrum fazowego anten GNSS metoda dynamicznego nachylania. Presented at Seminarium "Satelitarne metody wyznaczania pozycji we wspolczesnej geodezji i nawigacji" Poznan, 23–24 czerwca 2005
Dobrzycka M., Cisak J., (2001): Polska sie geodynamiczna 1997 epoka 0. Prace IGiK 2001, t. 48, z 103.
Görres B., Campbell J., Becker M., Siemes M., (2006): Absolute calibration of GPS antennas: laboratory results and comparison with field and robot techniques. GPS Solutions (2006) 10, pp. 136–145.
Grejner-Brzezinska D.A., Vázquez G.E. Hothem L.,(2006) Geodetic Antenna Calibration Test in the Antarctic Environment. ION GNSS 19th International Technical Meeting of the Satellite Division, 26–29 September 2006, Fort Worth, TX, pp, 2798–2806.
Krynski J., Zanimonskiy Y.(2002) Contribution of Data from Polar Regions to the Investigation of Short Term Geodynamics. First Results and Perspectives. Antarctic Geodesy Symposium 2001, 17–20 July, St Petersburg, Russia, SCAR Report, No 21 January 2002, Cambridge, UK, pp. 56–60.

Mader G., (1999): GPS Antenna Calibration at the National Geodetic Survey. GPS Solutions, Vol. 3, No. 1, 1999, pp. 50–58.

Meertens C., Rocken C., Braun J., Stephens B., Alber C., Ware R., Exner M., Kolesnikoff P., (1997): Antenna type, mount, height, mixing, and snow effects in high-accuracy GPS observations, The Global Positioning System for the Geosciences. Summary and Proceedings of a Workshop on Improving the GPS Reference Station Infrastructure for Earth, Oceanic, and Atmospheric Science Applications, NATIONAL ACADEMY PRESS Washington, D.C. 1997, pp. 211–218.

Niell A., (1997): Near-Field Problems for GPS Antennas. The Global Positioning System for the Geosciences. Summary and Proceedings of a Workshop on Improving the GPS Reference Station Infrastructure for Earth, Oceanic, and Atmospheric Science Applications, NATIONAL ACADEMY PRESS Washington D.C. 1997, pp. 219–222.

Ray J., (2006): Systematic Errors in GPS Position Estimates. IGS Workshop, Darmstadt, 11 May 2006 ftp://ftp.ngs.noaa.gov/dist/Jimr/igsframe-igs06.pdf

Rothacher M., Schmid R. (2002): GPS-Antennenkalibrierung aus nationaler und internationaler Sicht. 4. SAPOS-Symposium 21–23 May 2002, Hannover, pp 124–131.

Wübbena G., Schmitz M., Menge F., Seeber G., Völksen C., (1997): A New approach for field calibration of absolute antenna phase center variations, Inst Navigat 44: 2.

Wübbena, G., Schmitz, M., Boettcher, G. (2006): Near-field Effects on GNSS Sites: Analysis using Absolute Robot Calibrations and Procedures to Determine Corrections, Presented at IGS Workshop. The International GNSS Service (IGS): Perspectives and Visions for 2010 and beyond, 8–12 May 2006, Darmstadt, Germany.

Atmospheric Impact on GNSS Observations, Sea Level Change Investigations and GPS-Photogrammetry Ice Cap Survey at Vernadsky Station in Antarctic Peninsula

J. Cisak, G. Milinevsky, V. Danylevsky, V. Glotov, V. Chizhevsky, S. Kovalenok, A. Olijnyk and Y. Zanimonskiy

Abstract The state-of-art of research in the several fields of geodesy and geophysics investigations based on GPS observations in Ukrainian Antarctic Vernadsky station area are considered in the paper. The some results of the Atmospheric Impact on GNSS Observations project are discussed. The examples of the specific errors of the GNSS-solutions for Antarctic continent and of the troposphere-ionosphere coupling research are shown. The necessity of application of the new precision measurement methods is caused by complexity of the local geodynamic processes. This complication is illustrated by the examples of sea level data and results of the glaciers GPS-photogrammetry monitoring of the small ice caps dynamics of Argentine Islands Archipelago. The sea level changes as a consequence of climate change and geological processes are discussed. The present ice cap GPS-photogrammetry observations show a reduction in volume of around several percent of Galindez ice cap in eight years, suggesting that it could disappear within a century.

Keywords Sea level · climate change · tide gauge · GPS · ice cap · photogrammetry

J. Cisak
Institute of Geodesy and Cartography, Warsaw, Poland

G. Milinevsky
National Taras Shevchenko University of Kyiv, Kyiv, Ukraine,
e-mail: genmilinevsky@gmail.com

V. Danylevsky
National Taras Shevchenko University of Kyiv, Kyiv, Ukraine

V. Glotov
National University Lvivska Politechnica, Lviv, Ukraine

V. Chizhevsky
National University Lvivska Politechnica, Lviv, Ukraine

S. Kovalenok
Ministry Education and Science of Ukraine, Kyiv, Ukraine

A. Olijnyk
National Scientific Center "Institute of metrology", Kharkiv, Ukraine

Y. Zanimonskiy
Institute of Radio Astronomy National Academy of Sciences of Ukraine, Kharkiv, Ukraine

1 Introduction

The aim of the paper is to discuss the results of research, which based on GPS methods in the field of atmosphere and the climate change impact study at the Ukrainian Antarctic Vernadsky station. The upper atmosphere and climate research started in fifties at Faraday Base (see position in Fig. 1), which has been transferred by British Antarctic Survey (BAS) to Ukraine in February 1996. The three issues are discussed in the paper: (1) the atmospheric impact on the global navigation satellite system (GNSS) observations (Cisak 2005) and the troposphere-ionosphere coupling study which is based on the idea of energy exchange between neutral atmosphere and ionosphere plasma (Galushko et al. 2003, Yampolsky et al. 2004, Lisachenko et al. 2007); (2) the monitoring of sea level which is important due to the regional climate change (Krynski and Zanimonskiy 2004b, Turner et al. 2005); (3) the GPS-photogrammetry survey of the small ice caps (glaciers) in the purpose to monitor climate warming causes intensive melting of Antarctic Peninsula ice shelves, small islands ice caps and glaciers (Cook et al. 2005, Greku et al. 2006). The research was undertaken in the "Atmospheric impact on GPS observations in Antarctica" project, coordinated by Poland, and started at the XXVI SCAR meeting in Tokyo 2000, in the framework of the Geodetic Infrastructure of Antarctica (GIANT) Program (e.g. Krynski and Zanimonskiy 2001). At the XXVIII SCAR meeting 2004 in Bremen the project was prolonged under the Ukrainian leadership up to 2006 and renamed in "Atmospheric Impact on GNSS Observations in Antarctica in relation to Geophysical research" (AIGO) project. The GNSS provides the valuable data to study

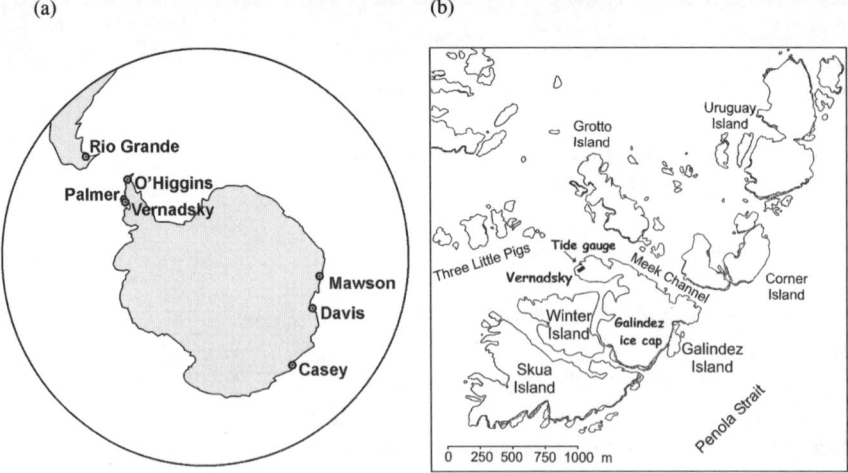

Fig. 1 The GPS stations in Antarctic Peninsula and East Antarctica regions which provided data used in the paper (**a**), Argentine Islands Archipelago: Faraday/Vernadsky station, tide gauge, and Galindez ice cap positions (**b**)

the troposphere-ionosphere coupling, sea level and climate change research (e.g. Yunck 2002, Krynski and Zanimonskiy 2004a). Antarctic Peninsula area is interesting because situated in an extremely active cyclonic region Drake Passage and by unprecedented regional climate warming (Turner et al. 2006).

The investigation of the atmospheric impact on the quality of GPS observations in Antarctica gives data for study the ionosphere disturbances on the planetary wave scale (Altadill et al. 2004, Cisak 2005) caused by energy transfer from troposphere. Results of analysis of the ionosphere disturbances caused by different sources were widely discussed (Stankov et al. 2006). Recently available global ionospheric maps (GIM) are useful instrument for detection and investigation the ionosphere response to planetary, regional and local disturbances from troposphere/Earth surface (Azpilicueta et al. 2006). The sources of disturbances from below are: strong weather fronts, typhoons, terminator, and orography structures (Yampolsky et al. 2004, Lizunov and Hayakawa 2004).

The second issue concerns "the enigma of 20th century sea level" which means that the rate of the observed sea level change is almost twice larger than it can be estimated from known climate change and other contributing processes (Munk 2002). It is known that global mean sea level has been rising in 20th century with rate increased nonlinearly (Cazenave and Nerem 2004, Jevrejeva et al. 2006, IPCC 2007). Global mean sea level change is the most significant indicator of Earth's global climate warming (IPCC 2007). Greenland and Antarctic ice sheets melting is considered as a main source of eustatic sea level rise (IPCC 2007). Tide gauge measurements are the principal data set used to determine rate of sea level change during last two centuries, because satellite altimeters operate during latest 15 years only (Woodworth and Player 2003, Cazenave and Nerem 2004, Krynski and Zanimonskiy 2004b). Data obtained by both techniques show that rate of sea level change is very different over the world's ocean (Cazenave and Nerem 2004).

Significant sea level rise is observed at Antarctic Peninsula (Thomas et al. 2004). There are many reasons of the sea level variations in that region, particularly, glaciers thawing and discharge, increasing annual mean temperature of atmosphere and sea water temperature (Rignot et al. 2004, Thomas et al. 2004). At the same time opposite processes as rising snowfall can result directly opposite effect on sea level (Vaughan et al. 2007).

The purpose of our study is to estimate sea level change at Argentine Islands (Fig. 1b) region caused by climate change. But vertical motion of the Earth's crust via effect of glacial isostatic adjustment is one of reasons of sea level changes relative to shoreline where Faraday/Vernadsky tide gauge have been located. The permanent GPS-measurements started at Faraday/Vernadsky station since 2005 will give the possibility to determine the vertical crustal motion and to obtain more realistic rate of sea level change in this region.

For the third issue, it well known, that regional climate change produces intensive melting of Antarctic Peninsula ice shelves and glaciers (Vaughan et al. 2003). In conditions of regional warming to 2.5°C annual mean according the measurements at the Faraday/Vernadsky station during last 50 years (Turner et al. 2005) the ice cover undergoes the considerable changes. The historical data of the Galindez Island ice cap and other glacier observations of the Antarctic Peninsula are used for our study. The precision GPS and photogrammetry observations started in 2002 enable to repeat the earliest ice cap measurement and provide new data. Several studies of the Galindez ice cap have been carried out during the last fifty years (Sadler 1968, Kovalenok et al. 2004). At the end of 1960s Sadler (1968) has published a report on the mass budget of the ice cap, which included details of the shape and size of the ice cap. GPS-based photogrammetry survey of small island ice caps in the Vernadsky station region (Fig. 1b) has been started in 2002 for monitoring of the ice caps changes and glaciers dynamics of Argentine Islands Archipelago in connection with the regional climate changes. The results of the small ice cap retreat are discussed in third part of the paper.

2 Atmospheric Impact on GNSS Observations

A considerable progress in the applications of GNSS methods and geodynamic research is the result of a development of GNSS measuring techniques. However, the qualitative results on Earth's crust movements are limited by the accuracy determination (e.g. Dietrich et al. 2004). The one of the goals of the GIANT project was the investigation of the atmospheric impact on the quality of GPS observations in Antarctica to improve their accuracy, and to develop recommendations for future Antarctic GPS campaigns and to model the GPS solutions. The investigation of atmospheric processes using GPS technique was another aim of the GIANT project and of the AIGO project as a part of the GIANT.

To investigate the impact of varying meteorological conditions on GPS solutions we used the time series of vector components obtained with the Bernese v.4.2 software for daily sessions of GPS observations from a number of permanent stations (Fig. 1a) (Cisak 2005). There are two interfering phenomena that are inseparable in GPS data if additional information is not available. First, the changing air mass and humidity, which cause the change of the tropospheric path delay, and second, the crust displacement due to the loading variations. Both phenomena produce the variations of cm-range that are within the accuracy of GPS observations. Variations in vertical components of the vectors are correlated with seasonal variations of atmospheric pressure. Total tropospheric zenith delay (TZD) data published by International GNSS Service (IGS) are correlated with variations of atmospheric pressure for Antarctic GPS stations because of low humidity in Antarctica (Cisak 2005). There is no significant troposphere variation impact on the final GPS solutions obtained with the use of the Bernese software. For commercial programs used for processing GPS data, our experiments indicate the dependence of GPS-derived vector components variations from TZD variations.

In the framework of the AIGO project and bi-lateral agreement of co-operation between the National Antarctic Scientific Center of the Ukraine and the Institute of

Fig. 2 GPS antenna of Ashtech Z-12 at Vernadsky station

Geodesy and Cartography (IGiK), Warsaw, Poland, the GPS dual frequency receiver Ashtech Z-12 of IGiK has been installed and set in permanent operation at Vernadsky station since April 2005 (Fig. 2). The data of two-year operation are now under processing according to the main tasks of the project. We used ionosphere total electron content (TEC) determinations from GPS data for study troposphere-ionosphere coupling processes as well (Mendillo 2006, Lisachenko et al. 2007).

The effect of the atmospheric pressure loading is well known in geodetic community. For example, the regression coefficients (about −0.5 mm/hPa) linking variations of pressure with the vertical movement of GPS sites were calculated using processed data of the European network GNSS stations (Kaniuth and Vetter 2006). But in that paper, however, the possible dependence of the regression coefficients on the period of pressure variations was not considered.

Within the framework of the AIGO project the frequency response of the regression coefficients was obtained. In the Fig. 3 the correlation between the atmospheric pressure variation difference on the Mawson and Davis Antarctic bases and the difference of heights of MAW1 and DAV1 GPS sites is shown as a result of spectral and

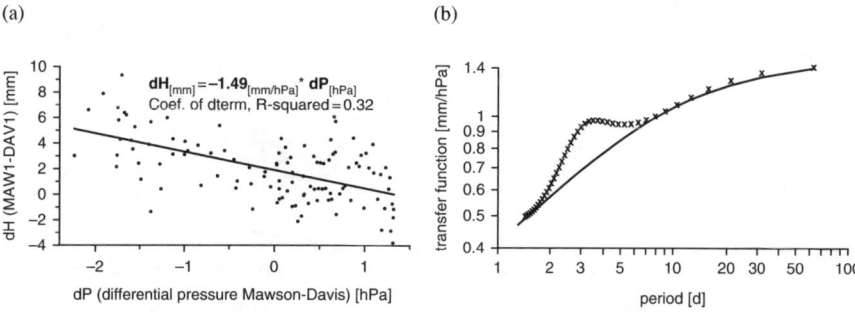

Fig. 3 The variations of dependence of height difference between the MAW1 and DAV1 GPS sites on the difference of pressures variations for the period larger then 20 days (**a**); transfer function of height-pressure difference versus period of pressure variations (**b**), crosses – from calculations, solid line – from the model

regressive analyses of the time series of atmospheric pressure and GPS-solutions. Dependence of regression coefficient on the period of variations (see Fig. 3b) was calculated on the basis of estimations of the atmospheric pressure variation spectrum and height difference.

The crosses in Fig. 3b mark the numerical analysis results while the curve corresponds to the analytical model with random spatial distribution of atmospheric pressure over investigated region. The distinction of two functions can be connected to the presence of the weather periodicity in the observed variations that were not taken into account in the model. In Fig. 3a the regression of height differences variations versus variations of pressure differences is shown for the long-period part of the frequency response in Fig 3b.

The results presented in Fig. 3 correspond to qualitative behaviour of atmospheric pressure loading. The approved method of quantitative estimation of pressure variations and Earth's crust vertical moving connections can be used for the pairs of the GPS sites located on a continent and on islands in an Antarctic region. In this sense it is difficult to over-estimate the importance of data, permanently accumulated on Vernadsky station, despite the fact that they are accessible only once a year.

At the current stage of the AIGO project our results indicate the dependence of geodetic GNSS solutions on the ionosphere conditions (Cisak 2005). The usage of the Bernese software-generated time series of GPS solutions based on overlapped sessions is of great importance for both quantitative and qualitative analysis (Cisak 2005). The relationship between the vector length (dD) and diurnal TEC average values in the long time series of GPS data allows for modelling the ionospheric impact on GPS solutions (see Fig. 4).

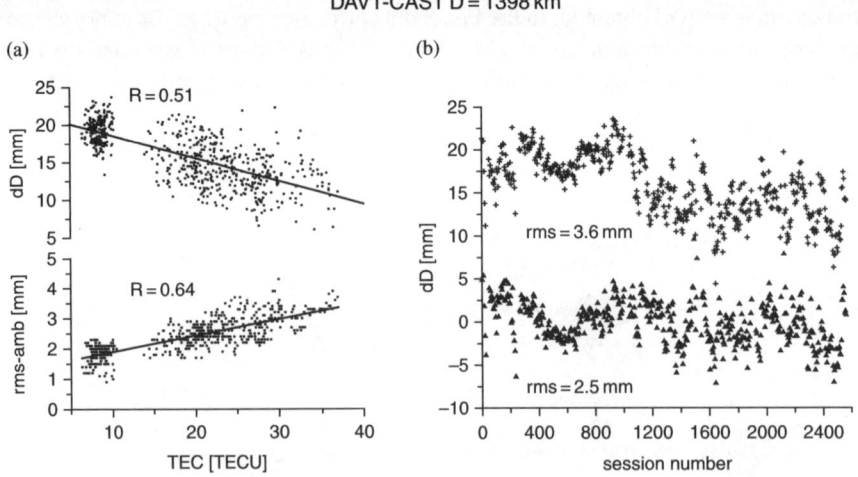

Fig. 4 Variations of vector length and the RMS of ambiguity resolution versus TEC (**a**), and time series of initial and corrected variations of the vector length (**b**)

The solutions for vector length (dD) could be improved using the simple regression model dD = k · (TEC), based on global ionospheric mapping (GIM) data. As an example for data from Davis and Casey Antarctic stations (see Fig. 1a) for the sites DAV1-CAS1 of 1398 km vector length the correction equals to −3 mm/10 TECU. Fig. 4a shows the relationship between DAV1-CAS1 vector length and diurnal average of TEC (upper graph) and the variations of the RMS of ambiguity resolution in GPS solution (lower graph) versus TEC diurnal average variations. The RMS of ambiguity resolution exhibits stronger correlation with TEC variations (correlation coefficient of 0.64) than with the vector length variations (correlation coefficient of 0.51). Generally, increasing of ionosphere electron content causes shortening of vector length obtained from GPS solution.

The corrected standardized to zero-TEC results (Fig. 4b, lower graph) show the smaller scattering and considerable systematic shift in comparison with GPS solutions obtained by the Bernese program (Fig. 4b, upper graph) without corrections. In further works it is necessary to prove the stability of the developed algorithm of the correction of the measuring distances in different regions. The correlations shown in Fig. 4 indicate the presence of a problem to be solved by the use of the improved model of signal propagation in ionosphere as it was done in the strategy of reprocessing of the global GPS data (Fritsche et al. 2006). The absence of systematic errors in GPS solutions arisen in the process of ambiguity resolution caused by ionosphere must be checked as well.

The spatial and temporal resolutions of GIM are sufficient to study ionosphere variations on planetary scale. To investigate regional and local ionosphere disturbances in the Drake Passage region the primary data from GNSS permanent sites at Rio-Grande and at O'Higgins, Palmer, Vernadsky Antarctic stations (Fig. 1a) – RIOG and OHI3, PALM, VER2 GPS points (http://www.tu-dresden.de/ipg/service/scargps/VER1.html) – were used. The local ionospheric maps provided in an ionospheric exchange (IONEX) data format in both calm and disturbed ionosphere and for severe weather front conditions (http://www.aiub-download.unibe.ch/ionex/) were applied as well. The TEC variations in spatial-temporal scale about 300 km and one-hour resolution were analyzed.

The several hundreds km variation spatial scale corresponds to mesoscale time-spatial ionosphere structures – well known traveling ionospheric disturbances (TID) of the electron density (Hocke and Schlegel 1996). The sources of the TIDs could be ionosphere storm and strong tropospheric disturbances (weather front, typhoon) (Yampolsky et al. 2004). It is suggested that the weather front or/and geomagnetic storm produce disturbances, which break up the regular TEC structure of the ionosphere and create TID.

From GPS phase data on two frequencies on each station and for every satellite the time series of geometry-free combination data were calculated (Cisak 2005). This combination data enables to determine slant total electron content (STEC) along the satellite-receiver line of sight. Here we discuss the GPS data processed from three stations PALM, OHI3, and VER2. To get the STEC variations we separate the regular part of TEC values using GIM data and receive the STEC variation difference time series (Fig. 5). The wave-like quasiperiodic variations are seen in

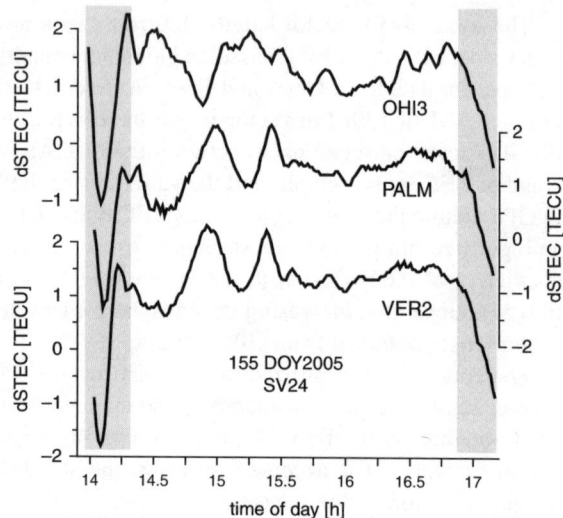

Fig. 5 Time series of STEC variations along the line of sight between SV24 and GPS receivers at three stations in Antarctic Peninsula

central part of the Fig. 5. We suppose those TEC variations appeared due to TID as the ionosphere response on troposphere weather front propagation (Lisachenko et al. 2007). The meteorological data (pressure variations, temperature and wind speed) from the Vernadsky station and TZD time series for nearest permanent stations were used to identify the front passage. serve as a proof of this assumption. The dissipation of the front along its path may cause the differences of STEC disturbances from lower to upper graph in Fig. 5. Note the similar dSTEC variations behaviour on VERN and PALM sites (distance ~50 km) and different behaviour at OHI2 site (~340 km from PALM) that reflects possible spatial weather front propagation (Yampolsky et al. 2004). Similar ionospheric response on the powerful tropospheric events was investigated in (Lisachenko et al. 2007).

In the analysis it is very important to take into account the higher order ionospheric refraction effects (Prokopov and Zanimonska 2003, Fritsche et al. 2005) that input the systematic errors in measured satellite-receiver distance. The errors depend on the TEC value and on the satellite elevation angle. The impact of higher-order ionospheric terms on GPS results caused by the small satellite elevation angle is sometimes difficult to be separated from the TID (Fig. 5, grey bars).

3 Sea Level Change Investigations in Faraday/Vernadsky Station Area

Tide level measurements that are performed at Faraday/Vernadsky station are used to research the sea level change at Antarctic Peninsula and particularly in Argentine Islands region (Fig. 1). The British Graham Land Expedition of 1934–1937 made the first tidal observations in the area from a ship frozen in Stella Creek. British

Antarctic Survey (BAS) at Galindez Island (Fig. 1b) installed the stilling well tide gauge connected to a standard Munro chart recorder in 1957, but continuous tidal measurements started in 1959 (Cartwright 1979). It is the longest set of tidal measurements in Antarctic. The gauge was operated by BAS, and the charts had been digitized at hourly intervals, checked for transcription errors and edited at the Institute of Oceanographic Sciences, UK.

Later the Proudman Oceanography Laboratory (POL) provided Faraday station and tide gauge operation and data processing. In 1992 the station was equipped with the pressure gauge. Ukrainian scientists provide the tide measurements at Faraday/Vernadsky station since 1996. Tide gauge at Galindez Island was included in Global Sea Level Observing System with number 188 to monitor long-term sea level variations due to climate change. Data are digitized at hourly intervals from chart recorders and 15 minutes intervals from pressure gauge. Sea temperature and barometric pressure are measured as well. These data are transferred to the POL and to Permanent Service for Mean Sea Level (PSMSL). Faraday/Vernadsky station has PSMSL code A/003, data are processed by Doodson X0 filter as provided by PSMSL procedure (Woodworth and Player 2003). The data are known with respect to the level of a tide gauge benchmark and have PSMSL status as Revised Local Reference (RLR) records. Monthly and annual mean sea level data obtained from Faraday/Vernadsky station hourly records are available on PSMSL web-site (http://www.pol.ac.uk/psmsl/psmsl_individual_station.html).

Figure 6a, b show monthly and annual mean sea level changes respectively, obtained from Faraday/Vernadsky tide gauge records (PSMSL website data). Monthly averages with the Doodson X0 filter eliminate main short-term tidal constituents, local seishes and surges from sea level measurements and annual averaging eliminates seasonal quasiperiodic sea surface level variations and long-period tidal constituents (Pugh 1996). According to Fig. 6b the sea level interannual changes are considerable at Argentine Islands in 1961–2006.

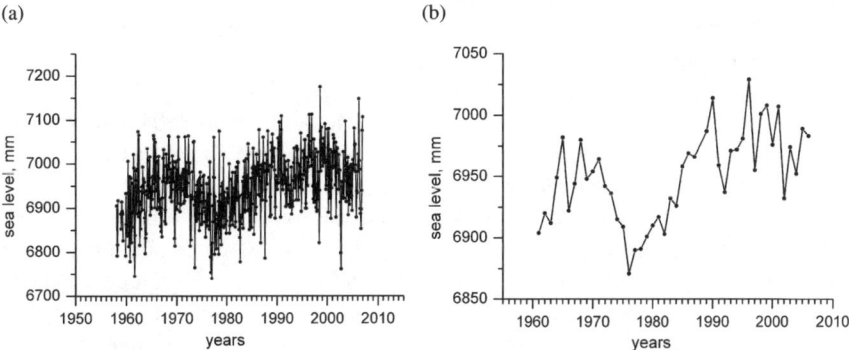

Fig. 6 Monthly mean sea level changes (**a**) and annual mean sea level changes by tide gauge at Faraday/Vernadsky station (**b**)

We tried to estimate an eperiogenic contribution to sea level change from land movement at Galindez Island. It is quite possible to expect the uplift the Earth's surface at this area by isostatic adjustment because of glaciers thawing and discharge, and it must be accompanied by sea level decreasing. Estimation of this kind of sea level change for Argentine Islands obtained by Peltier (2004) using ICE-5G (VM2 and VM4) models of glacial isostatic adjustment and postglacial variations in the sea level, both with a 90 km lithosphere. These estimations are equal to −2.36 mm/yr (VM2) and −2.31 mm/yr (VM4) at present, which means that sea level decreases in time span ±100 years relative to present (http://www.pol.ac.uk/psmsl/landmove.html). But Fig. 6b shows rising of the mean sea level during time span observations: linear trend is equal approximately +1.47 mm/yr. It means, if Peltier 2004) estimations of land uplift are correct, the non-eperiogenic processes increase sea level at this region on 3.8 mm/yr. It is noticeably greater than upper estimation of global sea level rate over the last century that is 2.0 mm/yr (IPCC 2007) and even exceeds global sea level rate estimation from TOPEX/Poseidon altimeter measurements over the period of 1993–2003, that is from 2.6 mm/yr (Jevrejeva et al. 2006) to 2.8 mm/yr (Cazenave and Nerem 2004).

But research of tectonic motions at different parts of Antarctic Peninsula region performed over 1995–1999 with GPS technique and recently published (Dietrich et al. 2004) suggests that realistic vertical land motion can be greater than Peltier's estimations. Real values of eperiogenic sea level changes at Argentine Islands can only be determined from Earth's crust motion measurements at this place. Because of the permanent GPS measurements started at Faraday/Vernadsky station recently, we expect to determine realistic rate of sea level change using GPS-measuremen of the vertical crust motion in this region.

Due to the general processes of Antarctic Peninsula ice sheet decreasing and glaciers discharge, and the ICE-5G (VM2 and VM4) models estimations and observational results (Dietrich et al. 2004) it is suggested that sea level has to decrease relatively to land. However, Faraday/Vernadsky sea level tide gauge data (Fig. 6) exhibit rapid and significant variations on different time scales: over period 1971–1976 sea level decreased continuously with mean rate of 19 mm/yr, and from 1976 to 1990 it increased with mean rate of 10.4 mm/yr, before 1976 and after 1990 it is changed significantly on smaller time spans. We suppose that climatic and meteorological factors and long-period tidal constituents basically form sea level changes at this region. Particularly, as it follows from (Meredith et al. 2004, Woodworth et al. 2006), interannual changes in ocean transport through Drake Passage forced variability in sea level at Faraday/Vernadsky station.

Tidal constituents analysis at Argentine Islands on 1959–1968 data set has been performed in (Cartwright 1979) by response analysis technique (Munk and Cartwright 1966). It was noted there that the relatively strong diurnal components of the tide are fairly typical of the Pacific Ocean, but the amplitudes of the principal lunar (M_2) and solar (S_2) semi-diurnal constituents are approximately equal each other. It implies a strong suppression, more than twice, of the lunar relative to the solar semi-diurnal tide. Also there is a slight suppression of the luni-solar principal diurnal tidal component K_1 relative to the principal lunar diurnal tide component O_1. These events imply selective absorption and dissipation of tidal energy at this region. Also it has been denoted that non-gravitational tidal effects caused by variable (diurnal and seasonal) solar radiation are generally rather small: seasonal (annual) tidal variation (Sa) was 37 mm, the semi-diurnal S_2 radiational tidal effect had amplitude 23 mm and main diurnal radiational tide S_1 amplitude was 3 mm.

There are tidal hights of daily high and low water on POL website (http://www.pol.ac.uk/) that are computed by POL with software POLTIPS on the basis of harmonic analysis technique. We have compared hourly tide heights measured at Vernadsky station in 2004 to max and min tidal heights predicted by POL for that period. In general the predicted results agree with the observed ones, however, some of them are different. It shows that the models still need adjustments.

We computed the power spectrum density of tide hights measured hourly at Faraday/Vernadsky station. The spectral analysis was used to estimate the relative impact of different modes of tide waves on resulting sea level changes amplitude and to separate possible non-tide components. The series length of the PSMSL hourly RLR data set of period 1984–2004 is rather long and allows performing spectral analysis with high resolution of period, up to 1–2 s. Power spectral density estimations (periodogram) of tides at Faraday/Vernadsky station can be seen in Fig. 7. These spectrums are obtained by analysis of hourly tide records measured on one-year time intervals and have not so high resolution as mentioned above. Fig. 7a shows that amplitude ratio of principal diurnal and semi-diurnal components in 2000 are similar to obtained in (Cartwright 1979). In Fig. 7a the symbols M_2 and S_2 are principal lunar and principal solar semi-diurnal constituents respectively, N_2 is lunar elliptic semi-diurnal constituent, K_1 and O_1 – diurnal luni-solar and principal lunar constituents respectively, Q_1 – lunar elliptic diurnal constituent.

The other diurnal and semi-diurnal components of tide spectra in 1984–2004 are all practically the same and are stable in their frequencies and amplitudes. On the contrary, the low-frequency tides spectrum (the periods of constituents are from 5 to more than 35 days) is rather complicated and shows very unstable properties

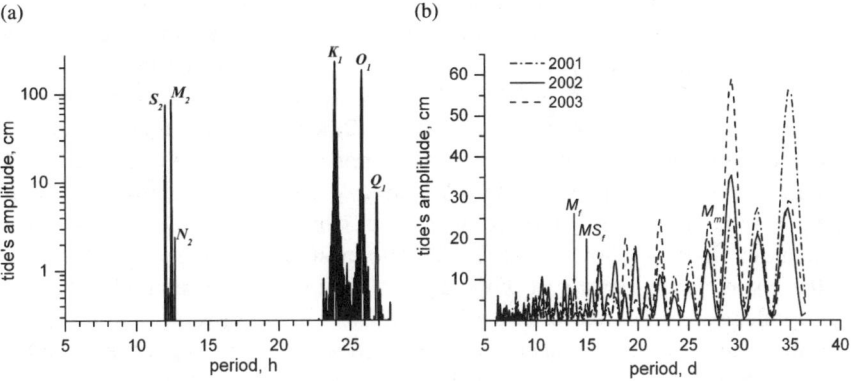

Fig. 7 The periodogram for diurnal and semi-diurnal tidal constituents for 2000 (**a**), the periodogram for low-frequency spectrum of tide hights for 2001–2003 years (**b**) computed

in time. Fig. 7b shows low-frequency spectra obtained for three successive years 2001–2003 which correspond to different annual averaged sea level values. Nevertheless, it can be seen from Figs. 6b to 7b that differences of yearly mean sea levels of these years correspond to differences of the tidal constituent amplitudes of 35 days periods and approximately 29 days (note: fortnightly M_f and MS_f constituents and monthly M_m constituent contributions are negligible). This interesting result needs more detailed study and correspond to conclusion made in (Pugh 1996) that the direct meteorological effects influence in the greatest extent on monthly and larger periods of mean sea level changes in higher latitudes and in the vicinity of wide continental shelves. We have both this conditions at Faraday/Vernadsky station and suppose that low-frequency tidal constituents 29 and 35 days are modulated by meteorological factors on long time scales and other quasiperiodic contributors of the sea level variation. These investigations will be continued in order to find out the causes of the observed low-frequency tidal harmonics of sea level changes.

4 GPS-Photogrammetry Ice Cap Survey

In this section we describe the results of the study of Argentine Islands Archipelago ice caps variations possibly caused by regional climate change using GPS-photogrammetry method (Dorozhynskyy et al. 2004, Gao and Liu 2001). The Argentine Islands are a small group of islands (Fig. 1b) lying seven kilometers west of the Antarctic Peninsula at 65°15'S, 64°15'W. Most of the islands of the group have small ice caps located in periglacial zone, which form in the region of exposed rocks at the north end, rise to a small summit and terminate at the southern end in a cliff. The ice caps are symmetrical about an axis parallel to the prevailing wind, which is from azimuth around 20° (Thomas 1963).

Several investigations of the Galindez ice cap have been carried out in 1947–2004 (Thomas 1963, Sadler 1968, Glotov et al. 2003, Shanklin et al. 2004). In the 1960s Thomas (1963) and Sadler (1968) published reports on the mass budget of the ice cap which included details of the shape and size of the ice cap: the contour intervals and spot heights at the top of the ice cap. It was noted that obvious changes were taking place and therefore the survey work to quantify the changes was established (Shanklin et al. 2004). The three types of measurements were introduced: (1) measurement of the 50 m contour, (2) measurement of a profile of the top the ice cap along the fixed sites, (3) measurement of the edge of the ice cliffs from stationary survey stations. The BAS surveys prior to 1987 show only small variation in the maximum height of the ice cap, which was between 55 and 56 m (Shanklin et al. 2004, Glotov et al. 2003). The most noticeable change of the ice cap has taken place in west part of the glacier. Here the ice lies on rock, which is close to mean sea level and is hence affected by seawater erosion. The precision GPS observations enable to repeat the earliest BAS investigations of the Galindez ice cap profile of the top the ice cap and 50 m height contour, which were based on aerial photography and theodolite observations (Thomas, 1963).

The GPS-based photogrammetry survey of small island ice caps in Antarctic Peninsula region has been started in 2002 (Dorozhynskyy et al. 2004). The mathematical model of photogrammetry of kinematics processes and complex technology of conducting digital terrestrial survey from a firm point and from a moving boat were elaborated and put into practice. During surveys in period of 2002–2004 using GPS-based digital phototheodolite about 1000 digital photograph pairs were made and were stereoscopically processed on the digital photogrammetry station Delta-2. The survey was done with different focal lengths of the cameras (Kodak-260C, Olympus E20p), which were determined by the angles of object coverage. According to the technology the coordinates of the survey points were determined using GPS in static state to improve the accuracy of the center locations. During the first campaign in 2002 the coordinate measurements of the survey point have been made by the GPS in the static regime for the determination of the exact center positions. The 2002 survey resulted the receiving of the 12 stereo pairs as normal as deflected to the left (right) with the different focus distance (92 and 280 mm). The measurement error of the initial calculation is 0.1% and mean accuracy is 0.3–0.5% (Dorozhynskyy et al. 2004). On the basis of survey data the large-scale digital maps 1:2000 and 1:1000 for Galindez Island were created (Litvinov et al. 2006).

The main objectives of the GPS-photogrammetry survey are producing the precise geodetic data for ice cap monitoring and the evolution model creation of the ice caps changes and glaciers dynamics of Argentine Islands Archipelago in connection with the region climate changes. The geodetic survey data give possibility to monitor the changes of size, shape, and deformation, moving velocity and the edge position of ice caps and use these data of the ice cap for the regional climate variability study and for the prognosis of the ice caps dynamic. The three digital GPS-photogrammetry survey campaigns were fulfilled in 2002, 2003, and 2004 (Glotov et al. 2003, Kovalenok et al. 2004, Greku et al. 2006).

The comparison of the longitudinal profiles and bedrock of the Galindez ice cap of 1961, 1998, 2002 surveys are shown in Fig. 8. Ice cap height is shown in meters

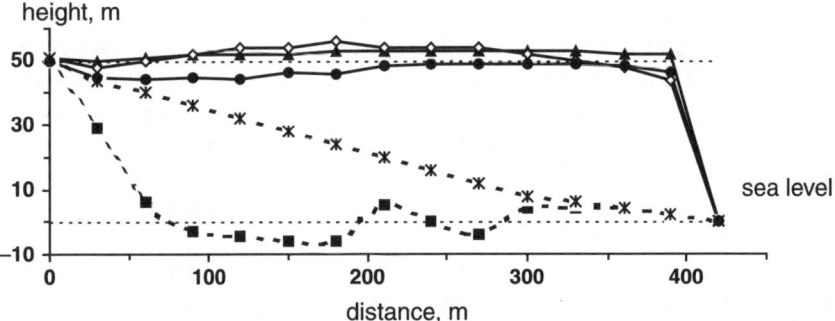

Fig. 8 The comparison of the longitudinal profiles survey and bedrock of the Galindez ice cap made in 1961 (*triangles*), 1998 (open rhombs) and 2002 (*circles*), dashed lines mark sea level and 50 m height. The bedrock profiles are shown below: asterisks – by Thomas (1963) modelling, squares by radio sounding data (Kovalenok et al., 2004)

versus distance along ice cap. The Galindez ice cap profiles show a notable reduction in height between the 1961 and 1998 surveys from around 3 m at the top of the ice cap to around 1 m close to the rocky outcrop (point "0" in Fig. 8). According 2002 survey data all profile points are below 50 m height, which show significant ice recession, then it means that the previous 50 m contour measurements method cannot be used anymore (Fig. 8).

The radio sounding research of the ice cap of the Galindez Island was fulfilled in 1998 (Kovalenok et al. 2004). The main task of sounding was the measuring of the ice thickness. The radio sounding measurements were carried out for 31 points on the 4 profiles. The measurements of the velocity of radio waves propagation were made using the method of the inclined sounding. The measured bedrock profile is shown in Fig. 8 (squares). The maximum ice thickness was determined as about 57 m. According the measurements 30 % of the ice bedrock is located under sea level. The radio sounding bedrock profile is more detailed than received from simple model of ice movement by Thomas (1963). That is confirmed by the latest electroresonance sounding measurements (Levashov et al. 2004, Bakhmutov et al. 2006).

The stereo photogrammetry and GPS continuous survey technique give the possibility to determine the ice volume changes. The measurements were made by the method of vertical network and via creation of the digital relief model. The changes of the ice volume between two surveys were calculated from the changes of the frontal (vertical) topography of the ice cliff. The accuracy of ice volume changes determination corresponds to RMS of 0.12% of measured volume. The observations show that the ice caps are rapidly retreating in 2002–2004. Since 2000, Galindez ice cap has lost 2–3% of its volume which correspond to a loss of about 20000 m^3/year. According 2003–2004 surveys the Winter Island ice cap nearby Galindez Island (Figs. 1b and 9a) has lost at least about 6000 m^3 of ice in a year.

Although the bulk of the Galindez ice cap is diminishing, visual observation shows that the change is not uniform over the entire ice cap (Glotov et al. 2003, Shanklin et al. 2004). Some areas have receded more than others (see Fig. 9a). The comparison of the Galindez ice cap boundaries from the BAS aerial photographs (Thomas 1963) and the Quick Bird satellite image (Fig. 9a) suggest that main receding area is the west part of ice cap. The decreasing of west side of the Galindez ice cap during 2002–2004 is about $\sim$50000 m^3. The seawater thermal state for the summer period in Argentine Islands region shows significant inter-annual oscillations which impacts on ice melting. For example, the seawater temperature maximum of $+4.9°C$ was recorded at summer 2000–2001. These conditions strongly influenced the ice cap melting for parts of glaciers below mean sea level, especially on its west part.

However, a few areas of island glaciers have actually increased or vary in ice cover, for example, the narrow neck between Marina Point and Galindez Island melted out completely in 1986, but was covered with ice some 2 m thick in 1995–1997, although crossed by two small crevasses (Shanklin et al. 2004), then in 2004 this isthmus was melted completely again (Fig. 9a). Local observation suggests that the major factor in driving accumulation is the direction of the wind during

(a) (b)

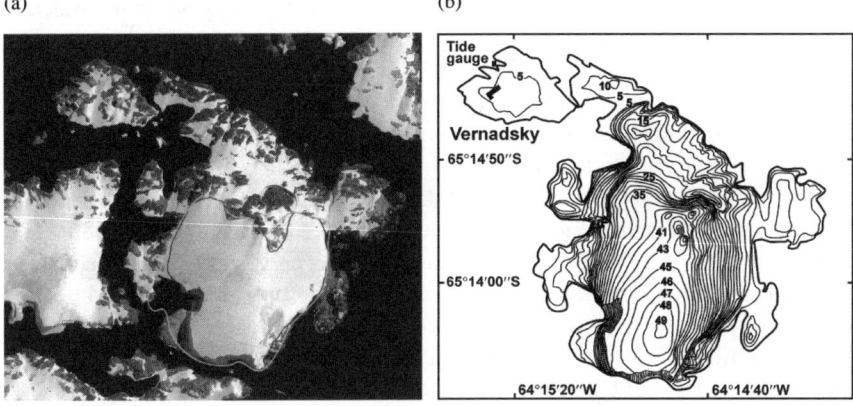

Fig. 9 Galindez ice cap boundaries: December 1956 (*yellow line*) and January 2004 (*red line*) (**a**); topographic map of the Galindez Island ice cap from 2002 GPS-photogrammetry survey (**b**)

major snowfalls, particularly the first snowfalls of each winter. Most ablation takes place in the summer during periods of heavy rain and strong winds when the temperature rise about 5°C. In general the results of 1956–2004 surveys show that the small ice caps Argentine Islands Archipelago as relicts of ice shelves are changing constant but those changes are not homogeneous. Recent observations in 2002–2004 show a reduction in volume of about several percent of Galindez ice cap in suggesting that in existent climate warming conditions it could disappear within a century.

5 Conclusions

The results in the fields of GPS-based investigations provided at Vernadsky station in Antarctic Peninsula are presented. The effect of atmospheric impact on GNSS observations was an object of research in several aspects as the analysis of the impact of ionospheric disturbances on variations of vector lengths obtained at high latitudes from GPS data, study the troposphere-ionosphere coupling, and atmosphere loading as the source of local geodynamics. The obtained results indicate the dependence of ambiguity resolution on the state of ionosphere. The use of time series of GPS solutions based on overlapped sessions is crucial in the performed quantitative and qualitative analysis. Generally, the unstable ionosphere causes shortening of vector length obtained from GPS solutions. Correlation analysis of data sets from chosen Antarctic stations shows a possibility to reduce GPS solutions errors induced by ionosphere disturbances using simple empirical models. The quasiperiodic variations on time series of STEC difference between the GPS on site observations and GIM data were observed. The variations very likely correspond to TID appearance in ionosphere as a sequence of atmospheric gravity waves generation due to troposphere weather front propagation.

Analysis of Faraday/Vernadsky sea level measurements suggests that the determination of real vertical crust motion is necessary to obtain an eperiogenic sea level change component, and GPS-technique will be used to solve this problem in next work. Besides, the spectral analysis of Faraday/Vernadsky tide gauge measurements allows estimating the relative impact of different modes of tide waves on resulting sea level changes amplitude and separate possible non-tide components. The tide periodogram shows that low frequency tidal constituents of 29 and 35-day periods could contribute noticeably to the observed sea level interannual variability. The search for other most effective contributors to sea level changes at Argentine Islands region will be the object of our next studies. Next steps are the studies of the mean and extreme sea level changes, which occur as a consequence of climate change, and investigations of geological processes, which affect sea level changes.

The results show the GPS-photogrammetry survey for monitoring of the ice cap size, shape and the edge position changes as an efficient and useful technique to study the impact of the rapid warming in Antarctic Peninsula region. These measurements suggest that major recession of the ice caps was taking place in the last decade. The Galindez ice cap volume is decreasing of about several percent of in last years. It indicates that permanent Argentine Islands ice caps monitoring would be important to predict their existence in the future.

Acknowledgments This research was partly made in the framework of the SCAR POLENET Project. The work was supported in part by the National Antarctic Scientific Center of the Ukraine, and National Taras Shevchenko University of Kyiv project 06BF051-12 and was partially supported by the Polish Committee for Scientific Research (Research Project No 8 T12 E045 20). Authors express their gratitude to Jan Krynski for his helpful suggestions that led to improve the manuscript of this paper. Authors thank anonymous referee for numerous important comments and useful advices.

References

Altadill D, Apostolov EM, Boska J, Lastovicka J, Sauli P (2004) Planetary and gravity wave signatures in the F-region ionosphere with impact on radio propagation predictions and variability. Annal Geophy Suppl 47(2/3):1109–1119

Azpilicueta F, Brunini C, Radicella SM (2006) Global ionospheric maps from GPS observations using modip latitude. Adv Space Res 38:2324–2331

Bakhmutov VG, Vaschenko VN, Grischenko VF, Korchagin IN, Levashov SP, Pishchany YM (2006) Methods and results of glaciers strength of small Wiggins (Antarctic Peninsula) and "Domashnij" (Galindez Island). Ukrainian Antarct J 4–5:47–51

Cartwright DE (1979) Analyses of British Antarctic Survey tidal Records. Br Antarct Surv Bull 49:167–179

Cazenave A, Nerem RS (2004) Present-day sea level change: Observations and causes. Rev Geophys 42:RG3001, doi: 10.1029/2003RG000139

Cisak J (2005) Overview of the research on the Atmospheric Impact on GPS Observation in Polar Regions. Report of the Fifth SCAR Antarctic Geodesy Symposium 23:15–23

Cook AJ, Fox AJ, Vaughan DG, Ferrigno JG (2005) Retreating Glacier Fronts on the Antarctic Peninsula over the Past Half-Century. Science 308:541–544

Dietrich R, Rülke A, Ihde J, Lindner K, Miller H, Niemeier W, Schenke H-W, Seeber G (2004) Plate kinematics and deformation status of the Antarctic Peninsula based on GPS. Glob Planet Change 42:313–321

Dorozhynskyy O, Milinevskyy H, Hlotov V (2004) Photogrammetric research conducted at the Antarctic station "Academician Vernadskyy". In: Proceedings Volume IAPRS XXXV B4 XXth ISPRS Congress: 642–644, http://www.isprs.org/istanbul2004/

Fritsche M, Dietrich R, Rülke A, Rothacher M, Steigenberger P (2006) Reprocessing of a Global GPS Network – Experiences and Results from a Joint Project at TU Dresden and TU Munich. Report IGS Workshop Darmstadt, 08–11 May 2006, Technische Universität Dresden, Institut für Planetare Geodäsie, http://nng.esoc.esa.de/ws2006/REPR4.pdf

Fritsche M, Dietrich R, Knofel C, Rulke A, Vey S, Rothacher M, Steigenberger P (2005) Impact of higher-order ionospheric terms on GPS estimates. Geophys Res Lett 32:L23311, doi:10.1029/2005GL024342

Galushko VG, Beley VS, Koloskov AV, Yampolski YM, Paznukhov VV, Reinisch BW, Foster JC, Erickson P (2003) Frequency-and-angular HF sounding and ISR diagnostics of TIDs. Radio Science 38(6):1102, doi:10.1029/2002rs002861:10–1–10–9

Gao J, Liu Y (2001) Applications of remote sensing, GIS and GPS in glaciology: a review. Progress in Physical Geography 25(4):520–540, doi:10.1177/030913330102500404

Glotov VM, Kovalenok SB, Milinevsky GP, Nakalov EF, Fulitka YV (2003) Monitoring of small ice caps as indicators of the Antarctic Peninsula region climate change. Ukrainian Antarct Journ 1:93–98

Greku R, Milinevsky G, Ladanovsky Y, Bakhmach P, Greku T (2006) Topographic and Geodetic Research by GPS, Echosounding and ERS Altimetric, and SAR Interferometric Surveys during Ukrainian Antarctic Expeditions in the West Antarctic. In: Fütterer DK, Damaske D, Kleinschmidt G, Miller H, Tessensohn F (eds) Antarctica, Contributions to Global Earth Sciences. Springer, Berlin, Heidelberg, doi:10.1007/3-540-32934-X, pp 383–390

Hocke K, Schlegel K (1996) A review of atmospheric gravity waves and travelling ionospheric disturbances: 1982–1995. Ann Geophys 14:917–940

IPCC 2007: Climate Change 2007: The Physical Science Basis. Contribution of Working Group I to the Fourth Assessment Report of the Intergovernmental Panel on Climate Change. In: Solomon S, Qin D, Manning M, Chen Z, Marquis M, Averyt KB, Tignor M and Miller HL (eds). Cambridge University Press, Cambridge New York

Jevrejeva S, Grinsted A, Moore JC, Holgate S (2006) Nonlinear trends and multiyear cycles in sea level records. J Geophys Res 111:C09012, doi:10.1029/2005JC003229

Kaniuth K, Vetter S (2006) Estimating atmospheric pressure loading regression coefficients from GPS observations. GPS Solutions 10(2):126–134, doi:10.1007/s10291-005-0014-4

Kovalenok SB, Milinevsky GP, Glotov VN, Tretjak KR, Chizhevsky VV, Greku RKh, Moskalevsky MYu (2004) Small ice cap dynamic in the Antarctic Peninsula rapid climate change conditions. In: XXVIII SCAR Open Science Conference, Bremen, 2004 – Abstracts Session 08: Climate history of the Antarctic from ice cores and meteorological reports S08/O10, pp 216–217

Krynski J, Zanimonskiy Y (2001) Contribution of data from Polar Regions to the investigation of short term geodynamics First results and perspectives. In: Antarctic Geodesy Symposium 2001, 17–20 July, St Petersburg, Russia, SCAR Report, No 21 January 2002, Publication of the Scientific Committee on Antarctic Research, Scott Polar Research Institute, Cambridge, UK, pp 56–60

Krynski J, Zanimonskiy Y (2004a) Investigation of regional troposphere processes using EPN data. In: Symposium of the IAG Subcommission for Europe (EUREF) held in Toledo, Spain, 4–7 June 2003. EUREF Publication No 13, Mitteilungen des Bundesamtes für Kartographie und Geodäsie, Band 33, Frankfurt am Main, pp 416–422

Krynski J, Zanimonskiy Y (2004b) Tide gauge records-derived variations of Baltic Sea level in terms of geodynamics. Geodesy and Cartography 53(2):85–98

Levashov SP, Yakymchuk NA, Usenko VP, Korchagin IN, Solovyov VD, Pishchany YM (2004) Determination of the Galindez Island ice cap thickness by the vertical electric-resonance sounding method. Ukrainian Antarct Journ 2:38–43

Lisachenko VN, Zanimonskiy YM, Yampolski YM, Wielgosz P (2007) Investigation of Ionospheric Total Electron Content Variations in the Antarctic Peninsula Region. Radio Physics and Radio Astronomy 12(1):20–32

Litvinov V, Glotov V, Kolb I, Chizhevsky V (2006) The analysis of the cartographic materials for the creation of GIS of Antarctic station "Academician Vernadsky". Ukrainian Antarct Journ 4–5:14–20

Lizunov G, Hayakawa M (2004) Atmospheric gravity waves and their role in lithosphere-troposphere-ionosphere interaction. IEEJ Transact. Fundament. Mater., 124(12), 1109–1120.

Mendillo M (2006) Storms in the ionosphere: patterns and processes for total electron content. Rev Geophys 44:1–47, RG4001

Meredith MP, Woodworth PL, Hughes CW, Stepanov V (2004) Changes in the ocean transport through Drake Passage during the 1980s and 1990s, forced by changes in the Southern Annular Mode. Geophys Res Lett 31:L21305, doi:10.1029/2004GL021169

Munk W (2002) Twentieth century sea level: An enigma. Proc Natl Acad Sci 99(10):6550–6555

Munk WH, Cartwright DE (1966) Tidal spectroscopy and predictions. Philos Transac Roy Soc, London A 259:533–581

Peltier WR (2004) Global glacial isostasy and the surface of the ice-age Earth: the ICE-5G (VM2) model and GRACE. Annu Rev Earth Planet Sci 32:111–149

Prokopov A, Zanimonska A (2003) The Second Order Refraction Effects for GPS Signals Propagation in Ionosphere. Report of the Fifth SCAR Antarctic Geodesy Symposium 23:25–28

Pugh DT (1996) Tides, surges and mean sea level. John Wiley and Sons, Chichester UK

Rignot E, Casassa G, Gogineni P, Krabill W, Rivera A, Thomas R (2004) Accelerated ice discharge from the Antarctic Peninsula following the collapse of Larsen B ice shelf. Geophys Res Lett 31: L18401, doi:10.1029/2004GL020697

Sadler I (1968) Observations on the Ice Caps of Galindez and Skua Islands, Argentine Islands, 1960–66. Br Antarct Surv Bull 17:21–49

Shanklin JD, Kovalenok SB, Milinevsky GP (2004) Impact of Antarctic Peninsula climate change on Galindez Island ice cap dynamics. In: Interdisciplinary Workshop "Antarctic Peninsula Climate Variability: History, Causes and Impacts", 16–18 September 2004, Cambridge UK. Abstracts, p 45

Stankov SM, Jakowski N, Tsybulya K, Wilken V (2006) Monitoring the generation and propagation of ionospheric disturbances and effects on Global Navigation Satellite System positioning. Radio Sci 41:1–14, doi:10.1029/2005rs003327

Thomas RH (1963) Studies on the Ice Cap of Galindez Island, Argentine Islands. Br Antarct Surv Bull 2:27–43

Thomas R, Rignot E, Casassa G, Kanagaratnam P, Acun C, Akins T, Brecher H, Frederick E, Gogineni P, Krabill W, Manizade S, Ramamoorthy H, Rivera A, Russell R, Sonntag J, Swift R, Yungel J, Zwally J (2004) Accelerated sea-level rise from West Antarctica. Science 306: 255–258

Turner J, Colwell SR, Marshall GJ, Lachlan-Cope TA, Carleton AM, Johnes PD, Lagun V, Reid PA, Iagovkina S (2005) Antarctic climate change during the last 50 years. Intern Journ Climatology 25:279–294

Turner J, Lachlan-Cope TA, Colwell S, Marshall GJ, Connolley WM (2006) Significant warming of the Antarctic winter troposphere. Science 311:1914–1917

Vaughan DG, Holt JW, Blankenship DD (2007) West Antarctic links to sea level estimation. EOS Transactions 88(46):485–486

Vaughan DG, Marshall GJ, Connolley WM, Parkinson C, Mulvaney R, Hodgson DA, King JC, Pudsey CJ, Turner J (2003) Recent rapid regional climate warming on the Antarctic Peninsula. Climatic Change 60:243–274

Woodworth PL, Hughes CW, Blackman DL, Stepanov VN, Holgate SJ, Foden PR, Pugh JP, Mack S, Hargreaves GW, Meredith MP, Milinevsky G, Fierro Contreras JJ (2006) Antarctic Peninsula sea levels: a real-time system for monitoring Drake Passage transport. Antarctic Science 18(3):429–436, doi:10.1017/S0954102006000472

Woodworth PL, Player R (2003) The permanent service for mean sea level: an update to the 21st century. Journ Coastal Research 19(2):287–295

Yampolsky YuM, Zalizovsky AV, Litvinenko LN, Lizunov GV, Groves K, Moldwin M (2004) Variations of the magnetic field in an Antarctic region and attended region (New England), stimulated by cyclonic activity. Radio Physics and Radio Astronomy 9(2):130–151

Yunck TP (2002) An Overview of Atmospheric Radio Occultation. Journ Global Positioning Systems 1(1):58–60

A Validation of Ocean Tide Models Around Antarctica Using GPS Measurements

Ian D. Thomas, Matt A. King and Peter J. Clarke

Abstract Ocean tide models around the coastline of Antarctica are often poorly constrained, due to sparse data input and poorly known bathymetry in the ice-shelf regions. Land-based measurements of Ocean Tide Loading Displacements (OTLD), such as those made by GPS, provide a means of assessment of ocean tide models in such regions. Up to 11 years of daily GPS data from 18 stations on the Antarctic continent were processed using an up-to date estimation strategy based upon a precise point positioning analysis. Carrier-phase ambiguities were fixed, and parameters representing harmonic ground displacements at 4 diurnal frequencies (M2, S2, N2 and K2) and 4 semi-diurnal frequencies (K1, O1, P1 and Q1) were estimated on a daily basis, and then combined to form the GPS estimates of OTLD. These were compared with estimates of OTLD computed by means of a convolution process with a Green's function from seven global ocean tide models: CSR4, FES99, FES2004, GOT00.2, NAO.99b, TPXO6.2, TPXO7.0, and four regional ocean tide models: CATS02.01, CADA00.10, MToS.05, AntPen04.01. Fixing of carrier phase ambiguities was, unexpectedly, found to result in a poorer agreement between GPS estimates and models. For Antarctica as a whole, the TPXO6.2 and TPXO7.0 global models offer very good agreement with the GPS estimates of OTLD in all regions, with CADA00.10, MToS.05, CATS02.01 also generally being in good agreement. In East Antarctica, where the models are well constrained and in good agreement, the GPS estimates offer good agreement with the models – often to a sub-millimetre level – particularly for the lunar N2 and Q1 constituents. In West Antarctica, there is greater divergence amongst the modelled estimates of OTLD due to the complex coastline and less well modelled ice sheet regions. Here, the TPXO6.2, TPXO7.0 and CADA00.10 models offer equally good agreement. In

Ian D. Thomas
School of Civil Engineering and Geosciences, Newcastle University, Newcastle upon Tyne, NE2 1JU, UK, e-mail: Ian.Thomas@newcastle.ac.uk

Matt A. King
School of Civil Engineering and Geosciences, Newcastle University, Newcastle upon Tyne, NE2 1JU, UK

Peter J. Clarke
School of Civil Engineering and Geosciences, Newcastle University, Newcastle upon Tyne, NE2 1JU, UK

summary, GPS measurements of OTLD are of sufficient accuracy to distinguish between the models in certain regions of Antarctica, although some systematic biases remain at solar frequencies.

1 Introduction

Various studies have shown that GPS is capable of measuring harmonic ground displacements at tidal frequencies (e.g. Schenewerk et al. (2001), Allinson et al. (2004) and King et al. (2005)). Recently, Thomas et al. (2006) have shown that such measurements are comparable in precision and accuracy with those made using Very Long Baseline Interferometry (e.g. Petrov and Ma (2003)). It has also been demonstrated by King et al. (2005) that such direct GPS measurement of Ocean Tide Loading Displacements (OTLD), at sites on the Antarctic continent, can be used as a means of validation of numerical ocean tide models around Antarctica.

As an alternative to direct measurement, OTLD can also be computed from numerical ocean tide models, by convolution with a solid Earth model (Farrell 1972). The limiting factor in accuracy of such OTLD computations is generally considered to be inaccuracies in the ocean tide models (e.g. Bos and Baker (2005)).

Whereas OTLD at any given location depends upon the global ocean mass distribution, it is the local ocean tides which are the principal contributor to the loading effect. For this reason a comparison of GPS estimates of OTLD and computed values can be useful as a regional assessment of the ocean tide models. This is of particular interest in places such as Antarctica, where the ocean tide models are often poorly constrained due to sparse data input, and the bathymetry under the ice-shelves remains poorly known.

In their study, King et al. (2005) compared GPS estimates of OTLD at 8 semidiurnal and diurnal frequencies with those computed from a selection of global and regional ocean tide models. King et al. (2005) noted that for much of East Antarctica, where the spatial pattern of ocean tides is relatively simple and the tidal models well constrained, the GPS and the various modelled estimates of OTLD are mostly in good agreement, often at sub-millimetre level. Around the coast of West Antarctica and the Antarctic Peninsula, the coastline is more complex and there are large floating ice shelves – notably the Filchner-Ronne, Ross and Larsen Ice Shelves – under which the bathymetry and tidal mechanisms remain relatively poorly constrained and still the subject of current study (Han et al. 2005). Therefore, unsurprisingly, there is greater uncertainty and divergence amongst the various ocean tide models here than is seen in East Antarctica, which translates directly into a greater spread amongst the computed OTLD estimates. For the oceans around West Antarctica, King et al. (2005) determined that the more recent global (e.g. TPXO6.2, FES99) and regional (e.g. CADA0.10, CATS02.01) models generally gave better agreement with the GPS estimates of OTLD (as well as gravity measurements) than did the older models (e.g. CSR3, TPXO2).

In this paper, we extend the study of King et al. (2005) by using additional GPS sites and time series of greater length. A refined GPS estimation strategy is employed (Thomas et al. 2006) which includes an investigation into the effect of fixing carrier phase ambiguities to integer values on GPS measurements of OTLD. We compare our GPS estimates of OTLD with estimates computed from the ocean tide models used in the study of King et al. (2005), plus additional global models CSR4, FES2004 and TPXO7.0 and regional models AntPen04.01 and MTOs.05.

2 GPS Estimation of OTLD in Antarctica

2.1 GPS data

We use daily data from 18 GPS receivers, located around the coastline of Antarctica (Fig. 1), with the exception of the Amundsen-Scott (South Pole) station. The GPS stations and their data availability are summarised in Table 1, grouped by longitude into four broad geographical regions: East Antarctica (5 stations), Ross Sea region (5 stations), the Antarctic Peninsula and Weddell Sea (7 stations) and the South Pole (1 station). The length of the GPS time series used in this study ranges from $\sim$2 years, up to $\sim$11 years for the continuously operating IGS stations; data were compared for the period up to mid-2006 where available.

Amundsen-Scott station (AMUN), at the South Pole, is located on an ice-sheet which is moving with a horizontal velocity of $\sim$10 m per year, roughly along the 40° west longitude line. The long-term velocity was calculated from a standard GPS

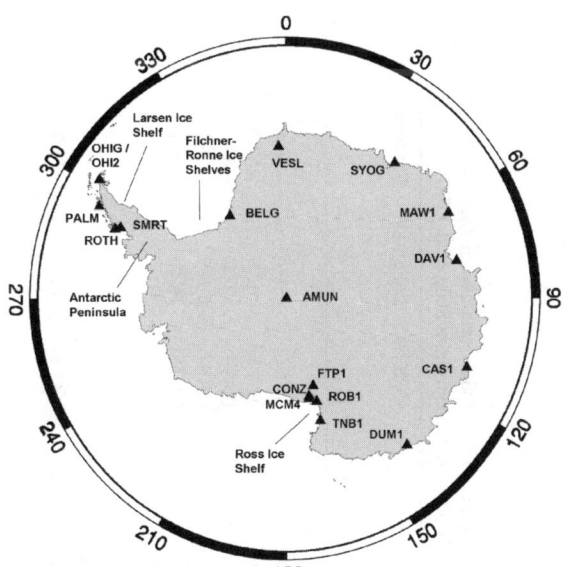

Fig. 1 Location map of the GPS sites used in this study (Polar-Stereographic projection)

Table 1 Summary of GPS data used in this study. Stations are listed in order of increasing Longitude, grouped into four geographical regions. Stations in bold are additional to those used in King et al. (2005)

Station	GPS ID	Longitude	Latitude	Start year	Start day	End year	End day	Number of days successfully processed	Data provider
East Antarctica									
Syowa	SYOG	39.6	−69.0	1995	74	2006	193	3605	IGS
Mawson	MAW1	62.9	−67.6	1995	5	2006	193	3534	IGS
Davis	DAV1	78.0	−68.6	1995	1	2006	193	3375	IGS
Casey Base	CAS1	110.5	−66.3	1996	2	2006	193	3359	IGS
Dumont d'Urville	DUM1	140.0	−66.6	1998	3	2005	365	2667	Laboratoire de Recherche en Géodésie (Bouin and Vigny 2000)
Ross Sea region									
Fishtail Point	**FTP1**	**162.6**	**−78.9**	**2002**	**9**	**2005**	**30**	**557**	**Ohio State University, TAMDEF network (Konfal et al. 2007)**
Cape Roberts	**ROB1**	**163.2**	**−77.0**	**2003**	**2**	**2005**	**38**	**547**	**Ohio State University, TAMDEF network**
Terra Nova Bay	**TNB1**	**164.1**	**−74.7**	**1999**	**1**	**2001**	**365**	**850**	**Italian Geodetic Project of PRNA, VLNDEF network (Casula et al. 2007)**
McMurdo	MCM4	166.7	−77.8	1995	25	2006	193	3576	IGS
Truncated Cones	CONZ	167.1	−77.5	2000	356	2006	365	1111	UNAVCO

Table 1 (continued)

Station	GPS ID	Longitude	Latitude	Start year	Start day	End year	End day	Number of days successfully processed	Data provider
Antarctic Peninsula									
Rothera	ROTH	291.9	−67.6	1999	55	2003	313	553	British Antarctic Survey
San Martin	**SMRT**	**292.9**	**−68.1**	**1999**	**112**	**2005**	**75**	**1378**	**Alfred Wegener Institute**
Palmer	PALM	295.9	−64.8	1998	189	2006	194	2798	IGS
O'Higgins	OHI2	302.1	−63.3	2002	5	2006	194	1286	IGS
O'Higgins	OHIG	302.1	−63.3	1995	70	2002	50	1527	IGS
Belgrano	BELG	325.4	−77.9	1998	34	2005	46	1581	Alfred Wegener Institute
Vesleskarvet	VESL	357.2	−71.7	1997	351	2006	154	1905	IGS
Amundsen-Scott (South Pole)	AMUN	139.2	−90.0	1998	189	2005	308	2298	IGS

precise point positioning analysis (Zumberge et al. 1997) of the full time-series. This computed velocity was removed from the daily GPS data at the observation level, using a procedure similar to that described by King et al. (2000), prior to the processing of the corrected GPS data using the same methodology as for the other stations, described below. All other sites are located on bedrock.

2.2 Estimation Strategy

We use the same GPS estimation strategy as Thomas et al. (2006) which is itself a refinement of the methodology employed by King et al. (2005) and Allinson et al. (2004). In summary, we use the GIPSY-OASIS II GPS analysis software from the NASA Jet Propulsion Laboratory (JPL) in its Precise Point Positioning (PPP) (Zumberge et al. 1997) mode, with an elevation cutoff of $7°$ and a sampling interval of 300 s. Solid Earth tide models are implemented according to the IERS Conventions 2003 (McCarthy and Petit 2004). Tropospheric zenith delays and gradients are modelled as random walk parameters with process noises of 10.2 and 0.3 mm $hr^{-1/2}$ respectively.

Within each 24 h PPP analysis, 51 constant parameters (three coordinate parameters, plus 48 additional OTLD parameters) were estimated once, along with their corresponding variance-covariance matrix. The 48 OTLD parameters represent harmonic motions (amplitudes of sine and cosine components), in each of the local co-ordinate directions (east, north, radial) at each of eight principal diurnal and semi-diurnal tidal frequencies (K1, O1, P1, Q1, M2, S2, N2 and K2).

The daily estimates of OTLD were stacked using a Kalman filter, effectively giving a weighted average, and nodal corrections were added to each constituent to account for the fact that we only estimate a subset of the tidal constituents. 15 mm^2 of process noise was added to the K2 constituent, in order to down-weight this constituent to account for its time variable behavior (King et al. 2005). Readers are referred to Thomas et al. (2006) for fuller details of the estimation strategy.

2.3 Ambiguity Resolution

In their study, King et al. (2005) did not fix carrier-phase ambiguities to integer values, due to limitations in the GIPSY software at the time of that study, although King et al. (2005) suggested that this may have a positive effect on the accuracy of the derived constituents, since radial and horizontal periodic signals have previously been shown to interact via the ambiguity terms in sub-daily position estimates (King et al. 2003).

However, in a subsequent global study, Thomas et al. (2006) considered ambiguity fixing using modified GIPSY routines provided by the Jet Propulsion Laboratory (JPL), which allowed harmonic parameters to be estimated from ambiguity-fixed daily solutions. Although it was shown that the fixing of ambiguities resulted in

A Validation of Ocean Tide Models

Table 2 Ambiguity resolution networks and mean percentage of carrier phase ambiguities fixed to integer values each day

Network	Stations					Mean daily fixing rate (%)
1	MAW1	CAS1	SYOG	DAV1		32.3
2	CONZ	MCM4	FTP1	ROB1	TNB1	28.1
3	PALM	OHIG	OHI2	ROTH	SMRT	53.4
4	AMUN	BELG	VESL			10.7
5	DUM1	DAV1	CAS1			25.6
6	BELG	SYOG	VESL			23.6

very marginal improvement in the GPS / model agreement in that global study (mainly K1 and K2), we decided that, for the most rigorous estimation of OTLD in Antarctica ambiguity resolution should be tested with more sites at high latitudes.

Ambiguity fixing was performed in regional networks of 3–4 stations per cluster (Table 2). For stations in more than one network, the solution from the network with the highest ambiguity fixing rate was used. On average, some ambiguities were successfully fixed on ∼90% of days. The percentage of ambiguities fixed on any given day was relatively low, with mean rates ranging within 10% and 53% (Table 2), lower than would be expected from a standard GIPSY PPP analysis. However, it is perhaps unsurprising, given the relatively large baselines and the significant reduction in redundancy (and corresponding increase in parameter formal errors) of the daily solutions due to the estimation of an additional 48 OTLD parameters over the standard solution.

The OTLD parameter estimates from the ambiguity-fixed daily solutions were combined by means of a Kalman filter, as described for the ambiguity floating solutions. For stations and days where no ambiguity fixing was achieved, the OTLD parameters from the ambiguity floating solution were used.

2.4 Modelled OTLD

We consider OTLD estimates computed from seven global barotropic ocean tide models (CSR4, FES99, FES2004, GOT00.2, NAO.99b, TPXO6.2, TPXO7.0) and four regional barotropic ocean tide models (CATS02.01, CADA00.10, MTOs.05, AntPen04.01). A summary of the ocean tide models is given in Table 3.

The models are, with the exception of CATS02.01 and AntPen04.01, inverse tidal models, assimilating Topex / Poseidon (T/P) altimetry data plus, for some of the models, one or more of the following additional data types: GPS, ERS altimetry, gravimetry, tide-gauge (TG) data.

The T/P data extend in latitude only as far as 66°S which, whilst achieving coverage of part of the Antarctic Peninsula and approaching the East Antarctic coastline, falls far short of much of the West Antarctic coastline. The TG data are also sparse in much of Antarctica (Padman et al. 2002). The other principal areas in which

Table 3 Summary details and references of the ocean tide models used in this study. The models additional to those used by King et al. (2005) are highlighted in bold (T/P – Topex / Poiseden; TG – Tide Gauge)

Model	Domain	Grid (degrees)	Data assimilated	Author / Reference
AntPen04.01	**Antarctic Peninsular region.**	**0.0333 × 0.0167**	None – Forward model, solves shallow-water wave equations.	**Willmott et al. (2007)**
CADA00.10	South of 58°S.	0.25 × 0.083	T/P plus 37 TG, GPS and gravimetric data.	Padman et al. (2002)
CATS02.01	South of 58°S.	0.25 × 0.083	None – Forward model.	Padman et al. (2002)
CSR4	**Global**	**0.5 × 0.5**	**T/P data.**	Updated version of CSR3 model (Eanes and Bettadpur 1995)
FES99	Global	0.25 × 0.25	T/P data plus 8 TG around Antarctica.	Lefevre et al. (2002)
FES2004	**Global**	**0.125 × 0.125**	**T/P data plus 8 TG around Antarctica.**	Update of FES99 model (Lefevre et al. 2002)
GOT00.2	Global	0.5 × 0.5	T/P plus ERS satellite data (not over ice shelves).	Update of GOT99.2 model (Ray 1999)
MTOs.05	**South of 30°S.**	**0.125 × 0.083**	**T/P data.**	K. Matsumoto (personal communication, 22 November 2005)
NAO.99b,	Global	0.5 × 0.5	T/P data.	Matsumoto et al. (2000)
TPXO6.2	Global	0.25 × 0.25	T/P data plus gravimetric data assimilated over Ross Sea (K1, O1 only).	Egbert and Erofeeva (2002)
TPXO7.0	**Global**	**0.25 × 0.25**	**T/P data.**	Updated TPXO6.2 model (Egbert and Erofeeva 2002)

the models are different from each other is in the bathymetry, particularly in the ice-shelf regions, and also the coastline definitions used, with there being several coastline definitions existing and in current use.

For these reasons, the output of the tide models differs considerably in some regions of Antarctica, most notably near the ice shelves of West Antarctica and around the Antarctic Peninsula. In these complicated regions, some of the models have local inverse solutions (e.g. TPXO6.2 has an inverse solution for the Ross Sea with assimilation of gravity data for K1 and O1 constituents). King et al. (2005) provide further background and references on the ocean tide modelling of the circum-Antarctic seas.

We compute OTLD at each of our 18 locations for the 8 main diurnal and semi-diurnal tidal frequencies: M2, S2, N2, K2, K1, O1, P1 and Q1, by convolving the selected ocean tide models with Green's functions derived from the Bullen-Gutenberg solid Earth model (Farrell 1972) using the SPOTL software (Agnew 1997). For the purposes of these global loading computations, the regional models are supplemented outside their boundaries by the TPX06.2 model.

Of the 11 models considered, CSR4, FES2004, TPXO7.0, AntPen04.01 and MTOs.05 are of particular interest here, since they are later versions of, or new models in addition to, the models tested by King et al. (2005).

3 Comparison of GPS and Modelled OTLD

3.1 GPS results

As pointed out before, a GPS data processing procedure based on a multi-day solution combination approach has been applied, in order to compute the estimations of OTLD and to compare these observables with the Oceanic Tidal Loading Displacements derived from different models.

We decided to omit the K1 and K2 constituents from the following analysis and discussion, owing to the well known systematic biases occurring in GPS estimation of OTLD at the K1 and K2 frequencies (e.g. King et al. (2005), Thomas et al. (2006)), moreover, we have to mention that, as expected, errors remain at these frequencies. These biases are considered to result from orbit mis-modelling and aliasing of errors, due largely to the orbital period of the GPS satellites corresponding to the K2 period, and the satellite constellation geometry repeat period corresponding closely to the K1 period. Systematic biases at these frequencies are likely to remain a limitation of the current GPS estimation strategy, perhaps until the European GNSS, the Galileo System, becomes available with its non-tidal orbital period. The reader is referred to King et al. (2005), King et al. (2005) and Thomas et al. (2006) for a fuller discussion and further references on the systematic errors in GPS derived OTLD at these periods.

The GPS estimates of OTLD for the M2, S2, N2, P1, O1 and Q1 constituents, from the unfixed ambiguity solution, are listed in Tables 4(a–c), expressed as an

Table 4 (a) GPS derived estimates of OTLD for M2, S2, N2, P1, O1, Q1 constituents plus 1-sigma formal errors – expressed as an amplitude and Greenwich phase lag. (East Antarctic sites). (b) as Table 4(a), for Ross Sea region. (c) – as Table 4(a), for Antarctic Peninsula region

(a)			SYOG	MAW1	DAV1	CAS1	DUM1
M2	E	Amplitude (mm)	1.2 ± 0.4	0.9 ± 0.3	1.4 ± 0.4	3.3 ± 0.3	2.5 ± 0.4
		Phase (deg)	−73.4 ± 18.1	−99.2 ± 21.1	5.4 ± 14.1	10.4 ± 6.0	−5.0 ± 9.2
	N	Amplitude (mm)	1.5 ± 0.3	1.0 ± 0.3	1.9 ± 0.3	0.9 ± 0.3	2.0 ± 0.4
		Phase (deg)	87.0 ± 13.3	−168.1 ± 17.5	−154.0 ± 9.7	161.1 ± 19.7	23.7 ± 10.7
	V	Amplitude (mm)	8.3 ± 0.7	4.3 ± 0.6	6.9 ± 0.5	8.6 ± 0.5	7.7 ± 0.7
		Phase (deg)	−91.2 ± 4.9	−18.4 ± 7.5	6.0 ± 4.5	−32.1 ± 3.3	−111.6 ± 4.9
S2	E	Amplitude (mm)	0.4 ± 1.0	0.8 ± 0.9	1.2 ± 1.0	0.8 ± 1.0	1.1 ± 1.1
		Phase (deg)	−164.0 ± 126.3	174.2 ± 71.2	165.2 ± 44.5	75.5 ± 67.0	24.9 ± 57.0
	N	Amplitude (mm)	0.8 ± 1.0	0.1 ± 0.9	0.5 ± 1.0	1.2 ± 1.0	1.0 ± 1.1
		Phase (deg)	85.1 ± 72.5	−69.4 ± 554.4	−156.1 ± 113.0	171.8 ± 44.1	125.3 ± 58.7
	V	Amplitude (mm)	5.0 ± 1.9	1.7 ± 1.9	4.1 ± 1.9	3.9 ± 1.8	3.6 ± 2.1
		Phase (deg)	−76.5 ± 22.1	177.6 ± 64.1	119.9 ± 25.6	43.1 ± 11.3	−57.1 ± 32.5
N2	E	Amplitude (mm)	0.2 ± 0.4	0.2 ± 0.3	0.5 ± 0.3	1.0 ± 0.3	1.0 ± 0.4
		Phase (deg)	124.1 ± 87.7	−161.0 ± 92.4	−20.5 ± 43.7	−24.8 ± 19.0	−43.4 ± 8.3
	N	Amplitude (mm)	0.6 ± 0.3	0.5 ± 0.3	0.7 ± 0.3	0.5 ± 0.3	0.6 ± 0.4
		Phase (deg)	83.1 ± 29.8	125.0 ± 30.6	142.8 ± 23.7	117.6 ± 36.9	21.3 ± 36.8
	V	Amplitude (mm)	1.7 ± 0.7	1.5 ± 0.5	2.1 ± 0.5	2.2 ± 0.5	2.1 ± 0.7
		Phase (deg)	−95.0 ± 23.6	−31.4 ± 20.1	−26.6 ± 14.5	−52.5 ± 11.5	−124.7 ± 17.0
O1	E	Amplitude (mm)	2.5 ± 0.3	2.7 ± 0.3	1.2 ± 0.3	0.8 ± 0.3	0.4 ± 0.4
		Phase (deg)	−122.4 ± 6.1	−103.1 ± 6.0	−106.7 ± 13.6	−86.8 ± 20.2	−85.0 ± 48.0
	N	Amplitude (mm)	0.9 ± 0.2	2.0 ± 0.2	2.5 ± 0.2	2.9 ± 0.2	3.0 ± 0.3
		Phase (deg)	−96.1 ± 13.6	−108.8 ± 6.1	−115.7 ± 5.1	−123.0 ± 4.3	−123.4 ± 5.3
	V	Amplitude (mm)	8.2 ± 0.6	8.1 ± 0.5	9.4 ± 0.5	11.1 ± 0.5	10.9 ± 0.6
		Phase (deg)	133.6 ± 1.4	96.7 ± 3.5	80.7 ± 2.8	49.6 ± 1.7	27.8 ± 3.1

Table 4 (continued)

(a)

			SYOG	MAW1	DAV1	CAS1	DUM1
P1	E	Amplitude (mm)	1.0±0.3	0.8±0.3	0.8±0.3	0.5±0.3	0.4±0.4
		Phase (deg)	−149.3±15.6	−113.4±20.6	−166.9±22.9	−22.1±35.2	−99.1±51.0
	N	Amplitude (mm)	0.5±0.2	0.6±0.2	0.9±0.2	1.5±0.2	1.5±0.3
		Phase (deg)	−2.7±27.8	−109.2±19.8	−125.6±14.5	−117.3±8.8	−118.9±11.3
	V	Amplitude (mm)	2.2±0.6	1.9±0.5	2.7±0.5	3.9±0.5	3.7±0.6
		Phase (deg)	148.9±16.3	106.6±15.4	76.2±9.9	53.9±6.2	36.3±8.5
Q1	E	Amplitude (mm)	0.5±0.3	0.4±0.3	0.1±0.3	0.0±0.3	0.2±0.4
		Phase (deg)	−138.6±21.8	−116.5±43.4	−157.3±211.9	126.8±1117.1	56.7±134.9
	N	Amplitude (mm)	0.2±0.2	0.4±0.2	0.5±0.2	0.5±0.2	0.6±0.3
		Phase (deg)	−92.9±50.6	−120.2±29.7	−126.2±25.5	−126.2±21.4	−125.0±26.2
	V	Amplitude (mm)	2.3±0.6	2.2±0.5	2.3±0.5	2.5±0.5	2.8±0.6
		Phase (deg)	121.7±15.6	91.4±13.0	75.0±11.8	39.3±8.1	10.0±12.2

(b)

			FTP1	ROB1	TNB1	MCM4	CONZ
M2	E	Amplitude (mm)	0.4±0.8	0.9±0.8	2.0±0.9	0.9±0.4	0.6±0.7
		Phase (deg)	4.9±127.5	24.3±50.7	0.1±25.1	−12.8±26.2	23.7±69.4
	N	Amplitude (mm)	1.1±0.8	1.4±0.8	1.0±0.9	0.9±0.4	1.6±0.7
		Phase (deg)	40.6±30.6	16.1±31.6	15.1±49.2	33.4±25.1	20.1±24.3
	V	Amplitude (mm)	1.3±1.3	1.2±1.2	3.4±1.5	1.7±0.7	1.1±1.2
		Phase (deg)	−38.0±48.3	−87.6±56.0	−114.1±24.7	−95.5±23.9	−61.4±64.4
S2	E	Amplitude (mm)	1.5±1.7	1.5±2.1	2.2±2.0	1.1±0.9	0.8±1.3
		Phase (deg)	39.1±53.1	56.8±77.3	11.7±53.3	3.7±47.3	33.3±94.6
	N	Amplitude (mm)	0.4±1.7	0.7±2.1	1.2±2.0	0.4±0.9	0.7±1.3
		Phase (deg)	80.0±269.7	42.9±75.8	−66.9±98.4	13.3±133.6	62.7±111.4
	V	Amplitude (mm)	3.6±3.1	1.3±4.7	3.6±3.8	1.6±2.0	0.7±2.8
		Phase (deg)	0.2±48.7	−13.4±214.3	−89.3±62.1	−51.7±61.2	−3.0±229.3

Table 4 (continued)

(b)

			FTP1	ROB1	TNB1	MCM4	CONZ
N2	E	Amplitude (mm)	0.5 ± 0.8	0.6 ± 0.8	0.7 ± 0.9	0.4 ± 0.4	0.6 ± 0.7
		Phase (deg)	−89.9 ± 91.7	−71.4 ± 69.9	−70.5 ± 75.9	−82.9 ± 53.8	−119.0 ± 72.0
	N	Amplitude (mm)	0.1 ± 0.8	0.5 ± 0.7	0.8 ± 0.8	0.4 ± 0.4	0.6 ± 0.7
		Phase (deg)	−2.6 ± 416.5	−35.8 ± 79.8	2.5 ± 63.1	−20.2 ± 51.0	−3.3 ± 61.4
	V	Amplitude (mm)	0.5 ± 1.2	0.4 ± 1.1	0.1 ± 1.5	0.6 ± 0.7	0.6 ± 1.2
		Phase (deg)	−16.7 ± 139.3	19.2 ± 164.5	−70.1 ± 1099.8	15.3 ± 70.1	15.4 ± 114.7
O1	E	Amplitude (mm)	1.7 ± 0.7	1.5 ± 0.6	0.7 ± 0.7	1.5 ± 0.3	2.1 ± 0.6
		Phase (deg)	94.3 ± 21.6	82.8 ± 21.5	118.1 ± 61.3	100.9 ± 11.9	76.4 ± 15.0
	N	Amplitude (mm)	2.3 ± 0.6	2.1 ± 0.5	1.8 ± 0.7	1.5 ± 0.3	2.2 ± 0.5
		Phase (deg)	−95.6 ± 15.6	−98.6 ± 14.5	−125.1 ± 20.6	−132.3 ± 5.5	−94.0 ± 13.7
	V	Amplitude (mm)	7.3 ± 1.2	7.6 ± 1.0	7.0 ± 1.3	8.8 ± 0.7	8.2 ± 1.2
		Phase (deg)	5.7 ± 9.2	−1.4 ± 7.9	−3.4 ± 11.2	−6.1 ± 4.5	−2.5 ± 8.2
P1	E	Amplitude (mm)	0.6 ± 0.7	1.2 ± 0.6	1.1 ± 0.8	0.7 ± 0.3	1.7 ± 0.6
		Phase (deg)	69.1 ± 60.0	129.8 ± 21.5	−161.7 ± 39.0	22.0 ± 27.1	110.9 ± 19.5
	N	Amplitude (mm)	1.3 ± 0.7	1.0 ± 0.6	0.4 ± 0.7	1.4 ± 0.3	0.9 ± 0.5
		Phase (deg)	−63.6 ± 29.9	−73.8 ± 31.9	−64.7 ± 93.8	−97.1 ± 11.6	−91.7 ± 34.9
	V	Amplitude (mm)	3.6 ± 1.2	4.6 ± 1.1	1.5 ± 1.4	2.9 ± 0.7	3.0 ± 1.2
		Phase (deg)	30.1 ± 18.5	37.0 ± 11.4	−28.3 ± 51.9	17.4 ± 13.8	56.4 ± 22.1
Q1	E	Amplitude (mm)	0.5 ± 0.6	0.8 ± 0.6	0.1 ± 0.7	0.5 ± 0.3	0.6 ± 0.5
		Phase (deg)	74.6 ± 79.8	67.8 ± 43.3	7.2 ± 444.9	90.6 ± 38.6	54.9 ± 52.1
	N	Amplitude (mm)	0.4 ± 0.6	0.5 ± 0.5	0.8 ± 0.7	0.3 ± 0.3	0.5 ± 0.5
		Phase (deg)	−122.4 ± 95.1	−92.8 ± 66.7	−120.3 ± 47.0	−115.1 ± 52.8	−86.0 ± 62.6
	V	Amplitude (mm)	1.4 ± 1.2	2.1 ± 1.1	1.8 ± 1.4	1.9 ± 0.7	1.9 ± 1.2
		Phase (deg)	−4.2 ± 46.7	−9.9 ± 28.0	−13.2 ± 44.5	−14.1 ± 20.9	−7.5 ± 35.9

Table 4 (continued)

(c)		ROTH	SMRT	PALM	OHIG	OHI2	BELG	VESL	AMUN
M2	E Amplitude (mm)	3.2 ± 1.1	4.2 ± 0.9	3.6 ± 0.4	2.1 ± 0.6	3.4 ± 0.5	1.5 ± 0.7	2.0 ± 0.5	1.3 ± 0.6
	Phase (deg)	−94.5 ± 20.1	−109.7 ± 12.8	−96.3 ± 5.8	−92.5 ± 16.1	−83.7 ± 8.0	145.5 ± 23.7	59.9 ± 14.1	43.2 ± 9.8
	N Amplitude (mm)	0.9 ± 1.0	0.3 ± 0.9	1.3 ± 0.3	2.7 ± 0.5	1.3 ± 0.4	4.7 ± 0.7	1.6 ± 0.5	0.6 ± 0.6
	Phase (deg)	−36.5 ± 56.3	−16.8 ± 153.6	51.3 ± 12.3	8.0 ± 10.9	40.6 ± 13.5	−127.9 ± 7.0	−165.4 ± 16.4	49.4 ± 44.2
	V Amplitude (mm)	14.1 ± 2.0	15.8 ± 1.5	17.9 ± 0.6	22.2 ± 1.0	23.5 ± 0.8	13.6 ± 1.1	4.6 ± 0.8	2.1 ± 0.9
	Phase (deg)	68.2 ± 8.1	67.0 ± 5.5	79.0 ± 1.8	85.6 ± 2.7	85.7 ± 1.8	81.0 ± 4.7	14.2 ± 10.3	39.3 ± 20.4
S2	E Amplitude (mm)	3.8 ± 1.7	4.3 ± 2.0	2.9 ± 1.4	2.1 ± 1.5	2.9 ± 1.5	0.7 ± 1.4	2.3 ± 1.1	0.5 ± 1.2
	Phase (deg)	−100.3 ± 25.3	−107.5 ± 26.6	−102.4 ± 26.3	−109.3 ± 41.4	−112.2 ± 29.3	139.3 ± 84.2	76.2 ± 28.0	0.5 ± 147.0
	N Amplitude (mm)	0.9 ± 1.6	1.0 ± 2.0	0.8 ± 1.3	1.5 ± 1.5	1.0 ± 1.5	3.3 ± 1.4	1.9 ± 1.1	1.1 ± 1.2
	Phase (deg)	74.7 ± 102.9	2.1 ± 112.3	26.2 ± 101.4	57.0 ± 55.3	47.7 ± 45.1	−97.7 ± 25.1	−72.2 ± 34.1	144.0 ± 57.7
	V Amplitude (mm)	2.6 ± 3.2	4.7 ± 2.8	6.2 ± 2.1	12.0 ± 2.8	11.6 ± 2.7	7.4 ± 2.5	4.7 ± 1.8	1.2 ± 2.3
	Phase (deg)	101.9 ± 71.3	106.6 ± 34.4	126.6 ± 17.8	122.2 ± 12.9	121.3 ± 13.3	95.8 ± 19.3	25.9 ± 22.2	47.2 ± 47.5
N2	E Amplitude (mm)	0.3 ± 1.1	0.8 ± 1.0	0.7 ± 0.4	0.6 ± 0.6	0.6 ± 0.5	0.7 ± 0.7	0.7 ± 0.5	0.3 ± 0.6
	Phase (deg)	−145.1 ± 239.5	−121.8 ± 67.2	−99.7 ± 30.3	−36.8 ± 46.3	−115.2 ± 46.1	126.1 ± 51.0	51.3 ± 33.4	−34.4 ± 115.1
	N Amplitude (mm)	0.4 ± 1.0	0.2 ± 0.9	0.3 ± 0.3	0.8 ± 0.5	0.4 ± 0.4	0.7 ± 0.7	0.3 ± 0.5	0.4 ± 0.6
	Phase (deg)	−64.6 ± 122.9	−80.9 ± 237.0	−55.4 ± 67.9	−46.5 ± 12.0	−39.6 ± 46.7	−139.3 ± 36.0	−148.5 ± 78.7	41.2 ± 52.2
	V Amplitude (mm)	3.3 ± 2.0	2.7 ± 1.5	2.7 ± 0.6	3.2 ± 1.0	3.9 ± 0.7	2.3 ± 1.1	0.8 ± 0.8	0.4 ± 0.9
	Phase (deg)	18.2 ± 34.1	25.4 ± 31.8	36.2 ± 10.8	53.1 ± 16.8	58.9 ± 10.5	68.0 ± 27.2	−44.3 ± 10.0	−51.5 ± 102.2
O1	E Amplitude (mm)	1.2 ± 0.8	1.4 ± 0.7	1.0 ± 0.3	1.3 ± 0.5	1.0 ± 0.3	2.7 ± 0.5	2.8 ± 0.3	1.1 ± 0.5
	Phase (deg)	−26.3 ± 39.6	−64.8 ± 29.5	−75.6 ± 14.7	−36.1 ± 19.1	−111.8 ± 18.6	−137.4 ± 4.8	−118.9 ± 6.9	68.8 ± 24.5
	N Amplitude (mm)	1.8 ± 0.6	2.2 ± 0.6	1.7 ± 0.2	0.9 ± 0.3	2.0 ± 0.2	3.2 ± 0.4	1.4 ± 0.3	2.8 ± 0.5
	Phase (deg)	34.5 ± 19.3	53.6 ± 13.3	69.6 ± 6.7	81.5 ± 21.5	92.3 ± 6.8	40.0 ± 5.7	33.4 ± 10.7	−132.5 ± 4.7
	V Amplitude (mm)	13.6 ± 1.6	13.4 ± 1.3	15.4 ± 0.5	16.8 ± 0.9	16.9 ± 0.6	10.5 ± 1.0	7.9 ± 0.7	1.3 ± 0.9
	Phase (deg)	−125.8 ± 6.4	−128.1 ± 4.9	−128.8 ± 1.4	−135.8 ± 0.6	−134.0 ± 0.5	−157.9 ± 5.6	178.4 ± 5.1	−107.6 ± 41.7

Table 4 (continued)

(c)			ROTH	SMRT	PALM	OHIG	OHI2	BELG	VESL	AMUN
P1	E	Amplitude (mm)	0.9 ± 0.9	0.7 ± 0.7	0.9 ± 0.3	0.7 ± 0.5	0.9 ± 0.3	1.5 ± 0.5	1.8 ± 0.4	1.4 ± 0.5
		Phase (deg)	−41.0 ± 36.8	−38.7 ± 49.5	−86.0 ± 17.5	7.2 ± 39.2	−93.3 ± 23.1	−146.6 ± 17.8	−107.0 ± 11.4	−39.4 ± 15.5
	N	Amplitude (mm)	0.4 ± 0.7	0.9 ± 0.6	0.5 ± 0.2	0.1 ± 0.3	0.6 ± 0.3	1.1 ± 0.4	0.8 ± 0.3	3.4 ± 0.5
		Phase (deg)	−86.5 ± 89.8	115.9 ± 39.0	104.1 ± 23.3	−176.9 ± 165.3	149.4 ± 23.5	50.9 ± 18.2	54.2 ± 20.0	−143.3 ± 7.2
	V	Amplitude (mm)	5.3 ± 1.7	4.8 ± 1.4	5.5 ± 0.5	5.3 ± 0.9	4.7 ± 0.6	3.6 ± 1.0	2.6 ± 0.7	1.3 ± 1.0
		Phase (deg)	−121.7 ± 17.5	−126.8 ± 14.4	−119.9 ± 4.9	−112.6 ± 9.9	−127.1 ± 6.7	−158.0 ± 16.7	−179.4 ± 15.5	−30.7 ± 40.8
Q1	E	Amplitude (mm)	0.6 ± 0.8	0.7 ± 0.7	0.2 ± 0.3	0.2 ± 0.5	0.2 ± 0.3	0.7 ± 0.5	0.7 ± 0.3	0.4 ± 0.5
		Phase (deg)	−33.6 ± 81.2	−17.0 ± 56.8	−24.0 ± 63.2	57.0 ± 164.2	−29.4 ± 94.9	−137.4 ± 17.6	−115.8 ± 28.2	77.3 ± 63.9
	N	Amplitude (mm)	0.6 ± 0.6	0.4 ± 0.6	0.3 ± 0.2	0.2 ± 0.3	0.5 ± 0.2	0.5 ± 0.4	0.3 ± 0.3	0.5 ± 0.5
		Phase (deg)	30.5 ± 55.1	17.6 ± 80.7	45.8 ± 7.2	43.4 ± 45.6	70.8 ± 28.7	24.6 ± 45.5	9.6 ± 53.9	−154.5 ± 50.6
	V	Amplitude (mm)	2.9 ± 1.6	3.0 ± 1.3	3.7 ± 0.5	3.8 ± 0.9	4.2 ± 0.6	2.9 ± 1.0	2.0 ± 0.7	0.6 ± 1.0
		Phase (deg)	−142.1 ± 27.6	−137.7 ± 13.5	−145.0 ± 6.8	−153.5 ± 13.4	−152.6 ± 8.3	−169.2 ± 20.1	172.0 ± 20.3	−172.0 ± 99.1

amplitude and Greenwich phase lag, along with their formal one-sigma errors. The convention adopted is positive Greenwich phase lags.

The GPS estimates are similar but slightly different to those of King et al. (2005), as is expected with our refined estimation strategy and longer time series. The formal one-sigma errors are of similar magnitude, or slightly bigger, than those obtained in the previous study.

It is noted that the uncertainties are relatively larger for the S2 constituents compared with the other constituents shown. This is largely due to the process noise introduced at the K2 constituent in the combination Kalman filter; since the K2 and S2 constituent frequencies are closely related, the process noise at K2 has the effect of modifying the relative uncertainties due to constituent correlations. Conversely, the uncertainties at the other constituents are smaller than would otherwise be the case (Thomas et al. 2006).

3.2 Ambiguity Resolution

To assess agreement between the GPS and modelled estimates of OTLD, we computed a misfit statistic (S^2), as calculated by King et al. (2005), defined as the sum of the squares of the magnitudes of the complex differences, across all constituents (j) and component directions (k):

$$S^2 = \sum_{k=1}^{3(1)} \sum_{j=1}^{6} (|Z_{\text{GPS}_{i,j}} - Z_{\text{model}_{i,j}}|^2) \quad (1)$$

The value of this statistic was summed regionally, for the three principal regions represented in Table 1 (excluding the South Pole), for each of the tidal models. In each case, we consider both the ambiguity-floating and the ambiguity-fixed GPS solutions, and for each solution we compute S^2 for the three-dimensional case (3 components) and for the vertical component alone. Table 5 summarises the results, with the model estimates ordered from best to worst agreement with GPS estimates in each case.

Comparing the ambiguity-floating and fixed solutions, it is apparent that for each of the three regions, the fixing of ambiguities has resulted in consistently poorer GPS / model agreement for all models and for both the three-dimensional and vertical cases. The increase of the misfit statistic due to the fixing of ambiguities is most notable in the case for East Antarctica, which is in fact the region for which we expect the modelled estimates to be the most reliable. This outcome of ambiguity fixing is surprising; particularly in light of Thomas et al. (2006) finding that ambiguity fixing resulted in no significant change to the GPS / model agreement for their global selection of stations (mainly at mid and low latitudes). We offer no explanation for this effect of fixing carrier-phase ambiguities at the present time, however for the remainder of this paper we revert to discussing the ambiguity-floating estimates of OTLD.

3.3 Regional Assessment of Models

In Table 5, we ordered the models according to their overall regional agreement with GPS. In Table 6 we extend this to identify the model that offers the best agreement at individual sites by listing the smallest S^2 misfit for each station, the formal 1-sigma error in S^2, and the corresponding model, for both the three-dimensional and vertical cases. We discuss the results from Tables 5 and 6 by region in the following sections.

3.3.1 East Antarctica

Figure 2 shows phasor plots, with 95% confidence ellipses, of modelled and GPS OTLD at Davis (DAV1) station, which is typical of the East Antarctica sites. The ocean tides in East Antarctica are well constrained, and this is reflected in the relatively close agreement of the modelled values at all constituents and components.

The IGS GPS sites in East Antarctica are some of the longest continuously recording sites used in this study, with a number of sites having >10 years of data, resulting in small formal errors and robust estimates. The GPS estimates are in good agreement with the models at most constituents / components; agreement is noted as particularly good at the small lunar constituents N2 and Q1, and also in the M2 and O1 constituents in the vertical component.

Of the previously un-tested models, MTOs.05, TPXO7.0 and FES2004 all offer good agreement with GPS for East Antarctica (Table 5), however the models are all in such close agreement that, with the possible exception of CSR4 and GOT00.2 which show slightly poorer agreement, we do not attempt to distinguish between the models.

3.3.2 Ross Sea region

The GPS stations in this region are all situated close to the east end of the Ross Ice Shelf. The GPS data series are of variable data length; IGS site MCM4 and site CONZ have longer time series than are available for the campaign sites in the TAMDEF (Konfal et al. 2007) and VLNDEF (Casula et al. 2007) networks. Figure 3 shows phasor plots of GPS and computed OTLD at TNB1; this site is chosen as an illustrative example, since it was not used by King et al. (2005). It also illustrates the reasonable GPS / model agreement that can be achieved with a relatively short time series, although the formal errors are larger than at sites with longer time series. The 850 days of data processed at this site is roughly the minimum data length of ∼900–1000 days determined by Allinson et al. (2004) as being required to achieve stabilisation of constituent estimates.

The spread of the models seen at TNB1, which is similar at all the locations in the region, is clearly seen to be larger than is the case for Davis and East Antarctica,

Table 5 Models ordered according to regionally summed S^2 misfit statistics, for the three main regions described in Section 2.1 and for each of 3-D and vertical only cases, excluding K1/K2 constituents in both cases. (Note that the AntPen04.01model is omitted for the East Antarctica and Ross Sea regions, since it is far outside its domain, and therefore the resulting OTLD are the same as TPXO6.2 in these regions)

S^2 GPS / model misfits by region (mm^2)

Ambiguity Floating				Ambiguity fixed			
Three-dimensional		Vertical		Three-dimensional		Vertical	
East Antarctica							
TPXO6.2	69.1	CATS02.01	21.8	FES2004	149.8	FES2004	85.9
MTOs.05	69.5	TPXO6.2	21.9	FES99	150.2	FES99	86.2
CATS02.01	70.6	MTOs.05	22.2	CADA00.10	151.4	TPXO6.2	87.2
TPXO7.0	71.6	NAO.99b	23.9	TPXO6.2	151.8	CADA00.10	87.5
NAO.99b	72.5	TPXO7.0	24.6	NAO.99b	154.6	NAO.99b	88.4
FES2004	72.7	CADA00.10	25.6	TPXO7.0	158.2	TPXO7.0	93.8
CADA00.10	72.8	FES2004	25.8	MTOs.05	159.8	MTOs.05	94.8
FES99	73.4	FES99	26.6	CSR4	160.6	CSR4	97.7
GOT00.2	84.0	GOT00.2	36.6	GOT00.2	165.5	GOT00.2	101.1
CSR4	85.5	CSR4	39.1	CATS02.01	170.3	CATS02.01	103.0
Ross Sea region							
FES2004	113.7	TPXO6.2	54.8	TPXO6.2	132.3	TPXO6.2	54.2
TPXO6.2	114.7	FES2004	60.0	FES2004	134.6	TPXO7.0	59.0
TPXO7.0	125.6	TPXO7.0	61.5	CATS02.01	140.9	FES2004	63.3
CATS02.01	126.7	CATS02.01	70.6	TPXO7.0	141.0	CATS02.01	67.3
GOT00.2	128.6	GOT00.2	73.4	CADA00.10	148.7	CADA00.10	70.9
CADA00.10	134.3	CADA00.10	74.0	GOT00.2	149.0	GOT00.2	76.4
MTOS.05	140.6	MTOS.05	91.4	MTOS.05	167.9	FES99	97.6
FES99	148.6	FES99	94.8	FES99	169.4	MTOS.05	100.0
CSR4	180.4	CSR4	124.7	CSR4	208.8	CSR4	137.4
NAO.99B	220.9	NAO.99B	153.6	NAO.99B	246.6	NAO.99B	161.2

Table 5 (continued)

S² GPS / model misfits by region (mm²)

Ambiguity Floating				Ambiguity fixed			
Three-dimensional		Vertical		Three-dimensional		Vertical	
Antarctic Peninsula							
TPXO7.0	115.1	CADA00.10	39.5	TPXO6.2	156.8	TPXO6.2	64.0
CADA00.10	115.6	TPXO6.2	40.2	MTOs.05	158.4	MTOs.05	65.8
TPXO6.2	116.0	TPXO7.0	41.6	CADA00.10	160.6	CADA00.10	67.6
AntPen04.01	122.3	MTOs.05	45.9	AntPen04.01	164.9	AntPen04.01	75.0
MTOs.05	122.4	AntPen04.01	48.5	GOT00.2	167.7	TPXO7.0	78.5
GOT00.2	125.3	CATS02.01	57.4	TPXO7.0	168.9	CATS02.01	81.6
CATS02.01	131.7	GOT00.2	58.9	CATS02.01	171.8	GOT00.2	84.1
FES2004	171.9	FES2004	99.9	FES2004	207.5	FES2004	119.0
FES99	182.4	FES99	112.0	FES99	216.7	FES99	129.3
CSR4	226.0	CSR4	144.0	CSR4	283.1	CSR4	186.1
NAO.99b	276.2	NAO.99b	185.5	NAO.99b	307.9	NAO.99b	211.2

Table 6 Smallest S^2 GPS / model misfit statistic for each station, along with sigma of S^2 and the corresponding model, for three dimensional and vertical cases (again, excluding K1 and K2 in both cases)

Site	Three-dimensional			Vertical		
	Smallest S^2 misfit (mm^2)	Sigma of S^2 misfit (mm^2)	Corresponding model	Smallest S^2 misfit (mm^2)	Sigma of S^2 misfit (mm^2)	Corresponding model
SYOG	16.1	4.4	FES99	3.4	2.8	CATS02.01
MAW1	12.5	3.6	FES2004	2.0	1.9	FES2004
DAV1	12.1	4.6	TPXO6.2	1.5	3.5	TPXO6.2
CAS1	9.3	4.9	NAO.99b	2.1	4.4	NAO.99b
DUM1	17.3	6.9	CATS02.01	10.7	6.6	CATS02.01
FTP1	27.8	23.7	FES2004	14.6	22.2	TPXO7.0
ROB1	20.3	17.9	TPXO6.2	7.5	16.5	TPXO6.2
TNB1	21.7	18.2	MTOs.05	9.4	15.6	MTOs.05
MCM4	11.5	6.5	TPXO6.2	4.3	6.1	TPXO6.2
CONZ	20.9	10.3	TPXO6.2	8.0	9.3	TPXO6.2
ROTH	10.9	11.4	TPXO7.0	5.3	10.1	TPXO7.0
SMRT	17.8	11.7	GOT00.2	6.8	9.8	TPXO6.2
PALM	15.4	7.6	GOT00.2	5.4	6.8	TPXO6.2
OHI2	20.0	10.6	CADA00.10	6.8	8.7	CADA00.10
OHIG	9.9	10.8	CADA00.10	5.2	9.8	CADA00.10
BELG	14.5	6.1	TPXO6.2	2.8	6.1	TPXO6.2
VESL	20.6	6.6	TPXO7.0	1.9	2.9	TPXO7.0
AMUN	23.0	7.5	TPXO6.2	3.7	5.9	TPXO6.2

in particular for the O1 constituent. MTOs.05 is the model with the smallest overall GPS / model S^2 statistic for TNB1, however the modelled estimates mostly fall within the 95% formal error ellipses for this region, and we do therefore distinguish between the models with significance at individual sites.

The remaining sites in this region of Antarctica show mainly similar patterns. A site of particular note is CONZ, where the GPS / model agreement is observed to be reasonable; the TPXO6.2 model offers the best agreement. This is in contrast to the anomalously poor GPS / model agreement observed here by King et al. (2005). It is now known that their earlier GPS estimates at this site were biased by an error in the GPS data used in that study. The latest GPS estimates of OTLD for site CONZ (Table 4(b)) are observed to be broadly similar to those for nearby site MCM4, although the S^2 misfits are somewhat larger, probably due to the shorter time series.

Of the models not tested in the previous study of King et al. (2005), FES2004 offers best agreement with GPS for the Ross Sea region of Antarctica, particularly in the three-dimensional measurments, followed by TPXO7.0 and MTOs.05 (Table 5). As noted by King et al. (2005), TPXO6.2 and CATS02.01 also agree well with the GPS estimates, with TPXO6.2 perhaps offering best overall agreement for the region as a whole.

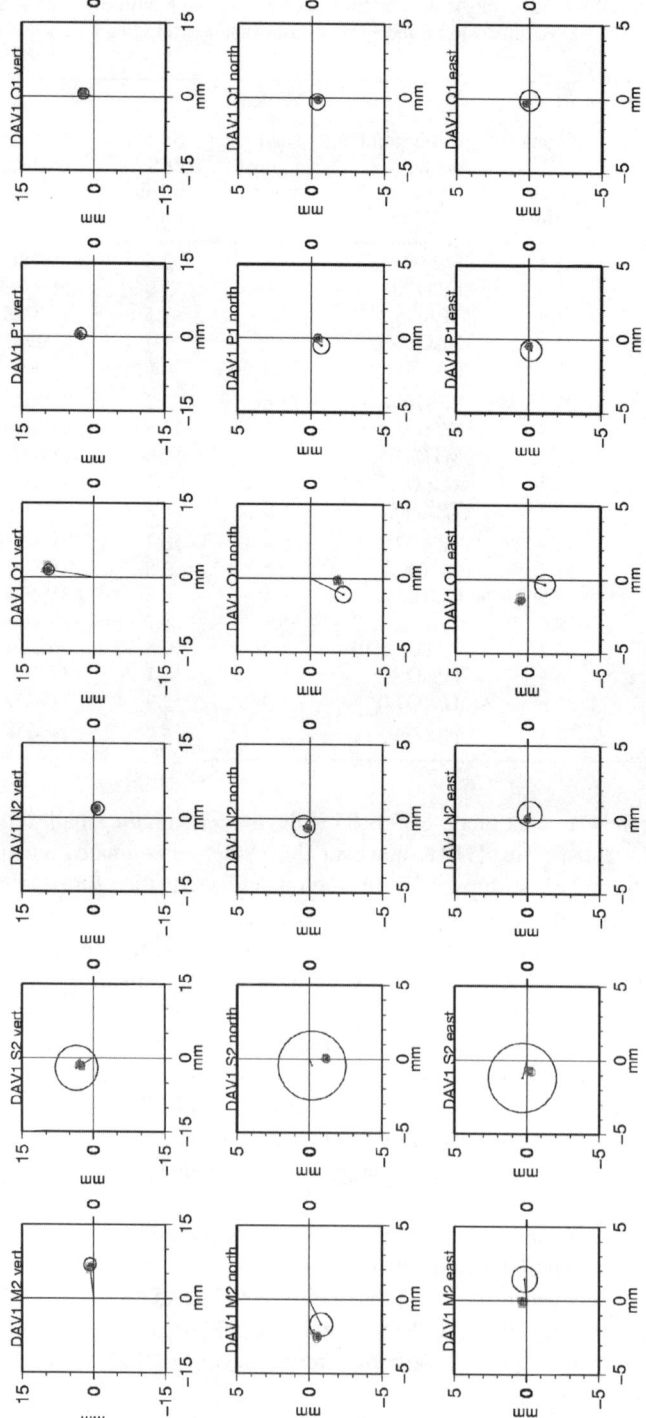

Fig. 2 GPS and model phasor plots of OTLD for station Davis (DAV1) for constituents M2, S2, N2, O1, P1, Q1. Symbols are as follows: Black vector – GPS estimate and 95% error ellipse; Triangles: red: CADA00.10 orange: CATS02.01 blue: AntPen04.01 green: MTOs.05; Squares: red: TPXO7.0 orange: GOT00.2 blue: FES2004 green: CSR4; Circles: red: TPX06.2 blue: FES99 green: NAO99b

A Validation of Ocean Tide Models

231

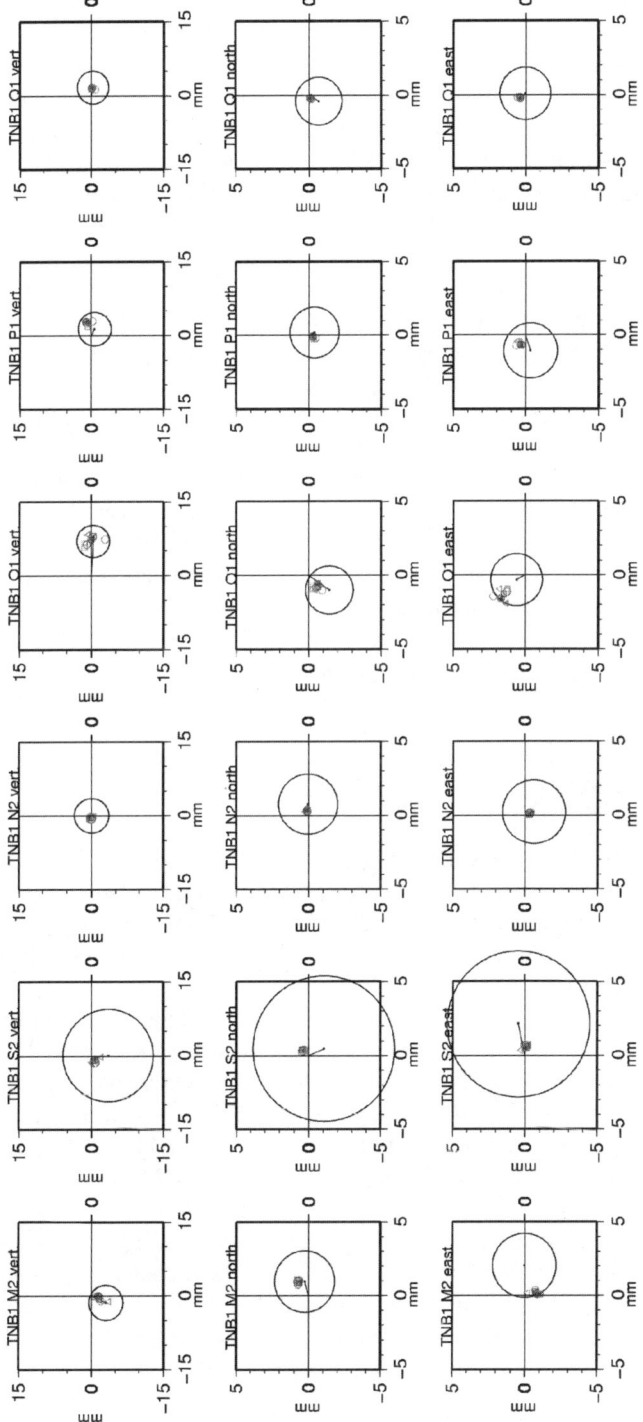

Fig. 3 Phasor plots for station Terra Nova Bay (TNB1). Note that symbols and scales used are the same as for Fig. 2

3.3.3 Antarctic Peninsula and Weddell Sea

Figure 4 shows phasor plots of GPS and modelled OTLD for San Martin (SMRT).

Along the Antarctic Peninsula and under the Filchner-Ronne Ice Shelves, the tides are predominantly semi-diurnal, with the M2 and S2 tidal constituents having the largest signals. Due to the ice shelves to the east of the peninsula, e.g. the Larsen Ice Shelf, and the topographically complex coastline to the west of the peninsula, the tides are not as well determined as in East Antarctica. This is apparent in the spread of modelled estimates in the phasor plots in Figure 4, and also explains why this region is one focus of attention in the ocean-modelling community, with a number of new regionally-tuned models, e.g., AntPen04.01.

The spread between the models at SMRT is most notable for the energetic M2 constituent, where the inter-model disagreement is of the order of several millimetres; for the vertical component the GPS estimate of OTLD disagrees with the CSR4 and NAO99b modelled estimates at the 95% level.

From Table 5, we see that the three models with the best agreement to GPS – TPXO7.0, CADA00.10 and TPXO6.2 – all offer very similar overall agreement for both three-dimensional and vertical. The AntPen04.01 and MTSo.05 models both offer good GPS /model agreement. The GOT00.2 also offers good agreement at San Martin and Palmer for three-dimensional OTLD.

We draw attention to FES2004, which offers surprisingly poor agreement in this region, both in three dimensions and vertically (Table 5).

4 Conclusions

This study has extended the work of King et al. (2005), using additional GPS sites, up-to date ocean models, and refined processing methods, including the resolving of carrier phase ambiguities to integer values. The resolving of ambiguities resulted in significantly poorer GPS / model agreement; the reason for this remains an open question for further investigation.

We have ordered the models region by region, based upon their agreement with the GPS estimates, coming to similar conclusions to those of King et al. (2005). For Antarctica as a whole, the TPXO6.2 model offers consistently good agreement with GPS estimates, with the CADA00.10 and CATS02.01 models also agreeing well. Of the newer models tested in this study, we suggest that TPXO7.0 model offers very good agreement at all sites / regions tested, with the MToS.05 also agreeing well. The FES2004 model offers very good agreement in East Antarctica and the Ross Sea region but is, surprisingly, one of the models offering poorest agreement in West Antarctica. FES99, CSR4 and NAO99b are the models that generally offer poorest agreement with GPS for Antarctica as a whole. There are still some regions that are poorly sampled by our present set of sites – the southern Ross and Filchner-Ronne ice shelves are where new data are most required.

A Validation of Ocean Tide Models

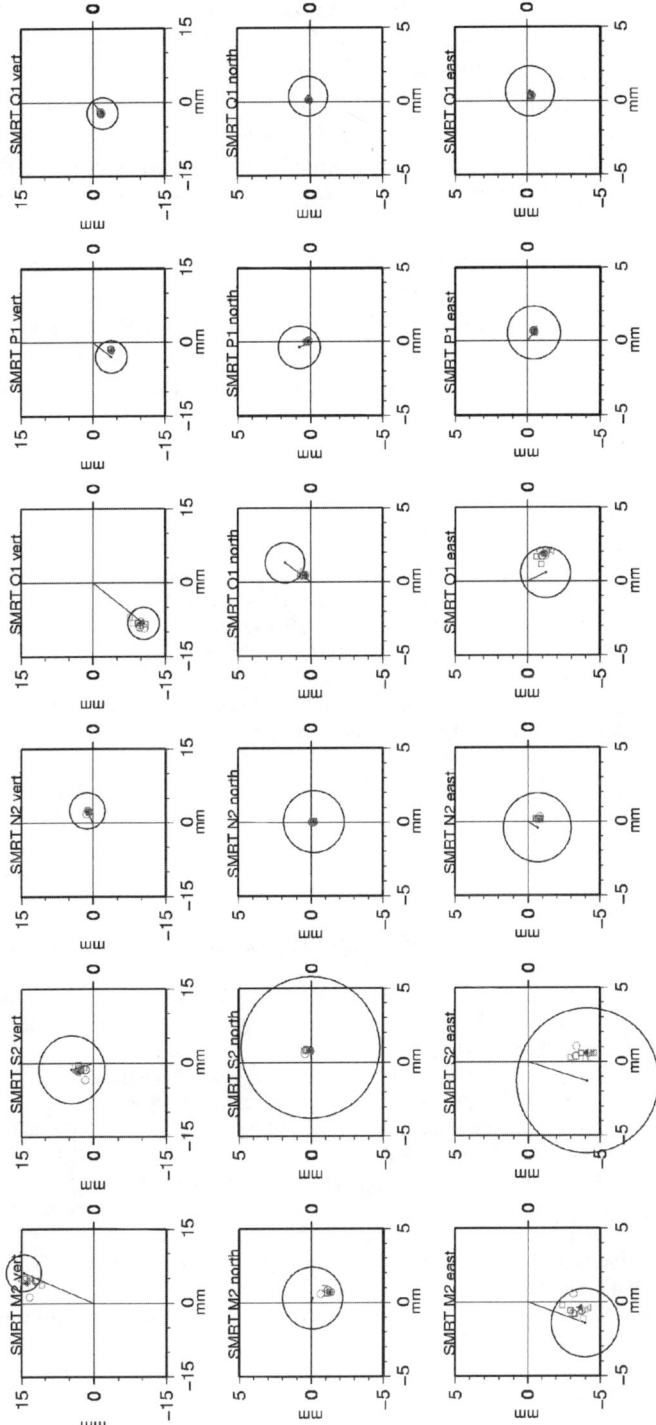

Fig. 4 Phasor plots for station San Martin (SMRT). Note that the symbols and scales used are the same as for Figs. 2 and 3

The refinements to the GPS estimation method made to the method used by King et al. (2005), have not significantly improved GPS / model agreement; we can conclude that the practical limit of accuracy achievable from the GPS strategy is being approached using present JPL satellite orbit, clock and earth orientation products. There still remain systematic GPS errors in the modelling at the solar frequencies S2, P1 as well as K1, K2. As previously pointed out by Thomas et al. (2006), the agreement between GPS and models at the lunar constituents, particularly the N2 and Q1 constituents, is generally very good, and this indicates the potential accuracies that could be achieved at the remaining constituents once the remaining, probably solar related, biases are eliminated. We await with interest the forthcoming global re-processing of IGS orbits, which will include additional improved models such as 2nd order ionosphere corrections. This will hopefully contribute towards a reduction in these biases, as should the forthcoming European GNNS, the Galileo System, with its non-solar orbital period.

Acknowledgments We would like to thank the following for making GPS data available to us: International GNNS Service (IGS) community; British Antarctic Survey (BAS); Laboratoire de Recherché en Géodésie, Ecole Nationale des Sciences Géographiques, Marne la Vallée, France; The Department of Geodetic Science and Surveying at The Ohio State University (OSU) (TAMDEF network); Alfred Wegener Institute, Bremerhaven; Italian Geodetic Project of PNRA (Programma Nazionale di Ricerche in Antartide) (VLNDEF network).

The NASA Jet Propulsion Laboratory (JPL) contributed by providing modified GIPSY routines, and their orbit, clock and Earth orientation products. The UK Engineering and Physical Science Research Council provided financial support to Ian Thomas through a Doctoral Training Account award to Newcastle University. Matt King was supported by a UK Natural Environment Research Council postdoctoral fellowship.

References

Agnew DC (1997). NLOADF: A program for computing ocean-tide loading. Journal of Geophysical Research-Solid Earth 102(B3):5109–5110.

Allinson CR, Clarke PJ, Edwards SJ, King MA, Baker TF, Cruddace PR (2004). Stability of direct GPS estimates of ocean tide loading. Geophys. Res. Lett. 31(15): doi:10.1029/2004GL020588, 2004.

Bos MS, Baker TF (2005). An estimate of the errors in gravity ocean tide loading computations. Journal of Geodesy 79(1–3):50–63, doi: 10.1007/s00190-005-0442-5.

Bouin MN, Vigny C (2000). New constraints on Antarctic plate motion and deformation from GPS data. Journal of Geophysical Research-Solid Earth 105(B12):28279–28293.

Casula G, Dubbini M, Galeandro A (2007). Modelling environmental bias and computing velocity field from data of Terra Nova Bay GPS network in Antarctica, by means of a quasi-observation processing approach. In 'Antarctica: A Keystone in a Changing World – Online Proceedings of the 10th ISAES USGS', Edited by Cooper, AK and Raymond, CR, Open-File Report 2007–1047, Short Research Paper 054; doi: 10.3133/of2007-1047.srp054.

Eanes RJ, Bettadpur SV (1995). The CSR 3.0 global ocean tide model. Center for Space Research, Technical Memorandum CSR-TM-96-05.

Egbert GD, Erofeeva L (2002). Efficient inverse modeling of barotropic ocean tides. Journal of Atmospheric and Oceanic Technolgy 19.

Farrell WE (1972). Deformation of Earth by Surface Loads. Reviews of Geophysics and Space Physics 10(3):761–797.
Han S, Shum CK, Matsumoto K (2005). GRACE observations of M2 and S2 ocean tides underneath the Filchner-Ronne and Larsen ice shelves, Antarctica. Geophysical Research Letters, VOL. 32, L20311, doi:10.1029/2005GL024296, 2005 32.
King M (2005). Kinematic and static GPS techniques for estimating tidal displacements with application to Antarctica. Journal of Geodynamics 41(1–3):77–86, doi:10.1016/j.jog.2005.08.019.
King M, Coleman R, Morgan P (2000). Treatment of horizontal and vertical tidal signals in GPS data: A case study on a floating ice shelf. Earth Planets Space 52:1043–1047.
King M, Coleman R, Nguyen LN (2003). Spurious periodic horizontal signals in sub-daily GPS position estimates. Journal of Geodesy 77(1–2):15–21, doi:10.1007/s00190-00002-00308-z.
King MA, Penna NT, Clarke PJ, King EC (2005). Validation of ocean tide models around Antarctica using onshore GPS and gravity data. Journal of Geophysical Research-Solid Earth 110(B8):B08401, doi: 10.1029/2004JB003390, 2005.
Konfal SA, Wilson TJ, Willis MJ (2007). GPS Surveys to detect active faulting in the Transantarctic Mountains, Antarctica. In 'Antarctica: A Keystone in a Changing World – Online Proceedings of the 10th ISAES', Edited by Cooper, AK and Raymond, CR, USGS Open-File Report 2007–1047, Extended Abstract 021.
Lefevre F, Yard FH, Le Provost C, Schrama EJO (2002). FES99: A global tide finite element solution assimilating tide gauge and altimetric information. Journal of Atmospheric and Oceanic Technology 19(9):1345–1356.
Matsumoto K, Takanezawa T, Ooe M (2000). Ocean Tide Models Developed by Assimilating TOPEX/POSEIDON Altimeter Data into Hydrodynamical Model: A Global Model and a Regional Model around Japan. Journal of Oceanography 56:567–581.
McCarthy DD, Petit Ge (2004). IERS Technical Note 32. IERS Conventions (2003), International Earth Rotation and Reference Systems Service (IERS), Frankfurt am Main, Germany: Verlag des Bundesamtes für Kartographie und Geodäsie.
Padman L, Fricker HA, Coleman R, Howard S, Erofeeva L (2002). A new tide model for the Antarctic ice shelves and seas. Annals of Glaciology 34:247–254.
Petrov L, Ma CP (2003). Study of harmonic site position variations determined by very long baseline interferometry. Journal of Geophysical Research 108(B4):2190, doi: 10,1029/2002JB001801, 2003.
Ray RD (1999). A Global Ocean Tide Model From TOPEX/POSEIDON Altimetry: GOT99.2. NASA Technical Memorandum 209478.
Schenewerk MS, Marshall J, Dillinger W (2001). Vertical ocean-loading deformations derived from a global GPS network. Journal of the Geodetic Society of Japan 47(1):237–242.
Thomas ID, King MA, Clarke PJ (2006). A comparison of GPS, VLBI and model estimates of ocean tide loading displacements. . Journal of Geodesy Online First.
Willmott V, Domack E, Padman L, Canals M (2007). Glacio-marine sediment drifts from Gerlache Strait, Antarctic Peninsula. Glacial Sedimentary Processes and Products, International Association of Sedimentologists (accepted).
Zumberge JF, Heflin MB, Jefferson DC, Watkins MM, Webb FH (1997). Precise point positioning for the efficient and robust analysis of GPS data from large networks. Journal of Geophysical Research 102(B3):5005–5017.

Continuous Gravity Observation with the Superconducting Gravimeter CT #043 at Syowa Station, Antarctica

K. Doi, K. Shibuya, H. Ikeda and Y. Fukuda

Abstract In April 2003, a new superconducting gravimeter CT #043 was installed at gravity observation hut in Syowa Station and TT-70 #016 was replaced with CT #043 at the end of 2003. Before the removal of the TT-70 #016, parallel observation with CT #043 and TT-70 #016 was carried out during the period from April 18 to November 4. A tidal analysis program BAYTAP-G was applied to the output signals from the both gravimeters for harmonic analysis, and tidal amplitude factors, phases and residuals were computed. Large instrumental drift over 200 µGal per year was observed in CT #043 record, but the drift could be eliminated accurately by fitting an exponential decaying function of first order. Tidal amplitudes and phases obtained from the both gravimeters data during the parallel observation period are in good agreement with each other. The detided gravity residuals agree both in variation pattern and amplitudes of the variation. This fact means reliability of the observation by CT #043 and adequacy of the manner to eliminate the instrumental drift. Tidal factors from CT #043 data after January 2004 were also compared with those obtained by previous studies using TT-70 #016 data, and they coincided with each other. From comparison of the gravity residuals from January through December 2004 with sea level trend, the obtained response coefficient to sea level change is about 0.07 µGal/cm.

K. Doi
National Institute of Polar Research, Kaga 1-9-10, Itabashi-ku, Tokyo 173-8515, Japan

K. Shibuya
National Institute of Polar Research, Kaga 1-9-10, Itabashi-ku, Tokyo 173-8515, Japan

H. Ikeda
Research Facility Center for Science and Technology Cryogenics Division,
University of Tsukuba, Tennodai 1-1-1, Tsukuba 305-8577, Japan

Y. Fukuda
Graduate School of Science, Kyoto University, Kyoto 606-8502, Japan

1 Introduction

Continuous observation with a superconducting gravimeter (SG) TT-70 #016 (hereinafter referred to as TT-70) started from March 22, 1993 to observe earth tides and earth's free oscillation (Sato et al. 1993) at a gravity observation hut (GOH) in Syowa Station (Fig. 1). Many results related to earth tides, earth's free oscillation and non-tidal signals were obtained by the observation. The overview is summarized in Shibuya et al. (2003).

In order to avoid accidents in making liquid helium and noise contamination to gravity signal during liquid helium transfers in about every six months, we decided to install a new SG of CT type with a 4-K cryocooler, which enable to liquefy gas helium. The new superconducting gravimeter CT #043 (hereinafter referred to as CT) was installed on a base for absolute gravity (AG) measurement in GOH at April 18, 2003 (Ikeda et al. 2005). Since then, parallel observation with the two SGs had been carried out until November 2003, when liquid helium in the dewar of TT-70 was exhausted and, therefore, observation by TT-70 terminated.

The CT was shifted to a base for SG observation at the end of 2003 from the AG base, and then AG measurements with two absolute gravimeters FG5 #210 and #203 were conducted for about two weeks in January 2004 on the AG basement by occupying the base alternatively in order to determine a scale factor of the CT as well as to monitor secular gravity change (Fukuda et al. 2005, Hiraoka et al. 2005).

In this manuscript, we show some results obtained from the observation with the CT for more than one year as well as those from the parallel observation by CT and TT-70. We aim to confirm the reliability of the observation with the newly installed SG by comparing the result of harmonic analysis from the one year observation with previously obtained results from TT-70. We also confirm the reliability of non-tidal

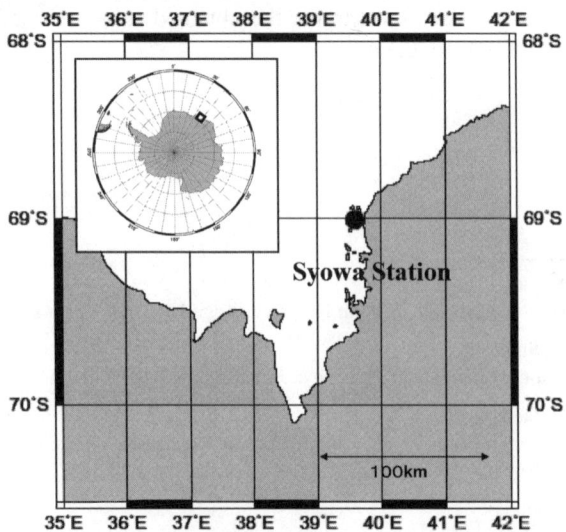

Fig. 1 Location of Syowa Station

signal by comparing residuals from CT with that from TT-70 using data during the parallel observation.

2 Tidal Analysis of the Data

2.1 Calibration of the SGs

The output of a SG gravimeter is the feedback voltage necessary to make a Niobium hollow sphere levitating in the magnetic field generated by a Niobium coil at the liquid helium temperature (4°K). A scale factor given in μGal/Volt is then needed to convert the output voltage of the feedback in units of acceleration of gravity. For this reason gravity measurements with an absolute gravimeter FG-5 were carried out three times in 1995 (Yamamoto 1996), 2001 (Kimura 2002), and 2004 (Fukuda et al. 2005, Hiraoka et al. 2005) on the base for AG measurement in GOH. The results from the measurements in 2001 and 2004 were used for determining the scale factors of TT-70 and CT SGs, respectively. The time series of simultaneous SG and AG observations are shown in Fig. 2 of Iwano et al. (2003) for SG TT-70 and in Fig. 1 of Fukuda et al. (2005) for SG CT. The evaluated calibration factors

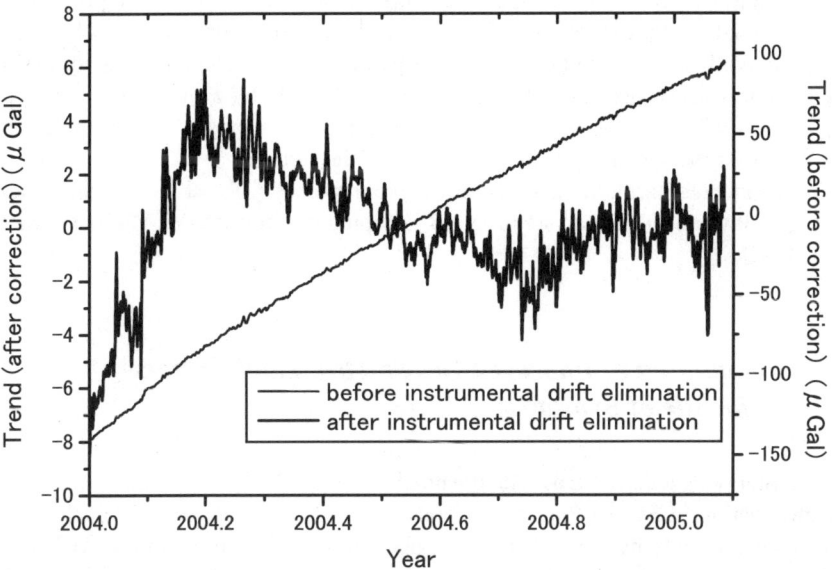

Fig. 2 Gravity residuals computed from data of the SG CT #043 in the period January 1, 2004–January 31, 2005. The residuals before eliminating the instrumental drift are shown by a thin line and those after subtracting it are shown by a thick line. The left vertical axis indicates scale for the trend after correcting the instrumental drift and the right vertical axis indicates that for the trend before the correction

Table 1 Scale factors of TT-70 #016 and CT #043

Gravimeter	2003.4.21–11.4	2004.1.1–2005.1.31
TT-70 #016	-58.168 ± 0.061	
CT #043	-59.453^a	-59.461 ± 0.079
unit: 10^{-8}m/s^2/Volt		

[a] Indirectly determined from the parallel observation by TT-70 #016 and CT #043.

by linear fittings are listed in Table 1. These values are employed for the following harmonic analysis. Incidentally, a scale factor value of -57.965 µGal/volt was used for the previous works (Sato et al. 1996, Tamura et al. 1997), which was determined by comparison of tidal amplitude of M2 tide observed by TT-70 with that observed by LaCoste & Romberg D-73 at GOH (Kanao and Sato 1995).

2.2 Preprocessing and Tidal Analysis

Using the scale factors in Table 1, hourly re-sampled output signals in voltage from CT and TT-70 SGs are converted to those in µGal (10^{-8}ms^{-2}). Since levitation of superconducting sphere of the SG was carried out several times during the installation procedure in 2003 and 2004, several steps induced by the levitation procedures are contained in the time series. To estimate magnitude of the steps, we, at first, removed synthesized earth tides and atmospheric pressure effect on gravity multiplying atmospheric pressure changes observed at GOH by a mean admittance of -0.3 (µGal/hPa), and then the differences between two averages of 10 samples before and after the step gave us the magnitude of the step.

We conducted harmonic analysis of the preprocessed CT and TT-70 data by applying a Baysian tidal analysis program grouping method (BAYTAP-G) (Tamura et al. 1991).

2.3 Computation of Improved Gravimetric Tidal Factors for the Site of Syowa

Syowa Station is located on an island named East Ongul Island and, therefore, gravity observation in Syowa Station is largely affected by ocean tides. We calculated ocean tides effects by a program for global ocean tidal correction 2 (GOTIC2) (Matsumoto et al. 2001) with global ocean tide models NAO99b (Matsumoto et al. 2000) and NAO99L (Takanezawa et al. 2001) and corrected the effects. We used latest version of GOTIC2, which contains fine mesh land–sea distribution data with a size of $2.25' \times 1.5'$ around Syowa Station. The gravimetric tidal factors corrected ocean tides effect are listed in Table 2 for the data period from January 1,

Table 2 Corrected tidal factors and ocean loading tides from NAO99b

Wave	Amplitude factor	Phase lag[a](°)	Amplitude (μGal)	Amplitude of loading tides (μGal)	Phase of loading tides[b](°)
Q1	1.155 ± 0.003	−0.152 ± 0.133	4.602 ± 0.012	0.594	20.771
O1	1.150 ± 0.001	−0.433 ± 0.026	23.923 ± 0.012	2.551	10.866
P1	1.144 ± 0.001	−0.255 ± 0.059	11.075 ± 0.012	0.676	5.996
K1	1.133 ± 0.000	−0.406 ± 0.018	33.163 ± 0.011	1.996	7.604
N2	1.186 ± 0.002	−2.273 ± 0.075	2.192 ± 0.003	0.464	14.901
M2	1.162 ± 0.000	−1.183 ± 0.016	11.221 ± 0.004	2.346	8.781
S2	1.140 ± 0.001	−1.372 ± 0.030	5.119 ± 0.004	1.622	−1.310
Mm	1.159 ± 0.013	0.61 ± 0.67	5.878 ± 0.072	0.242	−10.393
Mf	1.156 ± 0.004	0.37 ± 0.20	11.199 ± 0.040	0.421	−17.338
Mtm	1.148 ± 0.018	0.38 ± 0.94	2.148 ± 0.036	0.073	−23.424

[a] Local phase lag. Negative value means phase lag.
[b] Local phase. Negative value means phase lag.

2004 to January 31, 2005 as well as ocean tides effects at Syowa Station. In the Table 2, negative phase means lag and phases of ocean tides effects are shown in Greenwich phase.

BAYTAP-G (Tamura et al. 1991) can also evaluate response to atmospheric pressure changes simultaneously by inputting hourly atmospheric pressure time series measured at GOH as an associated data file. The mean real atmospheric pressure admittance estimated from the data of the whole period is -0.326 ± 0.005 μGal/hPa.

Tidal amplitude factors for diurnal and semi-diurnal tides derived from CT show good agreement with those from TT-70 indicated in Tamura et al. (1997) and Sato et al. (1996) after multiplying the previously obtained amplitude factors by a ratio of the two scale factors ($-58.168/-57.965$). Results for three long period tides (Mm, Mf, Mtm) are also compared with those obtained by Iwano et al. (2005). The amplitude factors and phase lags from CT coincide with those from TT-70 within their estimation errors.

2.4 The Residuals of SG CT #043

The BAYTAP-G package has an automated tool to output detided residuals, gravitational effect due to atmospheric pressure and other environmental effects. Long period tides, effect of polar motion on gravity, instrumental drift and non-tidal gravity change are contained in the output channel of residuals.

In order to retrieve the non-tidal gravity change, long period tides and effect of polar motion must be estimated. We computed long period tides using tidal factors obtained by Iwano et al. (2005) and effect of polar motion was estimated by the formula given in Wahr (1985) with IERS polar motion time series.

As mentioned above, output signal of CT contains large instrumental drift more than 200 µGal per year. Since the drift is not linear and it is gradually decreasing, it is difficult to discuss seasonal gravity change without carefully eliminating the drift. We employed an exponential decaying function of first order as a fitting function to the trend which is obtained by subtracting earth tides and effect of polar motion. The computed residuals from January 1, 2004 to January 31, 2005 is indicated by a thick line in Fig. 2 as well as a thin line indicates the residual before the elimination of the drift.

In Fig. 2, we can see the gravity rapidly increasing at the beginning of 2004 in the corrected residual channel. We think this is an irregular instrumental drift associated with migration of the SG from base for AG. In Fig. 2 we can also see local maximum of the trend at around mid February to March (2004.2) and local minimum at the beginning of August in 2004 (2004.7).

3 Discussion

3.1 Intercomparison Experiment Between CT #043 and TT-70 #016 SGs

In this section, we compare the tidal factors and trends from CT and TT-70 for the period of the parallel observation. We employed the data from April 21 to November 4, 2003 for the comparison.

The gravimetric tidal factor from TT-70 for the parallel observation period is also obtained by the BAYTAP-G (Tamura et al. 1991). The result is listed in Table 3 as well as the result from the CT data for the same period.

Tidal amplitude factors derived from CT for diurnal constituents agree with those from TT-70 within their estimation errors. Moreover, those for semidiurnal tides are also coincident with those from TT-70 except for M2 constituent. The difference is approximately 0.7 per cent and it can not be explained by errors in estimation and scale factor determination. Since the factor for M2 constituent from CT agrees with that derived from TT-70 in previous studies (Sato et al. 1996, Tamura et al. 1997) within 0.3 per cent, the difference may be induced by the degradation of data quality of TT-70 during the parallel observation period. The quality degradation was caused by increase of noise due to slight propagation of cold-head vibration induced by accretion of iced air in the cold-head neck.

Phase differences from theoretical earth tides observed by CT are coincident with those by TT-70 for diurnal and semidiurnal tides except for N2 tide. The phase difference for N2 tide is $1.716°$ and it is greater than the estimation errors. It may be also caused by the data quality degradation of TT-70 in the above mentioned period.

The trend of TT-70 from April 21 to November 4, 2003 is also obtained by the same procedure used for CT data. In the case of TT-70, linear instrumental drift was assumed and subtracted. The trend obtained from CT is shown in Fig. 3 by a thick line and that from TT-70 is represented by a thin line in the same figure.

Table 3 Tidal factors for the parallel measurement period (Apr. 21–Nov. 4, 2003)

Wave	CT #043			TT-70 #016		
	Amplitude factor	Phase lag[a] (°)	Amplitude (μGal)	Amplitude factor	Phase lag[a] (°)	Amplitude (μGal)
Q1	1.160 ± 0.011	−0.289 ± 0.492	4.620 ± 0.045	1.153 ± 0.018	−0.726 ± 0.808	4.592 ± 0.073
O1	1.148 ± 0.002	−0.446 ± 0.095	23.881 ± 0.044	1.149 ± 0.003	−0.465 ± 0.156	23.914 ± 0.072
P1	1.157 ± 0.005	−0.601 ± 0.213	11.199 ± 0.044	1.147 ± 0.007	−0.101 ± 0.353	11.105 ± 0.073
K1	1.134 ± 0.001	−0.316 ± 0.065	33.193 ± 0.040	1.134 ± 0.002	−0.488 ± 0.106	33.185 ± 0.065
N2	1.195 ± 0.006	−2.000 ± 0.252	2.208 ± 0.012	1.183 ± 0.010	−3.716 ± 0.422	2.187 ± 0.019
M2	1.163 ± 0.001	−1.256 ± 0.052	11.226 ± 0.012	1.155 ± 0.002	−1.393 ± 0.087	11.148 ± 0.020
S2	1.145 ± 0.003	−1.431 ± 0.099	5.141 ± 0.012	1.119 ± 0.004	−1.350 ± 0.168	5.027 ± 0.019

[a] Local phase lag. Negative value means phase lag.

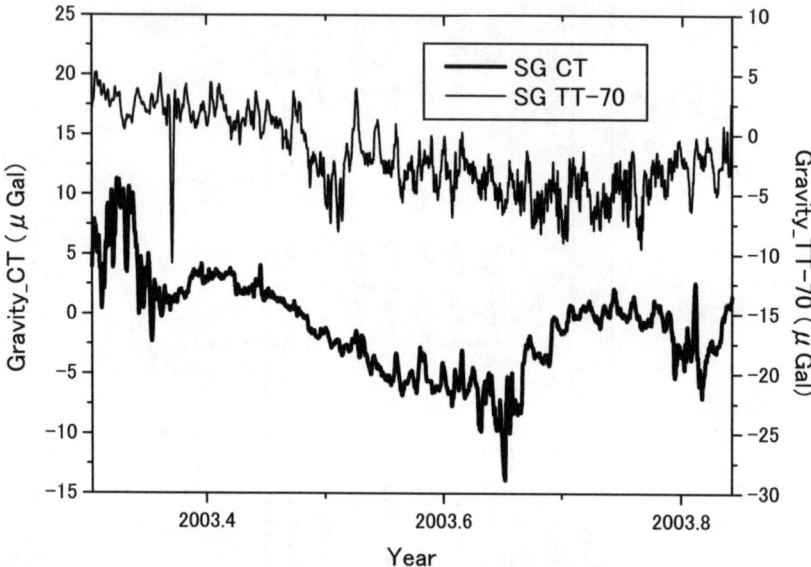

Fig. 3 Comparison of gravity residuals computed in the period April 21–November 4, 2003 using the data of the intercomparison experiment between the CT #043 and TT-70 #016 SGs. The left vertical axis indicates scale for SG CT #043 and the right vertical axis indicates that for SG TT-70 #016

The difference in gravity between raise and trough during the parallel observation is about 10 μGals for both trends. The trend curves from the two gravimeters are in phase each other in time scale of several months. They take relatively higher value at around June, 2003 (2003.40) and lower value at around September, 2003 (2003.65). The resemblance of the two trends means that the procedure used to estimate the instrumental drift was appropriate. Significant differences between the two trends are found at May (2003.33) and October (2003.75). The first difference at around May is considered to be the influence of startup procedure of CT. But the cause of the second difference is not clear until now.

3.2 Comparison with Sea Level Trend

Distance of GOH from the nearest coast is several hundreds meters. Hence seasonal sea level changes can also affect on the gravity significantly. We compared the gravity trend of CT for 12 months from January 1, 2004 with trend of sea level observed by a tide gauge at Nisi-no-ura Cove in Syowa Station. The sea level trend was derived using the package BAYTAP-G by almost the same procedure used for the gravity trend. The sea level trend is shown by a thick line in Fig. 4 as well as the gravity trend of CT by a thin line.

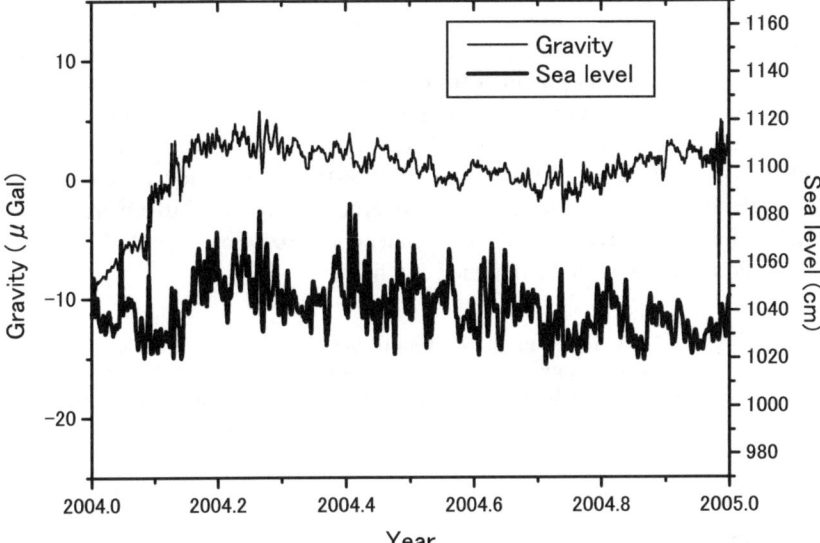

Fig. 4 Comparison of the gravity trend from CT #043 and sea level trend observed at Nisi-no-ura Cove in Syowa Station. The left vertical axis indicates scale for the gravity trend and the right vertical axis indicates that for the sea level trend

Response coefficient of gravity change to sea level change is 0.067 μGal/cm, which is derived by linear fitting using the whole data in Fig. 4. Standard deviation of gravity change time series is reduced to 2.62 μGal from 2.75 μGal by subtracting the influence of sea level trend.

4 Summary

A new superconducting gravimeter model CT number 043 was installed in April, 2003 on a base for absolute gravimetry in GOH at Syowa Station. After an intercomparison experiment lasted about seven months involving the two superconducting gravimeters CT-#043 and TT-70 #016, the TT-70 SG was switched off at the 4th of November 2003, and the CT one was moved to the base for SG at the end of the year.

We compared amplitude, amplitude factors and phases of the main tidal waves derived from the tidal analysis of data of the period of the two SGs parallel observation, and the residuals which were obtained by subtracting the modelled earth tides, which were obtained by subtracting earth tides, effect of polar motion and instrumental drifts, derived from data of the CT and TT-70 during the period of the parallel observation. The tidal factors from the two gravimeters indicate good agreement. The trends resemble each other in shape and the amounts of gravity change in the period are almost the same magnitude. The coincidence in the tidal factors and trends mean the reliability of observation with the gravimeter CT and the adequacy of the strategy used to eliminate instrumental drift.

Gravimetric tidal factors obtained from the CT data since January 2004 were also compared with those in previous studies using the TT-70 data. They were in good agreement. From the comparison of the gravity trend from January through December 2004 with trend of tide gauge data of the same period at Nisi-no-ura Cove, Syowa Station, the obtained response coefficient was 0.067 µGal/cm and the standard deviation of gravity change was reduced to 2.62 µGal from 2.75 µGal by correcting the influence with the obtained coefficient.

Acknowledgments We are grateful to all members of the JARE44 and 45 for their kind support in maintaining the SG observation in Syowa Station. This work was partially supported by a Grant-in-Aid for Scientific Research from the Ministry of Education, Culture, Sports, Science and Technology (No. 17540400).

References

Fukuda, Y., Iwano, S., Ikeda, H., Hiraoka, Y. and Doi, K. (2005): Calibration of the superconducting gravimeter CT #043 with an absolute gravimeter FG5 #210 at Syowa Station, Antarctica, Polar Geosci., 18, 41–48.

Hiraoka, Y., Kimura, I., Fukuda, Y., Doi, K. and Shibuya, K. (2005): Gravity measurements with the portable absolute gravimeter FG5 at Antarctica (III), Bull. Geogr. Surv. Inst., 108, 21–27 (in Japanese).

Ikeda, H., Doi K., Fukuda Y., Tamura Y. and Shibuya K. (2005): Installation of the superconducting gravimeter CT(#043) at Syowa Station, Antarctica, Polar Geosci., 18, 49–57.

Iwano, S., Fukuda, Y., Sato, T., Tamura, Y., Matsumoto, K. and Shibuya, K. (2005): Long-period tidal factors at Antarctica Syowa Station determined from 10 years of superconducting gravimeter data, J. Geophys. Res., 110, B10403, doi:10.1029/2004JB03551.

Iwano, S., Kimura, I. and Fukuda, Y. (2003): Calibration of the superconducting gravimeter TT70 #016 at Syowa Station by parallel observation with the absolute gravimeter FG5 #203, Polar Geosci., 16, 22–28.

Kanao, M. and Sato, T. (1995): Observation of tidal gravity and free oscillation of the Earth with a LaCoste & Romberg gravity meter at Syowa Station, East Antarctica, Proc. 12th Int. Symp. Earth Tides, ed. H. T. Hsu, Science Press, Beijing, 571–580.

Kimura, I. (2002): Gravity measurements with the portable absolute gravimeter FG5 at Antarctica (II), Bull. Geogr. Surv. Inst., 97, 17–23 (in Japanese).

Matsumoto, K., Sato, T., Takanezawa, T. and Ooe, M. (2001): GOTIC2: A Program for Computation of Oceanic Tidal Loading Effect, J. Geod. Soc. Jpn., 47, 243–248.

Matsumoto, K., Takanezawa, T. and Ooe, M. (2000): Ocean Tide Models Developed by Assimilating TOPEX/POSEIDON Altimeter Data into Hydrodynamical Model: A Global Model and a Regional Model around Japan, J. Oceanogr., 56, 567–581.

Sato, T., Shibuya, K., Nawa, K., Matsumoto, K. and Tamura, Y. (1996): On the diurnal and semidiurnal tidal factors at Syowa Station, Antarctica, J. Geod. Soc. Jpn., 42, 225–232.

Sato, T., Shibuya, K., Okano, K., Kaminuma, K. and Ooe, M. (1993): Observation of Earth tides and Earth's free oscillations with a superconducting gravimeter at Syowa Station (status report), Proc. NIPR Symp. Antarct. Geosci., 6, 17–25.

Shibuya, K., Doi, K. and Aoki, S. (2003): Ten years' progress of Syowa Station, Antarctica, as a global geodesy network site, Polar Geosci., 16, 29–52.

Takanezawa, T., Matsumoto, K., Ooe, M. and Naito I. (2001): Effects of the Long-period Ocean Tide on Earth Rotation, Gravity and Crustal Deformation Predicted by Global Barotropic Model – periods from Mtm to Sa-, J. Geod. Soc. Jpn., 47, 545–550.

Tamura, Y., Aoyama, Y. and Nawa, K. (1997): Gravimetric tidal factors at Syowa Station obtained from three-year observations with a superconducting gravimeter, Proc. NIPR Symp. Antarct. Geosci., 10, 1–10.

Tamura, Y. Sato, T., Ooe, M. and Ishiguro, M. (1991): A procedure for tidal analysis with a Baysian information criterion, Geophys. J. Int., 104, 507–516.

Wahr, J. M. (1985): Deformation induced by polar motion, J. Geophys. Res., 90, 9363–9368.

Yamamoto, H. (1996): Gravity measurements with the portable absolute gravimeter FG5 at Antarctica, Bull. Geogr. Surv. Inst., 85, 18–22 (in Japanese).

Tide Gauges in the Antarctic and sub-Antarctic: Instrumentation and Calibration Issues from an Australian Perspective

Christopher Watson, Roger Handsworth and Henk Brolsma

Abstract The measurement of sea level in the Antarctic and sub-Antarctic is of particular importance for a range of fundamental studies into the role of the region in the global climate system. As such, tide gauges remain a primary measurement tool for the oceanographic and geodetic communities, requiring accurate and uninterrupted estimates of sea level from a geographically diverse array of instrumentation. This paper provides a review of the Australian contribution to sea level observation in the Antarctic and sub-Antarctic, providing a detailed account of data availability and issues surrounding instrumentation and datum control. We also present a novel technique adopted by the Australian Government Antarctic Division (AAD) to achieve an in situ calibration of a tide gauge using a novel application of a GPS equipped buoy. First results from Davis station (68° 35′ S, 77° 58′ E) are presented as a case study with an emphasis on quantifying error sources within the gauge system. As the International Polar Year (IPY) 2007–2008 begins and as the temporal extent of Australian Antarctic and sub-Antarctic tide gauge data approaches nearly 15 years with near continuous operation, this provides a timely contribution to assist in maximising the scientific value of data acquired under the most demanding of conditions.

Keywords Antarctic sea level · pressure tide gauge · acoustic tide gauge · calibration · accuracy

---

Christopher Watson
Surveying and Spatial Science Group, School of Geography and Environmental Studies, University of Tasmania, Private Bag 76, Hobart, Australia 7001,
e-mail: cwatson@utas.edu.au

Roger Handsworth
Australian Antarctic Division, Channel Highway, Kingston, Tasmania,
Australia 7050, e-mail: Roger.Handsworth@aad.gov.au

Henk Brolsma
Australian Antarctic Division, Channel Highway, Kingston, Tasmania,
Australia 7050, e-mail: Henk.Brolsma@aad.gov.au

1 Introduction

Tide gauges remain the fundamental instrument used to observe and monitor changes in sea level caused by a range of phenomena varying in frequency from surface waves, tsunami, coastal seiche oscillations and tides through to secular changes in mean sea level. The analysis of these phenomena is often hampered by the irregular distribution of tide gauges (Douglas et al., 2001), clearly evident in the Southern Hemisphere. Records from the Permanent Service for Mean Sea level (PSMSL) highlight a significant bias towards the Northern hemisphere where some 91% of tide gauges with records spanning 40 years or more are located (PSMSL, 2004). In recent times, significant international effort has helped improve this distribution, with many tide gauges now operating throughout the sub-Antarctic islands and around the coastline of Antarctica.

It has long been recognised that the environmental conditions and sheer isolation that characterises Antarctica imposes several significant challenges for the successful operation of any form of tide gauge. Of primary difficulty, tide gauges must withstand significant seasonal variability in sea ice, often dictating that a passive observational technique is required to infer water level. As such, pressure gauges remain the gauge type of choice in Antarctic installations (see IOC, 1990; IOC, 1992 and Shih et al., 1995 for example). Active sensors, including acoustic gauges, and more recently radar style gauges have also been used with some success (see later), offering the advantage of direct sea level measurement yet incurring the obvious limitation of maintaining an ice free air/water interface and structural support fastened to the coastline and extending well above water level. Regardless of the observational technique, the demands placed on the data from the sparse Antarctic network are also exceptional. Issues of accuracy, continuity and datum stability are of fundamental importance for studies involving for example, water transport and the sea level response to the Antarctic Circumpolar Current and the Southern Annual Mode (Woodworth et al., 2006 and Aoki et al., 2002 for example), satellite altimeter calibration (Mitchum, 2000) and regional and global sea level change (Church et al., 2004). These demands will only increase given there is already evidence that sea level rise is accelerating (Church and White, 2006) and that current estimates of global sea level change are at the upper limits of previous Intergovernmental Panel for Climate Change (IPCC) predictions (Rahmstorf et al., 2007). Without doubt these issues increase the focus on Antarctica as a central and vital indicator of the Earth's response to a warming climate.

The provenance of Australian involvement in Antarctic sea level observation originates with Sir Douglas Mawson's 1911–1914 Australasian Antarctic Expedition (AAE). The expedition established gauges at Macquarie Island and Boat Harbour, Cape Denison observing for some ∼9 months and ∼4 months respectively (Doodson, 1939). Following the establishment of the Australian Antarctic Division (AAD) in 1947, various short periods of observations were undertaken throughout the 1940s and 1950s in continental Antarctica and the sub-Antarctic islands of Macquarie Island and Heard Island (see Summerson and Handsworth, 1995 for further details). The modern era of Australian involvement in continuous sea level observation in the region began in earnest in 1991 as part of a collaborative venture between the Australian Antarctic Division (AAD), the National Tidal Facility (NTF) at Flinders University and the Australian Surveying and Land Information Group (AUSLIG). The collaborative venture developed the Antarctic and Southern Ocean tide gauge network, with the primary objectives of defining a mean sea level datum for mapping and charting in addition to facilitating research into long term sea level change (Summerson and Handsworth, 1995). At the time the network was conceived, there was no "off the shelf" technology available that met the required specifications for monitoring secular changes in mean sea level without requiring major engineering works to avoid problems with rafted sea ice. This dictated that instrumentation be custom built to overcome these difficulties yet meet strict accuracy specifications. Sites for installation proved to be a compromise between oceanographic suitability and environmental guidelines, dictating that each installation is unique in many respects. The network remains in operation managed primarily by the AAD with assistance from the modern manifestations of the other founding bodies, the Bureau of Meteorology (BoM) – National Tidal Centre and Geoscience Australia. The network has remained in near continuous operation since the first gauge was installed at Mawson station in February 1992, and is the focus of this paper.

The paper is divided into four sections. First, tide gauge instrumentation utilised by the AAD is reviewed, with an emphasis on technical detail relating to long term stability. This is followed by a brief review of the AAD installations in the sub-Antarctic and Antarctic, with the third section focusing on the issue of datum control. Finally, a case study presented highlighting the development of a novel technique utilising GPS equipped buoys for tide gauge calibration.

2 Tide Gauge Instrumentation

The primary function of a tide gauge is to observe the separation distance between some fixed point (typically connected to ground based benchmarks) and the water surface. Many gauge types exist (see Pugh, 1987), all of which are constrained to some extent by the medium through which they observe or "sense" the water level. Such constraints include, for example, uncertainties in the water density or the speed of propagation of an acoustic pulse. Australian tide gauges in the Antarctic include bottom pressure gauges (BPGs) and acoustic systems, each presenting their own specific challenges in terms of accuracy, stability and cold environment operation. We focus largely on the BPG throughout this paper.

2.1 Bottom Pressure Gauges

The passive sampling operation of pressure gauges makes them most attractive for use in environmentally hostile areas where the accumulation of sea ice is a problem.

Sea level (h), relative to some fixed electronic centre within the pressure sensor is derived using the basic hydrostatic relationship (assuming fixed density and gravity along the water column):

$$h = \frac{(p - p_a)}{\rho g} \qquad (1)$$

where p is the total observed pressure at the submerged pressure point, p_a is the atmospheric pressure acting on the instantaneous water surface, ρ is the density of the water column and g is the local gravitational acceleration. The formula shown in (1) is an approximation given the water density is rarely constant throughout the water column, and the atmophseric pressure is practically not observed relative to the instantaneous water surface but at some fixed level above mean sea level. The range of pressure based tide gauges (see IOC, 1994; IOC, 2002 and Paroscientific, 2007 for example) can be categorised primarily as a function of the number of pressure sensors and their physical position with respect to the water surface. The so called "B gauge" system developed by the Proudman Oceanographic Laboratory utilises a series of three pressure transducers to provide precision datum control of the overall gauge system (Woodworth et al., 1996). This technique accommodates for effects such as instrumental drifts and stratifications in the density profile throughout the water column. Woodworth et al. (1996) conclude that the method is capable of providing at least subcentimetric, and potentially millimetric, precision datum control depending on the specifications of individual pressure sensors involved. The technique does however require the middle or "B" sensor to be installed within $\sim$0.1 m of the long term mean sea level which is often problematic in areas where sea ice accumulates. The system presented by Shih et al. (1995) provides a compromise and uses two pressure sensors located under water and separated by a fixed distance. The use of two sensors enables the determination of water density but does not rigorously reduce instrumental drift effects as discussed by Woodworth et al. (1996). A secondary category of pressure gauges include single transducer, BPGs, available for deployment at a range of depths. Single transducer systems offer a simplified self contained design, capable of running unattended and undisturbed by sea ice in the case of Antarctic installations. Single transducer BPGs however lack the ability to infer or observe accurate variations in water density.

The primary gauge system adopted by the AAD is a single transducer system utilising the Paroscientific PS-2 Digiquartz^{TM} sensor. The AAD sensor housing (signal processing, memory, communications and power supply) is a custom design described by Summerson and Handsworth (1995). The gauge is deployed within a concrete mooring (Fig. 1a) weighing approximately 700 kg, and typically situated in $\sim$7 m of water to avoid disturbance from floating sea ice.

Given the inherent difficulty in providing long term protection of telemetry cables at the sea/land interface, the AAD system is designed for autonomous use, with a battery design life of approximately 5 years and an internal ring buffer memory capable of storing two years of data. Communication with the gauge is achieved using an eletromagnetic induction interface that can be positioned over the head of the tide gauge (Fig. 1a and Fig. 2). In areas where sea ice is persistent over the winter

 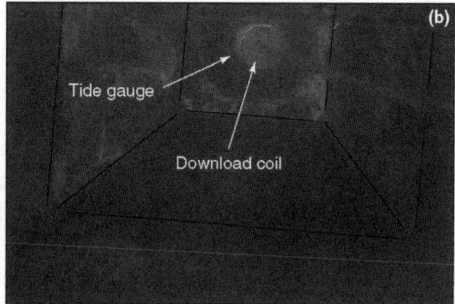

Fig. 1 Pressure gauge mooring used at the Australian Antarctic sites. (**a**) Shown before deployment, note the cylindicral hole for the tide gauge. (**b**) Shown submerged, the black lines have been added to improve definition of the mooring block. The download coil can be seen in position over the tide gauge

period, this induction coil may be left in place for regular download. In order to facilitate near real time download, the AAD are currently investigating the use of ultra low frequency radio communication that can simply be attached to the existing induction interface. Successful trials have been achieved using a pair of large diameter ($\sim$30 m) single turn loop antennae, one placed on the seafloor near the gauge, and the other on land as close as possible to the shoreline ($\sim$200 m). Bidirectional communication is achieved with a modulated carrier in the region of 76 kHz and antenna input power of 10–100 W.

The Paroscientific pressure sensor is a frequency output device. There are two outputs, one for pressure and one for temperature. Both outputs are pulse counted for 1 s and the resultant counts converted to a temperature compensated pressure value using the calibration data supplied with the sensor. This value is an integrated

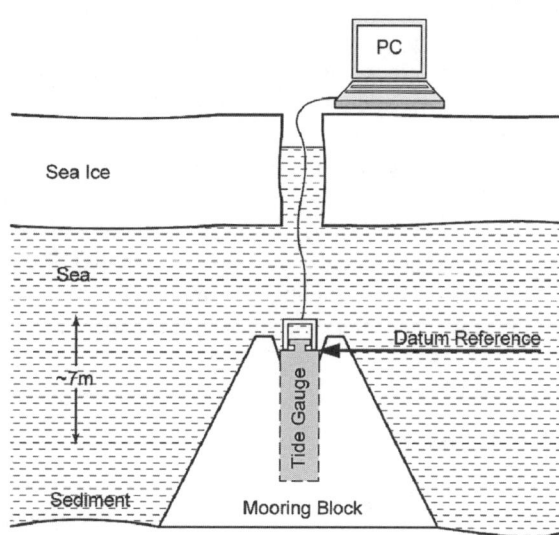

Fig. 2 Schematic illustration of the pressure gauge download configuration (not to scale). The induction coil used for the electronic interface is shown in place resting on the datum reference surface of the pressure gauge

pressure over the one second sample period. It is important to note that the electronic circuit is arranged such that no pulses are missed between successive 1 s integrations. For a non constant pressure input there will be a small error because the relationship between frequency and pressure is not linear. The resultant 1 s values are arithmetically averaged over the tide gauge sample period (typically 10 min). This 10 min average is stored in memory and represents a single sample. Because Antarctic gauges seldom experience large waves and are sufficiently deep that small surface waves are attenuated, the sampling technique avoids aliasing problems. After the gauge is downloaded, the 10 min pressure samples are converted to water pressure after subtracting interpolated barometric pressure observations, recorded relative to mean sea level at the meteorological station within 500 m of each gauge. These data are then converted to sea level above the gauge zero using a standard and fixed density of 1.028139 (i.e., density of water at 0°C and salinity 35 psu). Assuming a tidal range of 1 m, an error associated with using atmospheric pressure relative to mean sea level and not instantaneous level is at the ±0.5 mm level.

At this point it is important to distinguish between issues of precision and accuracy. A comprehensive treatment of the statistical behaviour of digital pressure water level gaues is provided by Carrera et al. (1996). The measurement precision is largely a function of the random errors associated with the sensor. Of arguably more interest to the oceanographic and geodetic communities is the system accuracy (relative to ground based benchmarks), which is significantly more difficult to determine due to contributions from systematic error sources within the sensor (sensor drift), the systematic nature of changes in water density (both temporally and spatially) and finally the differential motion of the gauge mooring with respect to fixed benchmarks. The PS-2 sensor used in the AAD pressure gauges has a stated pressure precision of 1 in 10,000 over the full temperature range (equating to ±2 mm water level). The output frequency of nominally 40 kHz changes by about 4 kHz over the full scale of 30 psi (2 bar). To maintain the claimed accuracy, the measuring clock needs to have a stability of somewhat better than 1 in 100,000. The measuring clock is derived from a carefully designed crystal oscillator that is also used to generate the real time clock. Drift in the clock can therefore be observed and quantified by observing drift in the real time clock. Gauges used in the Antarctic have generally kept time to better than 5 ppm. Drift within the quartz pressure sensor itself is difficult to quantify. Manufacturer tests from the 11–16 psi model show a median drift rate of −0.007 hPa/yr from trials of three sensors (Schaad, 2004). This translates to < 0.1 mm/yr effective sea level. Anecdotal evidence suggests higher drift rates are not uncommon and most likely at the level of the sea level change signal under observation. This highlights the importance of monitoring the gauge performance as discussed in a later case study in this paper.

2.2 Acoustic Gauges

The AAD also utilise acoustic reflection-time gauges at a number of installations in the sub-Antarctic and Antarctic. The principle relies on the measurement of the travel time of a pulse of sound to travel from a source fixed above the water level to the sea surface and return. In an Antarctic context, it must be noted that the acoustic pulse is reflected from the liquid water surface implying an ice free interface within the sounding tube. Despite the complexities surrounding the continuous heating of a sounding tube, the AAD have operated an experimental heated acoustic system at Mawson station since February 2003. Both the Mawson and Macquarie Island systems are installed on an incline through holes drilled in coastal bedrock to avoid damage by rafted ice at the shore. The angle of inclination is determined using terrestrial survey techniques. The dominant error contribution with an acoustic system is the uncertainty in the determination of the instantaneous velocity of sound in air. The Aquatrak sensors used by the AAD (Aquatrak, 2005) incorporate an automatic ratio based system of self calibration in an attempt to account for changes in the velocity of sound (as a function of temperature, atmospheric pressure and humidity). Hunter (2003) highlights that additional compensation is required in order to fully correct for non-linear thermal effects within the sounding tube and the sensor itself (a obvious requirement at the Mawson site given the heating required to maintain an ice free sounding tube). Temperature dependant errors are typically manifested by a range dependant error, as discovered at the sub-Antarctic installation at Macquarie Island. The calibration and accuracy assessment of this particular gauge is discussed in detail by Watson et al. (2008), using techniques that will also be applied to the Antarctic acoustic gauge installed at Mawson.

3 Australian Tide Gauge Installations

The AAD has permanent all year stations at Casey (66° 17′ S, 110° 32′ E), Davis (68° 35′ S, 77° 58′ E), Mawson (67° 36′ S, 62° 52′ E) and in the sub-Antarctic at Macquarie Island (54° 30′ S, 158° 57′ E). The AAD currently operates tide gauges at all permanently occupied Australian stations, with an additional gauge deployed in Nella Fjord (Larsemann Hills), about one kilometre from the Chinese station, Zhongshan (69° 23′ S, 76° 22′ E, also within Australian Antarctic Territory).

3.1 Sub Antarctic Installations

Macquarie Island is one of few sites in the Southern Ocean where tide gauge observations are possible. The earliest gauge installed on the island (1912, see Fig. 3) typifies the problematic conditions characterised by the harsh wave and atmospheric climatology, coupled with the jagged coastline.

The primary gauge at Macquarie Island is an Aquatrak acoustic sensor, with a backup single transducer Druck pressure sensor (model PDCR940), for details see Summerson and Handsworth (1995) and Tait et al. (1996). The calibration and accuracy assessment of the primary acoustic gauge was investigated in detail by Watson et al. (2008). This study showed that the acoustic system sea level data is biased

Fig. 3 Expeditioner Leslie Blake (cartographer) standing next to the original float tide gauge installed at Garden Bay, Macquarie Island, 1912. (Photograph by A. Sandell, from Doodson (1939))

both in offset and scale, with the offset error appearing to be sea state dependent indicating there is a potential draw-down effect within the gauge stilling well. The range dependent or scale error was found to be most probably related to temperature effects. The combined influence of these biases translates to the gauge over estimating the range to the sea surface by between ~33–115 mm depending on sea state (estimates determined at MSL, see Watson et al. (2008). Such systematic error is clearly significant for studies involving sea level change and absolute altimeter calibration, highlighting the importance of regular in situ calibration of unique installations such as that at Macquarie Island. The results from the Watson et al. (2008) study will be incorporated into future tide gauge processing and metadata records for future release (see metadata link in Brolsma, 1999).

The temporal availability of data from the Macquarie Island site is shown in Appendix A. Since the gauge began operation in December 1993, the record has been characterised by numerous small outages, most commonly caused by power failure. The site is remote from the station and uses solar power and lead acid storage batteries to power the data loggers. During storms it is not uncommon to have 1 to 2 m of water over the sensor and data logger installation. The solar panels are well clear of the water but suffer from being in a very corrosive environment. Some of the failures have occurred during unusually long periods of low sunlight levels. The AAD have recently tested a range of radio telemetry systems integrated into the station local area network which has enabled the remote operation of the gauge via the internet. Such operation facilitates the use of the gauge as part of the Australian tsunami early warning network.

3.2 Antarctic Installations

The first AAD bottom mounted pressure gauges were installed at Mawson and Casey (early 1992), followed by the gauge at Davis (early 1993). The first gauge at Casey was lost due to the sea ice freezing around a thin piece of rope attached to the mooring and shifting the mooring (a replacement was installed in early 1996). An identical BPG was also installed near Zhongshan in mid 2000. The temporal availability of data from these gauges is provided in Appendix B and shows only a small number of outages caused by a software related issues corrected in later gauge deployments.

Given the approximate battery life of five years, the operational plan at each station has been to download the gauge annually (and calibrate any clock offset in order to apply a linear drift correction), and install a second gauge system at each station prior to the initial gauge exhausting its battery. The simultaneous operation of two gauges enables the transfer of datum and the continuity of data (see for example the Mawson installation, Appendix B). The initial gauge can then be removed from it's mooring (either from a boat or preferably from the sea ice) and prepared for a future deployment. Before a re-deployment takes place, the mooring is checked for silt build up using an underwater video camera and any silt is removed using compressed air. The bottom of the gauge cavity in the mooring has a protruding peg designed to minimise the effects of silt build up, and contact with this peg is clearly felt during deployment. At Davis and Casey, logistical constraints prevented the secondary deployment in sufficient time before the primary gauge exhausted power (Appendix B). At Davis the gauge is 200 m from the shore and operation from a boat is very difficult. Conversely, when the ice is thick enough to operate from, light levels and visibility is often very low. There are therefore few opportunities when ice, wind and light conditions are favourable. Operation at Casey is made very difficult because the sea ice can break out at any time including midwinter, hence the ice is often not stable enough to operate from. The majority of tide gauge downloads at Casey have been made by AAD divers. This lack of data continuity is problematic and places the emphasis on accurate datum control transfer.

In addition to the BPG systems at each station, additional backup systems are in the process of being installed and connected to real time data interfaces. These secondary systems are designed purely as redundant systems to ensure a continuous record can be obtained into the future. The backup gauges at Mawson include a shore mounted Druck pressure gauge and an experimental shore based acoustic system, both installed through the same inclined hole drilled into the rock on East Arm (as per the Macquarie Island installation). The acoustic sounding tube is heated to ensure an ice free environment by heat traces and insulated using a double walled evacuated tube assembly. There are large temperature gradients down the tube, particularly where the bore is close to the rock surface. It will therefore most likely be necessary to install more sophisticated temperature control systems to minimise the errors due to temperature gradients, as discussed by Hunter (2003). The backup system at Casey is similar in principle to the system discussed by Shih et al. (1995) and will eventually replace the BPG as the primary system. The new Casey gauge has recently been installed in a 250 mm diameter vertical tube encased in the wharf with a short horizontal leg into the ocean. The construction of the wharf in 2005 has provided the first opportunity for a reliable shore based tide gauge system at Casey station. Each tube section in the gauge is lined with a double walled foam filled insulator and has a separate heat trace and pair of thermistors. In the vertical section are two Paroscientific PS-2 sensors separated vertically by 2 m and attached to a Monel rod linked to a reference point which is levelled to the local benchmarks. The two sensor installation allows the calculation of the water density. It is expected that the bore diameter of 150 mm is large enough to allow sufficient water exchange with the ocean to ensure that the water in the column is representative of the seawater at the inlet. However, the limitation caused by requiring the inlet well below the winter extent of sea ice is that the summer influx of surface melt-water may not enter the tube and hence the observed density may not reflect the actual average density, thereby biasing results if an exchange system is not developed.

4 Datum Control

One of the complexities of operating BPGs is the difficulty of monitoring the stability of the datum through regular and accurate survey connections to a surrounding network of tide gauge benchmarks (TGBMs). Changes in the datum may be caused by settlement of the gauge mooring and/or drift within the pressure sensor itself. A number of techniques have been trialled by the AAD, culminating in the GPS technique described in Watson et al. (2008) and applied for the first time to the Davis gauge below. Datum transfer via direct measurement is possible using a staff placed on the gauge reference point. This is complicated by the "blind" placement of the staff on the top surface of the gauge, the length of the levelling staff required ($\sim$7–9 m) and requirement to correct for thermal effects (expansion/contraction of the staff). A second technique involves timed water level measurements that can be made relative to local benchmarks. The technique requires exceptionally calm conditions and a staff that is temporally fixed in the water, close to the gauge location. The timed water level measurements (taken either manually or with the aid of a video system) can then be compared against the gauge observations. An extension of this technique is the application of kinematic GPS techniques. While this technique for water level measurement is not new, the application for the regular assessment of tide gauge performance is a novel application and has direct implications for monitoring the stability and accuracy of sub-Antarctic and Antarctic tide gauges. Early applications of the technique presented in the literature (for different purposes) include Aoki et al. (2002), who observed sea level using GPS at the Japanese station, Syowa, and achieved an RMS difference with respect to a BPG of $\sim$7 mm.

The extension of this technique (Watson et al., 2008), utilises a floating GPS buoy situated as close as possible to an inclined acoustic gauge, or in the case of the Antarctic gauges, directly above the BPG (the methodology is shown in schematic form in Fig. 4). The in situ technique relies on the closed loop between the GPS reference station (and associated TGBMs), the tide gauge, the water level and the GPS buoy in order to monitor the datum and error sources of the tide gauge as a complete system. The GPS buoy design was originally developed for satellite altimeter calibration as discussed in Watson et al. (2004).

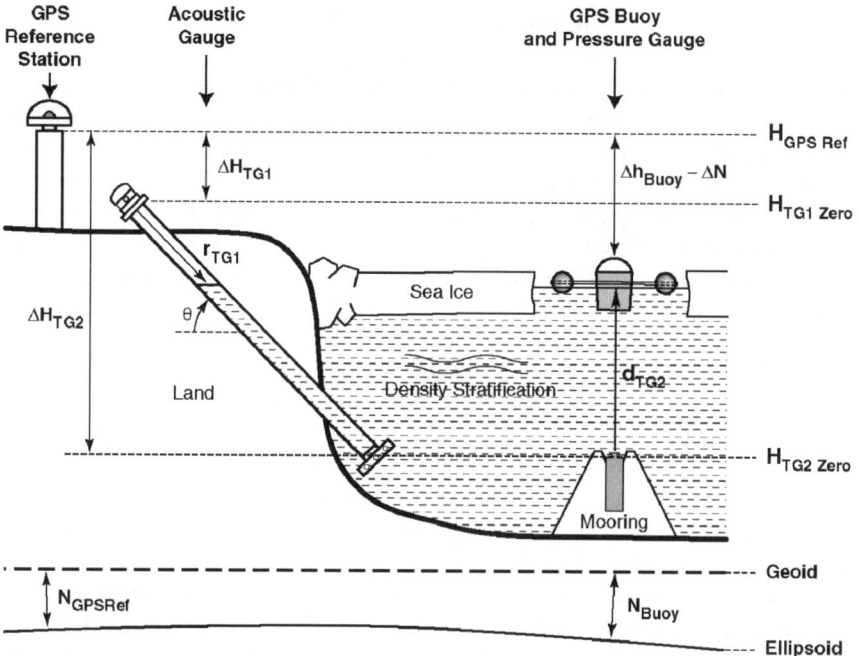

Fig. 4 Schematic illustration (not to scale) of the calibration methodology adopted at Australian Antarctic and sub-Antarctic tide gauge installations. Note it is important to distinguish between orthometric height (H), ellipsoidal height (h) and ellipsoid/geoid difference (N)

As shown in Fig. 4, raw tide gauge observations (r_{TG1} or d_{TG2} respectively) are observed relative to the electrical centre of the gauge instrument. Direct levelling connections to TGBMs and primary GPS monuments (ΔH_{TG1} and ΔH_{TG2} in Fig. 4, see Geoscience Australia, 2007) are then required to transform the tide gauge data to a fixed, absolute datum. In the case of the inclined acoustic gauges, the tide gauge zero is directly accessible, hence the GPS buoy technique is capable of determining range dependent and vertical offset errors within the tide gauge itself, solved for by forming a regression between the GPS buoy sea level data and the tide gauge data (see Watson et al. 2008 for further discussion). In the context of BPGs, the orthometric height connection to the tide gauge zero (ΔH_{TG2}) is not reliably known. Using the ellipsoid as the datum, the regression therefore becomes:

$$h_{Buoy} = \alpha \cdot d_{TG2} + \beta \qquad (2)$$

where h_{Buoy} is the filtered ellipsoidal height of the sea surface determined using the GPS buoy, d_{TG2} is the BPG raw observations (converted to depth), α is an (optional) range dependent term and β is a collective term referring to the offset of the tide gauge zero from the ellipsoid in addition to any residual offset error within the gauge. A component of β will include errors incurred due to any assumption made in reducing the pressure data using a fixed water density.

In order to monitor the performance of the AAD tide gauges, buoy deployments of at least 48 h duration are planned over each summer season in Antarctica. In order to achieve the highest accuracy, it is vital that utmost care is taken with respect to the determination of the GPS buoy antenna height above mean water level, and that identical antennae and radomes are used at either end of the GPS baseline (see Watson, 2005). Any variations in the density of the seawater from the nominal value used in the data reductions will affect the tide gauge observations and hence bias the datum connection. Studies such as Allison et al. (1985) show significant annual variation in salinity (and hence density) at Antarctic locations due to the interaction of the freeze/thaw cycle, in addition to influxes of freshwater melt draining from the Antarctic ice cap in the summer period. For this reason, a temperature and conductivity profiling instrument manufactured by Alex Electronics Co Ltd (Compact TD model) has been acquired to make in-situ density measurements at the time of each GPS buoy deployment. The device is claimed to have an in situ density precision of ± 40 ppm (given observational precisions of $\pm 0.02°C$ in temperature and ± 0.02 ms/cm in conductivity respectively), reducing the error associated with assuming a fixed density over a ~ 7 m gauge depth to within 0.5 mm. The seasonal density signal discussed by Allison et al. (1985) will however continue to bias sea level records from single transducer BPGs in Antarctic waters. CTD data acquired over a two year period from O'Gorman Rocks near Davis station area (A. Davison, pers. comm. 2007) shows a dynamic range of density estimates at the 1 part per 1000 level. This translates to a quasi seasonal error in sea level observations with a range at the 7 mm level.

4.1 Accuracy and Datum Control – A Case Study at Davis Station

The GPS buoy technique successfully applied at Macquarie Island was first tested in Antarctica at Davis station as a proof of concept trial. A 2 m square hole was cut into the sea ice allowing the GPS buoy to operate directly above the BPG (Fig. 5). Dual frequency geodetic quality GPS data was acquired at 10 s intervals at both the buoy (Leica MC500 receiver and AT504 antenna) and an inland reference station ~ 560 m away (Trimble 5700 receiver and Zephyr antenna). Note that logistical constraints at

Fig. 5 GPS buoy deployed over the Davis bottom mounted pressure gauge (Photograph by Richard Coleman)

the time prevented the necessary use of identical receivers and antennas at either end of the baseline, and given the use of the Trimble instrument, memory limitations dictated the use of 10 s data instead of the preferred 1 Hz data acquisition rate). Power limitations dictated only ~16 h of data was acquired during the trial deployment.

Processing of the GPS buoy data followed the procedure discussed in Watson et al. (2008), with the exception of the filtering strategy. To ensure concurrent and equivalent estimates of sea level, a 10 min moving average filter was used, matching the gauge sampling strategy as previously discussed. The GPS buoy time series (raw and filtered) and tide gauge time series (Fig. 6a), clearly shows the underlying tidal signal. Also evident is a higher frequency signal (period ~30 min, see 14:00 UTC for example) most likely representative of a seiche oscillation often observed within the tide gauge data. A number of spurious GPS results are also obvious (see 08:00

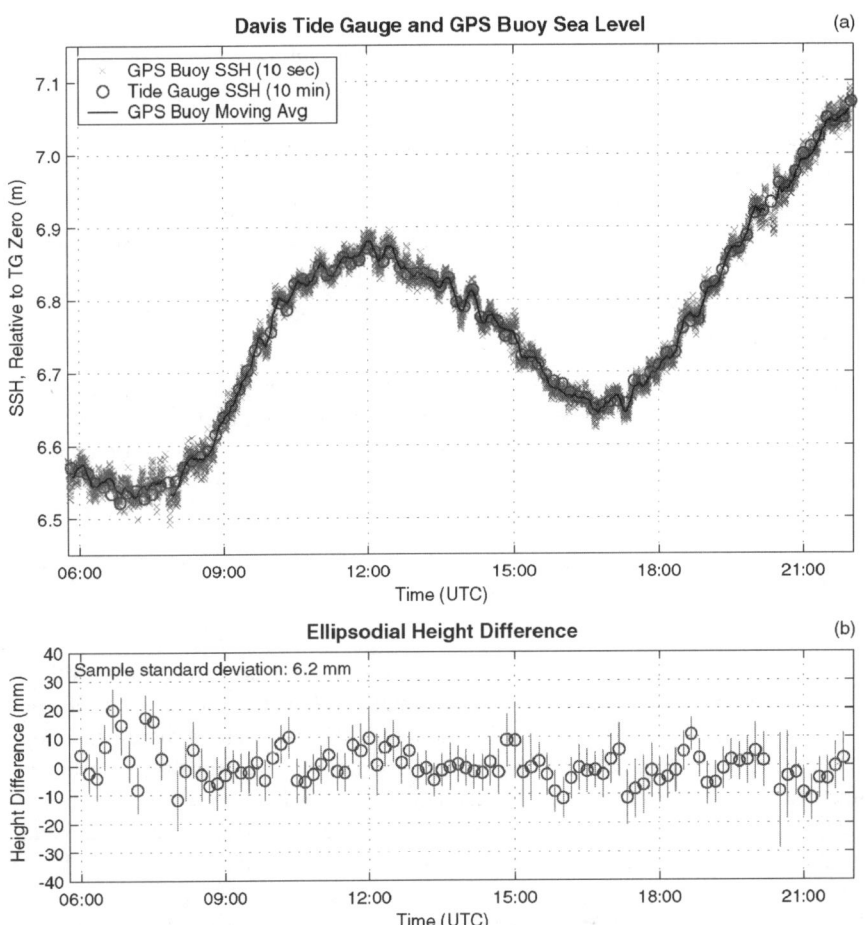

Fig. 6 (a) GPS buoy Sea Surface Height (SSH), filtered and unfiltered, and tide gauge SSH. (b) Difference time series (GPS buoy – tide gauge) with mean offset, β, removed

UTC for example). These are indicative of marginal satellite constellation geometry at the time.

As the temporal extent of the collected data in this trial did not span multiple cycles of high and low water, the range dependent term, α, in (2) was not solved. The residuals following the removal of the mean offset, β, are shown in (Fig. 6b). The RMS of these residuals is 6.2 mm, reducing to 5.1 mm with the exclusion of the spurious data between 07:00 and 08:15 UTC. Some level of serial correlation within the residual time series is clear and requires further invesigation. The low variability of this residual time series highlights the potential of the technique for monitoring the Antarctic BPGs. The absolute accuracy of the technique is difficult to quantify given the constraints introduced due to logistics in this trial, but are influenced by the following preferred approaches to the technique:

- Baseline length: Preferably < 100 m allowing L1/L2 solutions without the need to form the ionosphere free LC observable or solve for any tropospheric delay.
- Data duration: Preferably a minimum of 48 h of contiguous GPS and tide gauge data.
- Data rate: Preferably 1 Hz.
- Observation environment: Base station to be situated in ideal non-multipath environment as close as possible to the tide gauge.
- Antennae/radomes: Must be identical at both ends of the baseline.
- GPS buoy antenna height: Must be determined in situ by floating the buoy in water of identical density to that over the tide gauge.
- Water density: Must be determined in situ to avoid contribution to the error budget > 0.5 mm.

Given each of these criteria, the standard error about the mean offset over an entire deployment is likely to be at the 2–3 mm level. Given the limitations of other techniques, this technique provides an additional and superior practical methodology to monitor the combined gauge datum and performance (drift) over time.

5 Conclusions

The operation of tide gauges in the sub-Antarctic and Antarctic is clearly difficult and problematic due to a host of environmental and logistical constraints. The acquisition of some ~15 years of near continuous sea level data is therefore a considerable achievement, bettered only by the operation of the real-time or near-real time installations currently being tested. The need to fully understand the accuracy and stability of these gauges does however remain, particularly given the demanding scientific applications such as long term mean sea level change currently being investigated.

Single transducer, bottom mounted pressure gauges have proved highly successful in Antarctic installations, yet are limited in terms of their ability to accurately connect the tide gauge zero to inland benchmarks. The gauges also lack the ability to infer or observe in situ water density; hence quasi seasonal signals in water density will bias the sea level record periodically at the < 10 mm level. The long term drift rate of the Paroscientific sensor is also an issue given it has the potential to mask the underlying sea level signal under investigation.

Given the difficultly of applying traditional survey techniques, GPS buoy deployments coupled with accurate estimates of water density provide an in situ methodology capable of monitoring the datum and drift of the tide gauge system. The AAD have adopted a regular program of GPS buoy deployments that will require careful analysis over time in order to place realistic error bounds on the rate of change of the emerging sea level record.

Appendix A

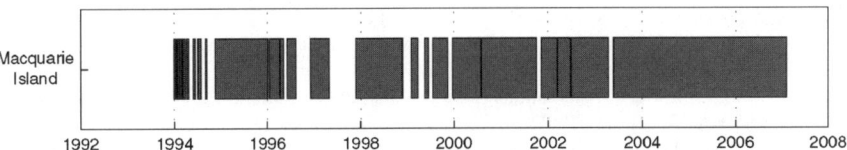

Appendix A. Availability of modern era acoustic sea level data at Macquarie Island

Appendix B

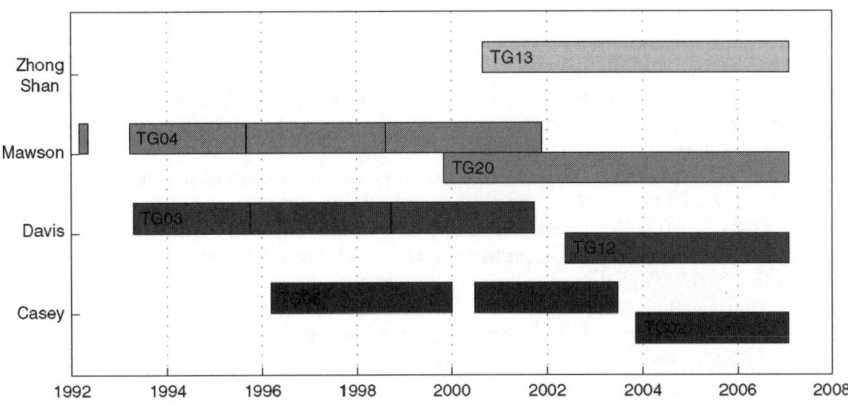

Appendix B. Availability of modern era bottom mounted pressure tide gauge data from Antartic sites

References

Allison, I., Tivendale, C. M., and Copson, G. R. (1985). "Annual Salt and Energy Budget Beneath an Antarctic Fast Ice Cover." *Annals of Glaciology*, 6, pp 182–186.

Aoki, S. (2002). Coherent sea level response to the Antarctic Oscillation. *Geophysical Research Letters*, 29(20), doi:10.1029/2002GL015733.

Aoki, S., Shibuya, K., Masuyama, A., Ozawa, T., and Doi, K. (2002). Evaluation of Seasonal Sea Level Variation at Syowa Station, Antarctica, Using GPS Observations. *Journal of Oceanography*, 58(3), pp 519–523.

Aquatrak (2005). Aquatrak Acoustic Liquid Level Sensors. Aquatrak Corporation, Available at: http://www.aquatrak.com.

Brolsma, H. (1999, updated 2006). Tide Gauge Records, Macquarie Island, Australian Antarctic Data Centre, CAASM Metadata (/aadc/metadata/).

Carrera, G., Tessier, B., and O'Reilly, C.T. (1996). Statistical Behaviour of Digital Pressure Water Level Gauges. *Marine Geodesy*, 19, pp 137–163.

Church, J.A. and White, N.J. (2006). A 20th Century Acceleration in Global Sea Level Rise. *Geophysical Research Letters*, 33, L01602, doi:10.1029/2005GL024826.

Church, J.A., White, N.J., Coleman, R., Lambeck, K., and Mitrovica, J.X. (2004). Estimates of the Regional Distribution of Sea Level Rise over the 1950 to 2000 Period. *Journal of Climate*, 17(13), pp 2609–2625.

Doodson, D. (1939). Tidal Observations, Part II, Australian Antarctic Expedition Scientific Reports, Series A, Vol. II, pp 61–85.

Douglas, B.C., Kearney, M.S., and Leatherman, S.P. (2001). Sea Level Rise: History and Consequences. International Geophysics Series, 75, Academic Press, San Diego.

Geoscience Australia (2007). Connections to Antarctic Tide Gauge Benchmarks, Available at: http://www.ga.gov.au/geodesy/antarc/antgauge.jsp.

Hunter, J.R. (2003). On the Temperature Correction of the Aquatrak Acoustic Tide Gauge. *Journal of Atmospheric and Oceanic Technology*, 20(8), pp 1230–1235.

IOC (1990). IOC Workshop on Sea level Measurements in Antarctica. Workshop Report No. 69, Intergovernmental Oceanographic Commission, Leningrad, USSR, 14 pp.

IOC (1992). IOC Workshop on Sea level Measurements in Antarctica. Workshop Report No. 69 (Supplement), Intergovernmental Oceanographic Commission, Leningrad, USSR, 102 pp.

IOC (1994). Manual on Sea level Measurement and Interpretation, Volume 2 – Emerging Technologies. Manuals and Guides No. 14, Intergovernmental Oceanographic Commission, 52 pp.

IOC (2002). Manual on Sea level Measurement and Interpretation, Volume 3 – Reappraisals and Recommendations as of the Year 2000. Manuals and Guides No. 14, Intergovernmental Oceanographic Commission, 55 pp.

IPCC (2007). Climate Change 2007: The Physical Science Basis, Summary for Policymakers. Contribution of Working Group I to the Fourth Assessment Report of the Intergovernmental Panel on Climate Change. 18 pp. Available at: http://www.ipcc.ch/.

Mitchum, G.T. (2000). An Improved Calibration of Satellite Altimetric Heights Using Tide Gauge Sea levels with Adjustment for Land Motion. *Marine Geodesy*, 23(3), pp 145–166.

Paroscientific (2007). Digital Quartz Pressure Transmitters for Accurate Water Level Measurements. Paroscientific Inc. Available at: http://www.paroscientific.com/waterlevel.htm.

PSMSL (2004). Geographical Distribution of PSMSL Stations. Permanent Service for Mean Sea level, Proudman Oceanographic Laboratory, Available at: http://www.pol.ac.uk/psmsl/images/allpsmsl.gif.

Pugh, D. (1987). Tides, Surges and Mean Sea level: A Handbook for Engineers and Scientists. John Wiley and Sons, Chichester, 472 pp.

Rahmstorf, S., Cazenave, A., Church, J.A., Hansen, J.E., Keeling, R.F., Parker, D.E., and Somerville, R.C.J. (2007). Recent Climate Observations Compared to Projections. *Science*, Published online 1 February 2007 [DOI: 10.1126/science.1136843] (in *Science Express Brevia*).

Schaad, T. (2004). Fifteen-Year Test of Barometers Long-Term Stability. Paroscientific Inc. Available at: http://www.paroscientific.com/longtermstab.htm.

Shih, H.H., Moss, M.K., and Dixon, J.C. (1995), A New Water Level Gauge for Cold Region Application in Offshore Mechanics and Arctic Engineering, The American Society of Mechanical Engineers, Book No. H00938, pp 65–76.

Summerson, R.M.V. and Handsworth, R.J. (1995). "Instrumentation for Sea level Measurement in Antarctica and the Southern Ocean." in PACON '94, Recent Advances in Marine Science and Technology, (O. Bellwood, H. Choat, and N. Saxena, eds.), James Cook University of North Queensland, Australia, pp 375–384.

Tait, M.E., Summerson, R.M.V., Lennon, G.W., and Handsworth, R.J. (1996). "Macquarie Island – a Platform for Monitoring Sea level and its Role in Climate Variability." in Ocean Atmosphere Pacific Conference, 23–27 October, Adelaide, Australia, pp 345–351.

Watson, C. (2005). Satellite Altimeter Calibration and Validation Using GPS Buoy Technology. Thesis for Doctor of Philosophy, University of Tasmania, Hobart, Australia, 264 pp. Available on-line at: http://adt.lib.utas.edu.au/public/adt-TU20060110.152656/.

Watson, C., Coleman, R., and Handsworth, R. (2008). Coastal Tide Gauge Calibration: A Case Study at Macquarie Island Using GPS Buoy Techniques. *Journal of Coastal Research*, 24(4), pp 1071–1079.

Watson, C., White, N., Coleman, R., Church, J., Morgan, P., and Govind, R. (2004). TOPEX/Poseidon and Jason-1: Absolute Calibration in Bass Strait, Australia. *Marine Geodesy*, 27(1), pp 285–304.

Woodworth, P.L., Hughes, C.W., Blackman, D.L., Stepanov, V.N., Holgate, S.J., Foden, P.R., Pugh, J.P., Mack, S., Hargreaves, G.W., Meredith, M.P., Milinevsky, G., and Contreras, J.J.F. (2006). Antarctic Peninsula Sea levels: A Real-Time System for Monitoring Drake Passage Transport. *Antarctic Science*, 18(3), pp 429–436.

Woodworth, P.L., Vassie, J.M., Spencer, R., and Smith, D.E. (1996). Precise Datum Control for Pressure Tide Gauges. *Marine Geodesy*, 19(1), pp 1–20.

Tidal Gravimetry in Polar Regions: An Observation Tool Complementary to Continuous GPS for the Validation of Ocean Tide Models

Mirko Scheinert, Andrés F. Zakrajsek, Sergio A. Marenssi,
Reinhard Dietrich and Lutz Eberlein

Abstract Ocean tides are an important phenomenon that has to be considered in a variety of geoscientific investigations in polar regions. Using ocean tide models, appropriate corrections can be calculated and applied in the different analyses, e.g. to infer temporal mass variations from satellite data, to investigate ice-shelf dynamics and deformation, to come up with precise station positions and velocities within a terrestrial reference system. However, those models show larger uncertainties in the polar seas and at areas covered by ice shelves and, therefore, need to be improved. Within this context, tidal gravimetry provides a powerful tool to validate ocean tide models. Sensing the cumulative signal of different effects (solid earth tides, ocean tidal loading, atmospheric loading) tidal gravimetry can be regarded as an independent observation technique and thus complements continuous GPS measurements. This paper reviews the background and the work done so far by the authors in Greenland and in Antarctica. Within a current project tidal gravimetry is applied at the Argentine Antarctic stations Belgrano II and San Martín. This joint Argentine-German project is a contribution to the IPY project POLENET. The set-up of the instruments in Belgrano II as well as first results of the ongoing observations, started February 2007, are presented. We discuss the problem of observing ocean tides in the area of the glacier near Belgrano II, which is – at least in parts – freely floating.

Mirko Scheinert
Technische Universität Dresden, Institut für Planetare Geodäsie, 01062 Dresden, Germany

Andrés F. Zakrajsek
Dirección Nacional del Antártico, Instituto Antártico Argentino, Cerrito 1248 (C1010AAZ), Buenos Aires, Argentina

Sergio A. Marenssi
Dirección Nacional del Antártico, Instituto Antártico Argentino, Cerrito 1248 (C1010AAZ), Buenos Aires, Argentina

Reinhard Dietrich
Technische Universität Dresden, Institut für Planetare Geodäsie, 01062 Dresden, Germany

Lutz Eberlein
Technische Universität Dresden, Institut für Planetare Geodäsie, 01062 Dresden, Germany,
e-mail: scheinert@ipg.geo.tu-dresden.de

Keywords Tidal gravimetry · GPS · ocean models · Antarctica · Greenland

1 Introduction

One of the major goals of the International Polar Year 2007/2008 (IPY), which has been started 01 March 2007, is to collect "a broad range of measurements that provide a snapshot in time of the state of the polar regions" and "to improve our understanding of global processes in these important areas"[a]. With regard to geodesy and geophysics the IPY project 185 "Polar Earth Observing Network" (POLENET) pursues exactly this goal with aiming on deploying autonomous observatories at remote polar sites including GPS, seismics, gravity and tide gauges. Closing observational gaps in polar regions POLENET will substantially help to investigate polar geodynamics and to gain deeper insight into the interactions between cryosphere, solid earth, oceans and atmosphere. Within this framework, tidal gravimetry is one technique contributing to the goals of POLENET. Tidal gravimetry through its observations provides a connection between the aforementioned disciplines. Sensing the tides of the solid earth as the main contribution, gravimetric time series observations are sensitive to ocean loading, to ice mass changes and the viscoelastic reaction of the earth linked to these as well as to atmospheric mass changes and loading. In the following, we will exemplarily review the results of gravimetric time series observations carried out by the Dresden University of Technology group, especially with regard to the different kind of location and geodynamic setting. A joint Argentine-German project will be presented carrying out gravimetric time series observations at Argentine Antarctic stations.

2 Using and Validating Ocean Tide Models

While solid earth tides can be observed, modelled and applied for corrections with sufficient accuracy, the treatment of the ocean tide impact largely depends on the reliability and accuracy of the ocean tide model in the area of investigation. Ocean tides have to be considered for the separation of ice mass changes and oceanic mass transport in Antarctica using temporal mass variations inferred from the GRACE mission (Horwath and Dietrich 2006), for the investigation of ice-shelf dynamics and deformation (Horwath et al. 2006), for the determination of precise offshore heights and height changes using dedicated satellite missions like ICESat and CRYOSAT (Padman and Fricker 2005), and to infer station positions and velocities from GNSS observations in the frame of a precise and reliable terrestrial reference frame (Dietrich et al. 2005; Steigenberger et al. 2006). In the polar regions, satellite altimetry – due to the limited latitude range – may only partly contribute to the

---

[a] www.ipy.org

determination of ocean tide models. In situ data have to be assimilated in order to improve the models. Extended regions of poorly known ocean tides still exist especially in Antarctica with its large ice shelves. King et al. (2005); King and Padman (2005) carried out a validation and assessment of ocean tide models around Antarctica. Using a point positioning strategy (King and Aoki 2003) it was demonstrated how GPS offshore data can be used to test ocean tide models. GNSS turned out to be feasible to overcome limitations of other techniques like coastal tide gauges or bottom pressure gauges, which cannot be deployed at regions with ice shelves or heavy sea ice coverage like the Filchner-Ronne ice shelf (FRIS) and the Weddell sea or the Ross ice shelf (RIS) and the Ross sea. Tidal gravimetry is a further technique to be applied in ocean tide investigations (for a detailed discussion of principles and applied techniques see e.g. (Melchior and Francis 1996) for world-wide validation or e.g. (Bos et al. 2002) for a regional investigation in the North Atlantic/Arctic ocean). With regard to Antarctica, most stations with long-term tidal gravimetry observations were located far from the large ice shelves, like Amundsen-Scott (Bos et al. 2000) or Syowa (Sato et al. 1997). Nevertheless, GNSS and gravimetry provide independent data to test ocean tide models through a comparison with the respective load tide vector. For instance, gravimetry may give better estimates for the K1 and K2 constituents, which are systematically biased in GPS observations (King et al. 2005). King et al. (2005) also showed that discrepancies between different ocean tide models are largest for the semidiurnal constituents (M2, S2) for FRIS, and for the diurnal constituents (O1, K1) for RIS.

3 Tidal Gravimetry in Polar Regions

Among a variety of other observations, the TU Dresden group recorded gravimetric time series in polar regions. The different setting of these observations and their results will now be reviewed.

3.1 Greenland Inland Station: Kangerlussuaq

In Kangerlussuaq, West Greenland (see Fig. 1), a gravimetric time series was recorded from June 1996 to March 1997 (Scheinert et al. 1998) using LaCoste & Romberg gravity meter D-193 with a L&R digital feedback system DFB-144, see Fig. 2. Special attention was given to a stabilized temperature regime. Additionally, air pressure values were obtained for the entire time period. The tidal analysis was carried out using the software package ETERNA v3.21 (Wenzel 1995). The least-squares adjustment of 13 tidal constituents in the diurnal and semidiurnal bands resulted in a standard deviation of 0.28 µGal. Correlating the residual gravity signal with air pressure variation (Fig. 3) resulted in an atmospheric pressure admittance factor of −0.43 µGal/hPa. For further details of the analysis see (Scheinert

Fig. 1 Tidal gravimetry stations beyond 60 degree north held in the database of the International Centre of Earth Tides (www.astro.oma.be/ICET). Tidal gravimetry is discussed for Kangerlussuaq (section 3.1), situating in the ice-free region of West Greenland

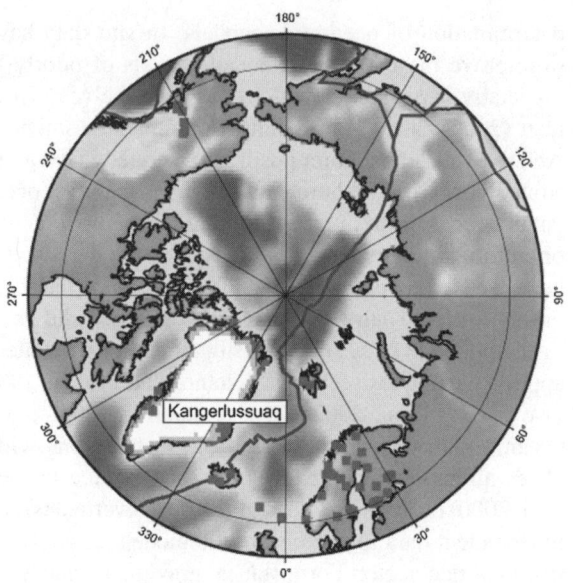

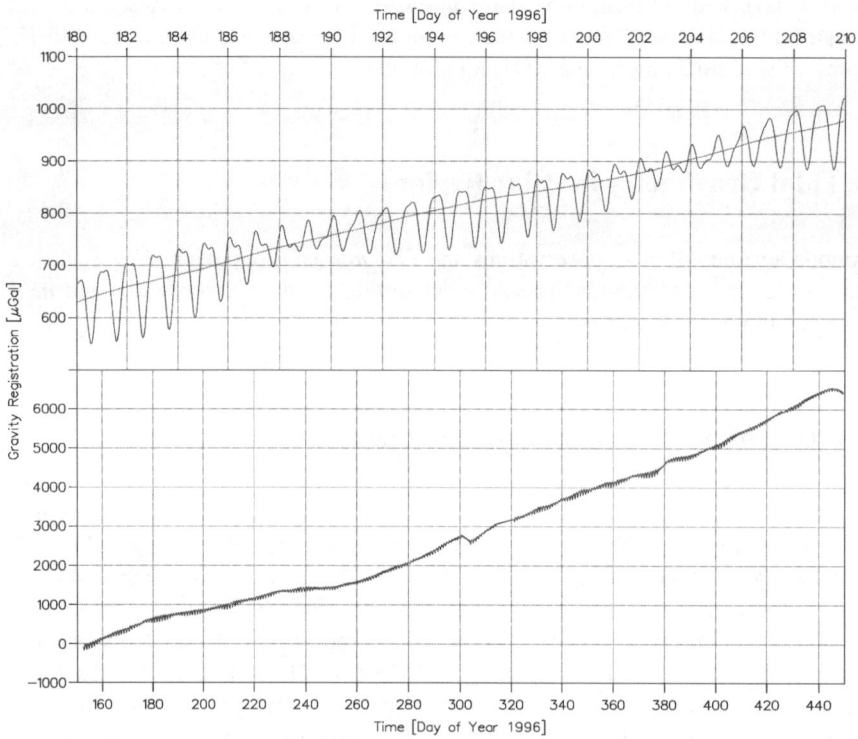

Fig. 2 Gravimetric time series at Kangerlussuaq, June 1996 to March 1997 (lower panel) and detail for July 1996 (upper panel)

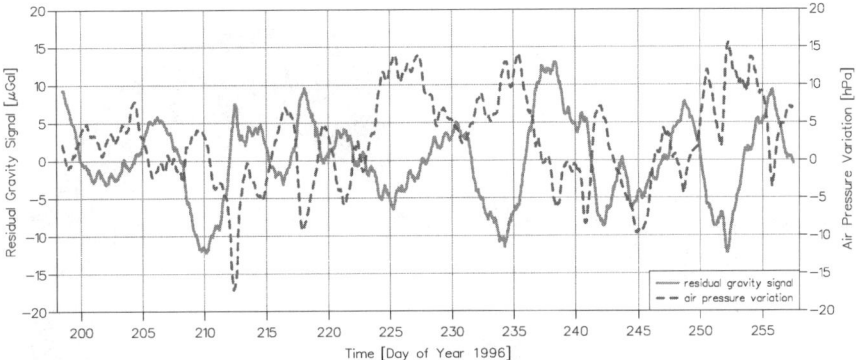

Fig. 3 Gravimetric time series of station Kangerlussuaq Comparison of residual gravity signal (red curve) and air pressure variation (green curve) for about two months

et al. 1998). The residual vectors of the four main constituents M2, S2, K1 and O1 were compared with the ocean load vector calculated using the model FES95.2 (Le Provost et al. 1995, 1998). Now, about ten years later, three newer models are used to compute the ocean load vector for comparison. The results are given in Table 1. Estimating the results a good agreement of the residual gravity vector with the ocean tide load vector for all four tide models can be stated. This is not that much surprising, since the gravimetric station Kangerlussuaq is situated in a distance of approximately 130 km from the coast, which leads to a considerable weakening of the ocean load signal. Furthermore, at a latitude of about 66.5°N (close to the northern polar

Table 1 Comparison of residual gravity vector and ocean tide load vector[b] for the gravimetric time series Kangerlussuaq (June 1996–March 1997) (units: amplitudes [μGal], phases [°]; phases are local phases)

Tide	Adjusted parameters		Residual vector	Ocean tide load vector			
				FES95.2	GOT00.2	FES2004	TPXO6.2
	Amplitude phase	Amplitude factor	Amplitude phase	Amplitude phase			
M2	16.01	1.40	2.91	2.58	2.61	2.70	2.71
	−3.75		−21.07	−14.32	−14.93	−16.53	−14.33
S2	6.86	1.29	1.14	1.09	1.03	0.95	1.10
	−7.96		−56.71	−58.69	−53.53	−50.63	−44.63
K1	36.77	1.17	1.34	1.22	1.06	1.01	1.12
	0.84		23.80	32.11	39.49	29.89	37.19
O1	25.92	1.16	0.35	0.41	0.38	0.42	0.37
	0.66		58.95	78.44	81.39	79.39	82.79

[b]Ocean tide load vector estimates provided by Olivier Francis (FES95.2) and the Free Ocean Tide Loading Provider (Bos and Scherneck 2007) (other models).

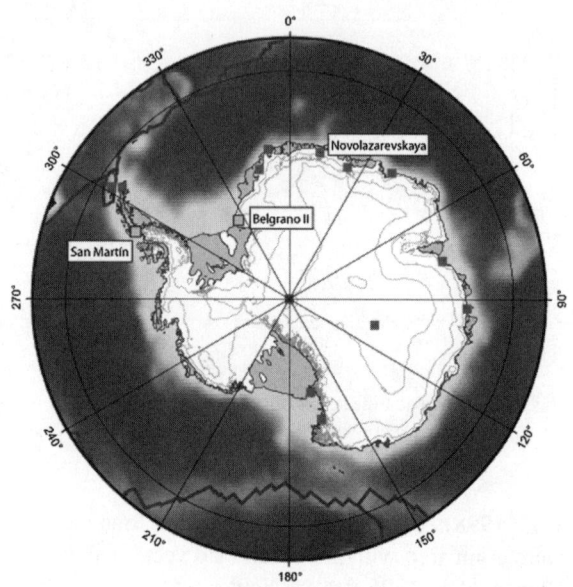

Fig. 4 Tidal gravimetric stations beyond 60 degree south held in the database of the International Centre of Earth Tides (www.astro.oma.be/ICET). Tidal gravimetry is discussed for Novolazarevskaya (section 3.2) as well as for Belgrano II (running 2007) and San Martín (planned 2008, section 3.3)

circle) the ocean tide models could make use of the satellite altimeter data. Remaining differences can be explained by the local tides of the nearby fjord system of the Søndre Strømfjord, which exhibits with a length of about 150 km a tidal range of about 3 m even in the inner part of the fjord (Fig. 4).

3.2 Antarctic Coastal Station: Novolazarevskaya

In Antarctica, gravimetric tide observations are constrained to manned stations. Therefore, most gravimetric time series were observed at coastal stations. In the 1990s our institute's research in Antarctica was concentrated on the region of Schirmacher Oasis, central Dronning Maud Land. A first gravimetric tide series using an Askania GS11 gravimeter was observed in 1991, but was of limited accuracy (Dietrich et al. 1995). A second gravimetric time series could be observed from February to April 1995 for about 54 days. As in the Greenland case, we used the L&R gravimeter D-193 together with the L&R digital feedback DFB-144 (Dietrich et al. 1998). At Novolazarevskaya, the gravimeter station was set up in a small hut on a concrete pillar adopting some thermal control. Again, we may compare the residual vector for the main tidal waves with the corresponding ocean tide load vector. The results are given in Table 2. Complementary to the model predictions by O. Francis as used in the paper by Dietrich et al. (1998), now additional estimates by two further software packages shall be given (Free Ocean Tide Loading Provider (Bos and Scherneck 2007) and SPOTL (Agnew 1996)). All three predictions are based on the same ocean tide model (FES95.2, (Le Provost et al. 1995)). For a

Table 2 Comparison of residual gravity vector and ocean tide load vector[c] for the gravimetric time series Novolazarevskaya (February–April 1995) (units: amplitudes [μGal], phases [°]; phases are local phases; relative errors [%]). Relative errors are given with respect to the values of the residual vectors

Tide	Adjusted parameters		Residual vector	Ocean Tide Load Vector					
				FES95.2 (1)		FES95.2 (2)		FES95.2 (3)	
	Amplitude phase	Amplitude factor	Amplitude phase	Amplitude phase		Amplitude phase		Amplitude phase	
				Value	*Relative error*	Value	*Relative error*	Value	*Relative error*
M2	10.97	1.35	1.56	1.55	*0.8*	1.08	*30.8*	1.94	*24.4*
	2.42		17.08	16.46	*3.6*	14.13	*17.3*	17.65	*3.3*
S2	5.63	1.48	1.24	1.47	*18.8*	0.97	*21.8*	1.75	*41.1*
	−2.46		−10.64	−8.48	*20.3*	−12.47	*17.2*	−7.72	*27.4*
K1	32.81	1.21	2.08	2.07	*0.4*	1.55	*25.5*	2.43	*16.8*
	−0.12		−1.55	1.74	*212.1*	−3.94	*153.9*	−8.43	*443.6*
O1	24.50	1.27	2.25	2.62	*16.4*	2.07	*8.0*	2.93	*30.2*
	0.20		1.84	11.07	*501.8*	6.97	*278.5*	1.76	*4.6*

better overview Table 2 gives also the relative errors of the ocean tide load vector predictions with respect to the residual vectors. In terms of the amplitudes, the estimates by Francis agree best, while the values by Scherneck/Bos underestimate the residual amplitudes, and the values by Agnew overestimate them. In terms of the phases, all three model predictions perform equally while the deviations are generally larger than for the amplitudes. Since the same ocean tide model was used, the differences originate in the different software realizations, which brings some problems of ocean load tide prediction into the focus. Most important is how the land-sea distribution is adopted by the respective software. Despite the original coverage and resolution of the ocean tide model, the software may introduce some interpolation algorithms at finer grids in the vicinity of the point of interest. Since for the present case the gravimeter station is situated close to the coast and at a height of about 130 m above sea level, cf. Fig. 5, the different treatment of the direct effect caused by the ocean cells close to the point of interest may easily explain the differences in amplitude. Furthermore, for Antarctica the case of a coastal station gets even more complicated, since in the presence of an ice shelf the coastline is defined by the grounding line (which marks the transition of the grounded glacier or inland ice to the swimming ice shelf). Figure (5) adopts the grounding line of the Antarctic Digital Database 4.0 (ADD 4.0) (ADD Consortium 2000). A more precise location of the grounding line can be determined e.g. by interferometric SAR (InSAR) using ERS-1/2 data, which has been done by Baessler et al. (2002, 2003) for the region of the Schirmacher Oasis. Therefore, in order to improve the ocean tide predictions not

---

[c] Ocean tide load vector estimates provided by Olivier Francis (1), the Free Ocean Tide Loading Provider (Bos and Scherneck 2007) (2) and SPOTL (3) (Agnew 1996).

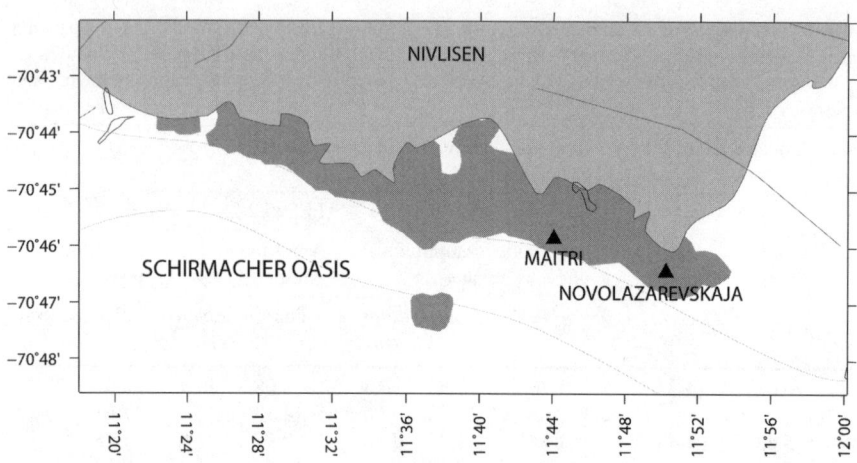

Fig. 5 Overview map of Schirmacher Oasis, central Dronning Maud Land, East Antarctica. The gravimeter observations were conducted at the Russian station Novolazarevskaya. Map source: Antarctic Digital Database 4.0 (ADD Consortium 2000)

only the tide model itself has to be improved but also the land-sea mask has to be given at grids with appropriate resolution, which is a challenging task especially for ice shelf covered regions. For such regions, the investigation of ice-shelf dynamics might help to solve such questions, see e.g. (Horwath et al. 2006).

3.3 Antarctic Coastal Station: Belgrano II

Acknowledging the success of the gravimeter observations described above, a joint Argentine-German project was initiated to record gravimetric time series at the Argentine Antarctic stations Belgrano II (East Antarctica) and San Martín (Antarctic Peninsula). This project is a joint effort of Dirección Nacional del Antártico – Instituto Antártico Argentino (IAA), Buenos Aires, and TU Dresden (TUD), and contributes to the project "Polar Earth Observing Network" (POLENET) within the framework of the International Polar Year 2007/2008. During 2007, the observations are being carried out at Belgrano II. An overview of the area of investigation in the vicinity of Belgrano II is given in Fig. 6, where an ASTER image (acquisition time 2006/01/16) showing ice surface topography is used as background. Belgrano II is situated on top of the Bertrab Nunatak at the southern bank of the Vahsel Bay in the south-eastern part of the Weddell Sea (comp. also Fig. 4). Vahsel Bay is covered by a glacier moving in north-western direction with a flow velocity of about 150...200 m/yr (preliminary result from an analysis of a pair of ASTER images covering a time span of one year). Comparing the present coastline given by the ASTER image with the coastline given by the ADD 4.0 (blue line), the largest offset is visible in the area of the Vahsel Bay for the edge of the glacier flowing into

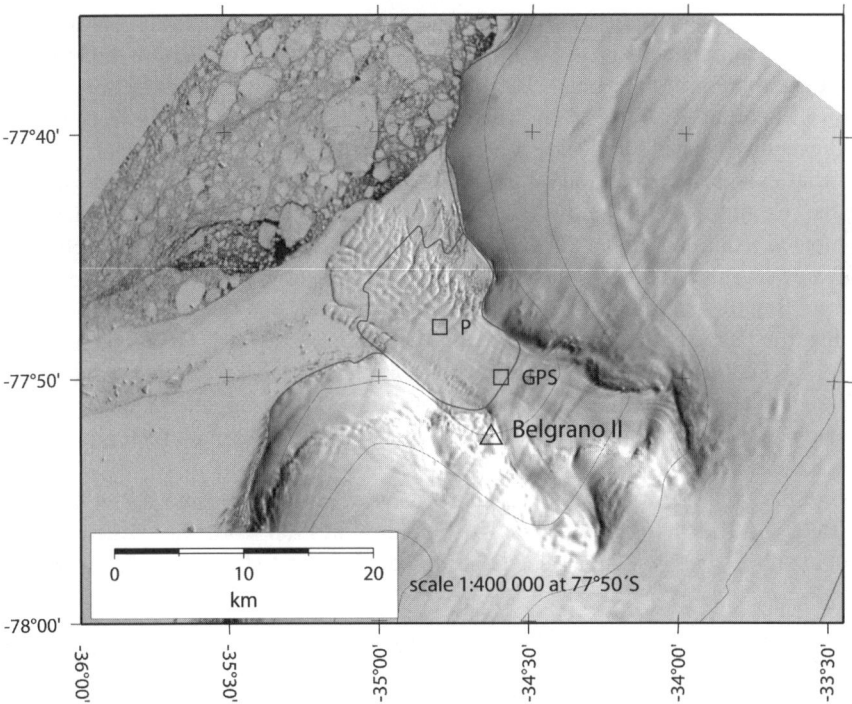

Fig. 6 Overview of the surroundings of Belgrano II. Satellite image: ASTER-L1B-VNIR 2032694470, acquisition time 2006/01/16. Map source: Antarctic Digital Database 4.0 (ADD Consortium 2000). Blue line: ADD coastline. Red lines: 200 m height contours. Polar stereographic projection

the Weddell Sea. A rough estimation of this offset yields a value of about 3.5 km. The time of surveying of the coastline entering the ADD 4.0 can be assumed to be at least 15 years before the acquisition time of the ASTER image (2006), which would – assuming linear behaviour – result in a flow velocity of about 230 m/yr. Thus, this consideration can only give a preliminary estimation of the order of magnitude of the horizontal flow velocity of the Vahsel Bay glacier. Furthermore, the dynamics of the floating glacier is influenced by the ocean tides. A freely floating ice plate would follow the vertical movement of the tidal water masses. This was proven for the realistic case of ice shelves. Only within a limited transition zone the deformation of the ice shelf changes from zero (grounded ice) to one (freely floating ice). This zone, also denoted as grounding zone, has a width of 1 to 10 km, depending from the mechanical and geometrical properties of the ice plate as well as from the bathymetry (Vaughan 1994). In the case of the Vahsel Bay glacier a response to ocean tides similar to that of ice shelves can be expected. The geometry and extent makes the great difference: The location of the grounding line given by the ADD (blue line further south-east of the glacier edge) is uncertain. Additionally, the bay has a width of only 10 km at maximum. Hence, in order to investigate the ocean tides in the area of Belgrano II GNSS (GPS/GLONASS) provides an observation technique complementary to tidal gravimetry, giving the advantage to directly observe the motion of the glacier. Therefore, a GPS station was set up for about three days in February 2007 (Fig. 6). The analysis of the data has still to be done and will eventually enable to decide if at that location the glacier is floating or not, and to what extent the excitation of the ocean tides is damped. To get an impression of the magnitude of the ocean tides in the area of investigation, a tide series was computed using the model TPXO6.2 (Egbert and Erofeeva 2002) and the SPOTL software (Agnew 1996). To meet that cell of the model which is located closest to Vahsel Bay the coordinates 77.8°S and 34.8°W were found (location denoted by P in Fig. 6). The expected amplitude of the ocean tides is about 1.7 m at that location (Fig. 7). The corresponding ocean tidal loading signal at Belgrano II would have an amplitude of about 10 µGal as shown in Fig. 8. To investigate this geodynamic phenomena and to validate the ocean tide model(s), the presented tidal gravimetry project was initiated (Fig. 9).

For the tidal gravimetry, an especially constructed and prepared hut in the area of Belgrano II is being used (Fig. 10). A hard excavation work was carried out removing more than two cubic meters of intricated granitic stones in order to erect a

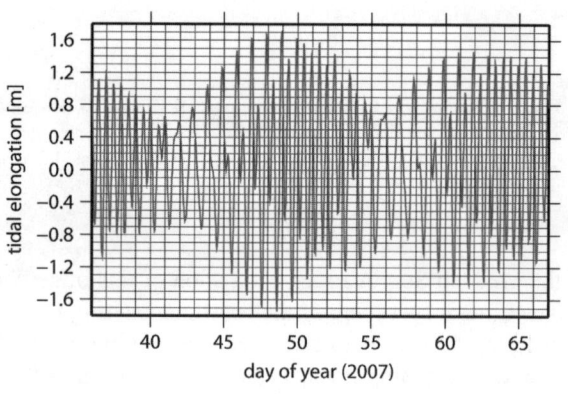

Fig. 7 Prediction of ocean tides for the location denoted by P (Fig. 6) and a time interval of one month, using model TPXO6.2 (Egbert and Erofeeva 2002) and the SPOTL software (Agnew 1996)

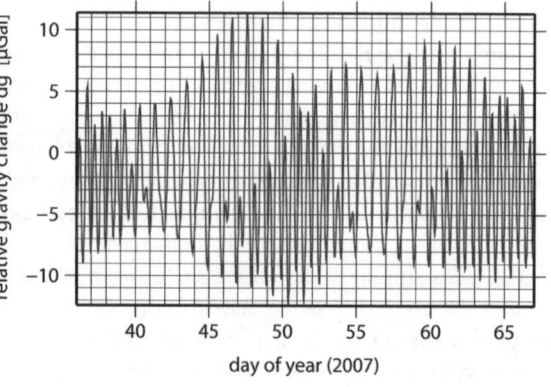

Fig. 8 Prediction of relative gravity change caused by ocean tidal loading for the gravity station at Belgrano II (Fig. 6) and a time interval of one month, using model TPXO6.2 (Egbert and Erofeeva 2002) and the SPOTL software (Agnew 1996)

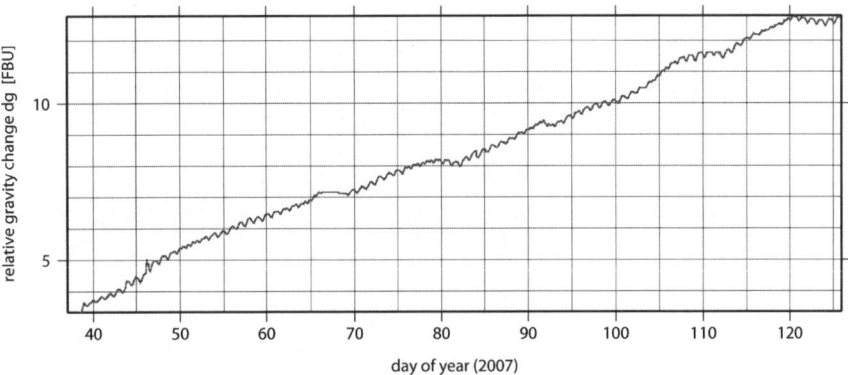

Fig. 9 Time series of the gravity registration by the L&R D-193 gravity meter at Belgrano II (preliminary raw data, feedback units), February 6 to May 5, 2007

concrete pillar aiming at a mechanically uncoupling from the surrounding ground and foundations as much as possible. Small grain stones were used to fill up the space between the pillar and the hut's foundations. At this pillar, the TUD gravimeter L&R D-193 together with the feedback DFB-144 as well as the IAA gravimeter G-748 were set up at the beginning of February 2007 (Fig. 11). The simultaneous observations with two gravimeters allow to cross-check the performance of the instruments as well as to achieve redundancy. The preliminary raw time series of gravimeter L&R D-193 is shown in Fig. 9. This raw time series has still to be corrected for offsets, spikes and the gravimeter drift. Nevertheless, already now the tidal signal is clearly visible. Once the time series observation will be finished, it is subject to the processing of the entire data record to estimate and to separate the contributions from solid earth tide, from ocean tidal loading as well as from atmospheric loading. For the latter part, we will make use of meteorological data being

Fig. 10 The gravimeter hut in the area of the Argentine Antarctic station Belgrano II. In the background, the Vahsel bay glacier can be seen

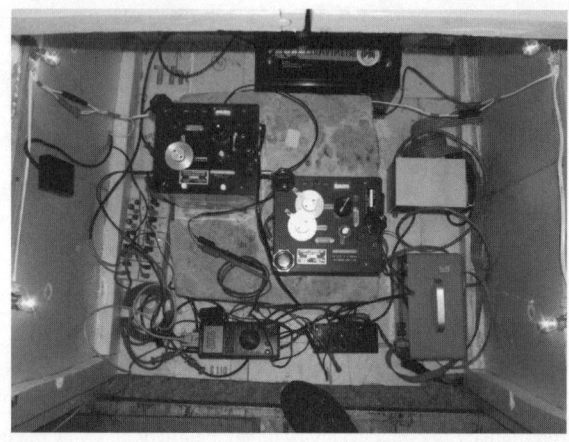

Fig. 11 Set-up of the instruments at a grounded pillar within the gravimeter hut, Belgrano II: L&R G-748 (left), L&R D-193 (right)

recorded at Belgrano II. After the successful completion of the observations, the instruments should be moved to San Martín during the upcoming Antarctic season 2007/2008.

4 Conclusions

The feasability of conducting gravimetric tide series has been demonstrated. Especially in polar regions this technique provides independent observations complementary to GPS. For stations close to the coast the reliability of estimating ocean tide effects from models depends not only from the used ocean tide model itself, but also from the realization of the land-ocean mask. This has been exemplarily demonstrated by the results from the inland station Kangerlussuaq (Greenland) and the coastal station Novolazarevskaya (East Antarctica). Carrying out a joint Argentine-German project, we can state that the location of the current gravimeter station in Belgrano II at the south-east coast of the Weddell Sea, close to the Filchner ice shelf, gives us the great opportunity to test the ocean tide models in that region, especially with regard to the poorly determined constituents M2 and S2. Additional GPS measurements on the floating ice will enable us to directly observe the ocean tides despite a certain damping of the tidal amplitudes can be expected due to the location of the grounding zone and the mechanical properties of the ice. Combining gravimetry and GPS enables to better discriminate the different effects of ocean tides, ocean tidal loading, solid earth tides and atmospheric loading.

Acknowledgments The investigations were and are supported by many colleagues and institutions, among others: Danish Polar Center (DPC), Copenhagen (Denmark), and Kangerlussuaq International Science Support (KISS) (Greenland); Russian Antarctic Expedition (RAE), St. Petersburg (Russia); Alfred Wegener Institute for Polar-and Marine Research (AWI), Bremerhaven (Germany); Comando Conjunto Antártico, Comando Antártico de Ejército (CAE) and Dirección

Nacional del Antártico (DNA), all Buenos Aires (Argenina). We would like to thank the people of Ejército Argentino and of Laboratorio Belgrano (LaBel), overwintering at Belgrano II during 2006 and 2007, resp., for their great support in setting up the gravimeter hut and controlling the gravimeter observations. Parts of the work were supported by research funds granted by the German Research Foundation (DFG) and the Federal Ministry of Education and Research (BMBF). The joint Argentine-German project is supported by grant ARG-01/Z08 of the scientific-technological cooperation programme of BMBF. Finally, we like to thank the editors for inviting to contribute this paper to this collection.

References

ADD Consortium (2000) Antarctic Digital Database, Version 3.0. Database, manual and bibliography. Scientific Commitee on Antarctic Research, Cambridge, manual and bibliography, 93 pp.
Agnew DC (1996) SPOTL: Some Programs for Ocean-Tide Loading. SIO Ref. Ser. 96–8, 35 pp., Scripps Institution of Oceanography, La Jolla, CA (version 3.2.1 of 26 July 2005).
Baessler M, Dietrich R, Shum C (2002) Investigations of Ice Dynamics at the Grounding Zone of an Antarctic Ice Shelf Utilizing SAR-Interferometry. In: Jekeli C (ed) Proceedings of the Weikko A. Heiskanen Symposium in Geodesy, 1–4 Oct 2002, Columbus, Ohio, USA.
Baessler M, Dietrich R, Shum C (2003) Horizontal velocity field, strain analysis and grounding zone location for the area of the Nivlisen ice shelf, Dronning Maud Land. Poster presentation at FRINGE Workshop, Frascati, Italy, December 1–5, 2003.
Bos MS, Scherneck HG (2007) Free ocean tide loading provider. URL http://www.oso.chalmers.se/loading, http://www.oso.chalmers.se/loading.
Bos MS, Baker TF, Lyard FH, Zürn WE, Rydelek PA (2000) Long-period lunar Earth tides at the geographic South Pole and recent models of ocean tides. Geophysical Journal International 143(2):490–494.
Bos MS, Baker TF, Røthing K, Plag HP (2002) Testing ocean tide models in the Nordic seas with tidal gravity observations. Geophys J Int 150:687–694.
Dietrich R, Liebsch G, Dittfeld H, Noack G (1995) Ocean tide and attempt of Earth tide recordings at Schirmacher Oasis/Dronning Maud Land, Antarctica. In: Hsu HT (ed) Proc. 12th International Symposium on Earth Tides, Science Press, Beijing – New York, pp 563–569.
Dietrich R, Dach R, Korth W, Polzin J, Scheinert M (1998) Gravimetric Earth Tide Observations in Dronning Maud Land/Antarctica to Verify Ocean Tidal Loading. In: Ducarme B, Paquet P (eds) Proc. 13th International Symposium on Earth Tides, Brussels, July 22–25, 1997, Obs. Royal de Belgique, Brussels, pp 529–536.
Dietrich R, Rülke A, Scheinert M (2005) Present-Day Vertical Crustal Deformations in West Greenland from Repeated GPS Observations. Geophys J Int 163(3):865–874, doi: 10.1111/j.1365-246X.2005.02766.x.
Egbert GD, Erofeeva SY (2002) Efficient inverse modeling of barotropic ocean tides. J Atmos Oceanic Technol 19(2):183–204.
Horwath M, Dietrich R (2006) Errors of regional mass variations inferred from GRACE monthly solutions. Geophys Res Lett 33:L07, 502, doi:10.1029/2005GL025, 550.
Horwath M, Dietrich R, Baessler M, Nixdorf U, Steinhage D, Fritzsche D, Damm V, Reitmayr G (2006) Nivlisen, an Antarctic ice shelf in Dronning Maud Land: Geodetic-glaciological results from a combined analysis of ice thickness, ice surface height and ice flow observations. J Glac 52(176):17–30.
King M, Aoki S (2003) Tidal observations on floating ice using a single GPS receiver. Geophys Res Lett 30(3):1138, doi:10.1029/2002GL016, 182.
King M, Penna NT, Clarke PJ (2005) Validation of ocean tide models around Antarctica using onshore GPS and gravity data. J Geophys Res 110:B08, 401, doi:10.1029/2004JB003, 390.

King MA, Padman L (2005) Accuracy assessment of ocean tide models around Antarctica. Geophys Res Lett 32:L23, 608, doi:10.1029/2005GL023, 901.

Le Provost C, Bennett AF, Cartwright DE (1995) Ocean Tides for and from TOPEX/POSEIDON. Science 267:639–642.

Le Provost C, Lyard FH, Molines JM, Genco ML, Rabilloud F (1998) A hydrodynamic ocean tide model improved by assimilating a satellite altimeter-derived data set. J Geophys Res 103(C3):5513–5529.

Melchior P, Francis O (1996) Comparison of recent ocean tide models using ground-based tidal gravity measurements. Mar Geod. 19:291–330.

Padman L, Fricker HA (2005) Tides on the ross ice shelf observed with ICESat. Geophys Res Lett 32:L14, 503, doi:10.1029/2005GL023, 214.

Sato T, Ooe M, Nawa K, Shibuya K, Tamura Y, Kaminuma K (1997) Long-period tides observed with a superconducting gravimeter at Syowa Station, Antarctica, and their implication to global ocean tide modeling. Phys Earth Plan Int 103(1):39–53.

Scheinert M, Dietrich R, Schneider W (1998) One Year of Gravimetric Earth Tide Observations in Kangerlussuaq/West Greenland. In: Ducarme B, Paquet P (eds) Proc. 13th International Symposium on Earth Tides, Brussels, July 22–25, 1997, Obs. Royal de Belgique, Brussels, pp 201–208.

Steigenberger P, Rothacher M, Dietrich R, Fritsche M, Rülke A, Vey S (2006) Reprocessing of a global GPS network. J Geophys Res 111, B05402, doi:10.1029/2005JB003747.

Vaughan DG (1994) Investigating tidal flexure on an ice shelf using kinematic GPS. Ann Glaciol 20:372–376.

Wenzel H (1995) ETERNA 3.20 Earth Tide Data Processing Package. (Software manual).

Joint Geophysical Observations of Ice Stream Dynamics

S. Danesi, M. Dubbini, A. Morelli, L. Vittuari and S. Bannister

Abstract Ice streams play a major role in the ice mass balance and in the reckoning of the global sea level; they have therefore been object of wide scientific interest in the last three decades. During the 21st Italian Antarctic Expedition, in the austral summer 2005–06, we deployed a joint seismographic and geodetic network in the area of the David Glacier, Southern Victoria Land. This campaign followed a similar experiment carried out in the same area during the austral summer 2003–04 with the deployment of a seismographic network that recorded significant microseismicity beneath the David Glacier, primarily occurring as a few small clusters. In the latest 2005–06 deployment, 7 seismographic stations and 3 GPS geodetic receivers operated continuously for a period of 3 months (November 2005–early February 2006) in an area of about 100×150 km^2 around the David Glacier. We have carried out several analyses using the combined data sets. These included the examination of the temporal evolution in earthquake magnitude and location and also the contemporaneous observation of both seismic activity and surface kinematics of the ice stream to possibly correlate the recorded microseismicity with the movement of the glacier, affected by the Ross Sea tides. Unfortunately, a clear correlation between the occurrence of seismic events and the movement of the glacier is not evident. Here

S. Danesi
Istituto Nazionale di Geofisica e Vulcanologia, Sezione Bologna, Via Donato Creti, 12 – 40128 Bologna IT, e-mail: danesi@bo.ingv.it

M. Dubbini
Università di Modena e Reggio Emilia, DiMec, Via Vignolese 905 – 41100 Modena IT, e-mail: marco.dubbini@unimo.it

A. Morelli
Istituto Nazionale di Geofisica e Vulcanologia, Sezione Bologna, Via Donato Creti, 12 – 40128 Bologna ITe-mail: morelli@bo.ingv.it

L. Vittuari
Università di Bologna, DISTART, Viale del Risorgimento, 2 – 40136 Bologna IT, e-mail: luca.vittuari@mail.ing.unibo.it

S. Bannister
GNS Science, Fairway Drive, PO 30368 Lower Hutt 5040, New Zealand, e-mail: s.bannister@gns.cri.nz

we present some details of the two temporary networks and preliminary results and implications.

1 Introduction

Ice streams are the fast flowing channels through which polar sheets pour ice into surrounding oceans and influence their mass balance, thermal regime and circulation with evident consequences on large-scale climate changes. Global sea level is basically regulated by thermal expansion and glaciers mass budget, with growing impact of Greenland and Antarctic ice sheet mass variation. From mass balance studies it has been inferred and generally accepted that the complete melt of polar ice would potentially result in a dramatic sea level rise (about 70 m) (Rignot and Jacobs 2002; Alley et al. 2005).

One fundamental grasp of the past 30 years is that the response of polar ice sheets to climate changes is far from being slow and steady; on the contrary ice sheets are extremely sensitive to global warming, with rapid dynamic feedback on ice flow and consequent pronounced variations in mass balance. For this reason, the understanding of ice stream mechanics and the measurement of flow variations are valid gateways towards the study of ice sheet (and climatic) stability.

The David Glacier and the Drygalski Ice Tongue (its 100 km-long floating seaward extension) represent the main structure of the very complex drainage system of the Victoria Land area. The glacier has two main tributaries, flowing from DomeC and Talos Dome, which converge after a remarkable icefall due to a subglacial ridge crossing the ice flow and after a large basin called David Cauldron (Fig. 1).

Several glaciological studies have been carried out over the last decades to detect the grounding line location and to study the floating ice tongue profile. The grounding line was initially thought to be detected at the mouth of the glacier bay, but InSAR observations now indicate that it is about 15 km upstream (Fig. 1), just downstream of the David Cauldron (Frezzotti 1993; Frezzotti et al. 2000; Rignot 2002).

The glacier base loses about 68% of its mass from bottom melting within 20 km of the grounding line, driving an up-welling of "super-cooled" water that allows the deposition of enough marine ice to preserve the length of the Drygalski Ice Tongue (Frezzotti et al. 2000, Rignot and Jacobs 2002).

During the 2005–06 Austral Summer, we carried out an observational campaign in the area of the David Glacier with the aim of collecting simultaneous time series of geodetic and seismological data.

The target of the experiment is the study of possible correlations between vertical/horizontal displacements of the glacier and the weak typical seismic activity inland beneath the glacier. Location of seismographic and geodetic stations operating in the David area are listed in Table 1 and shown in Fig. 1 (stars and circles respectively).

The Global Navigation Satellite System (GNSS) takes advantage of satellite positioning techniques to provide accurate positioning information with reliable time reference.

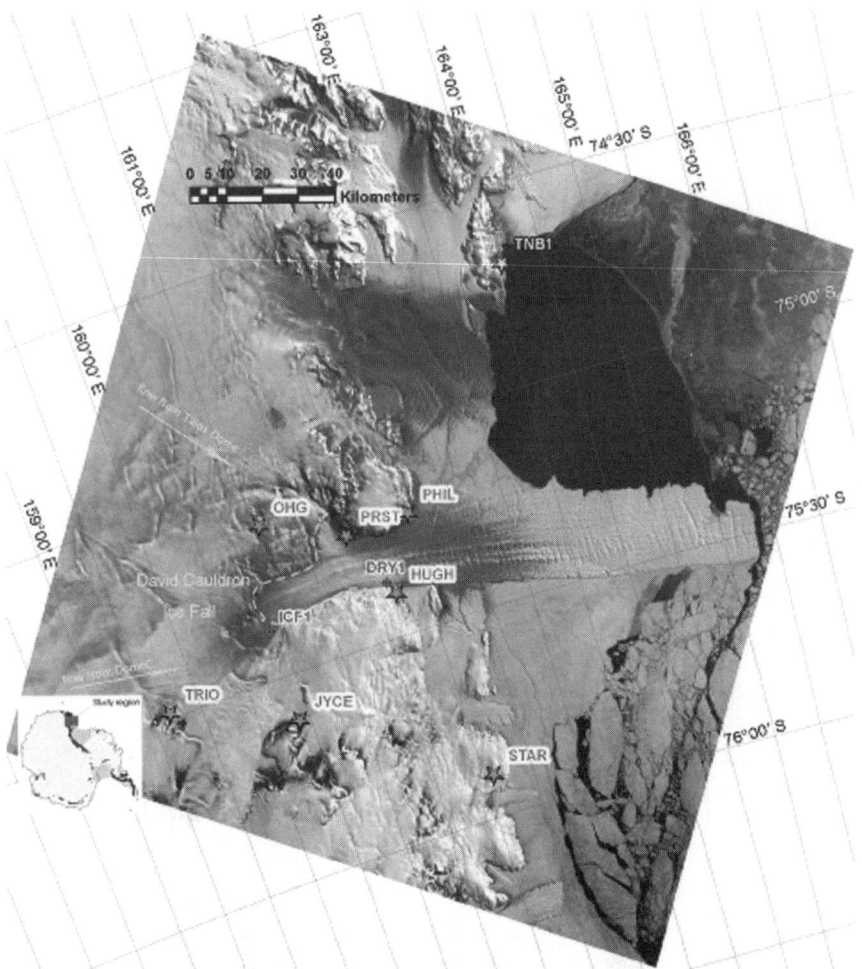

Fig. 1 Reference map. A satellite image of the David Glacier (ASTER_VNIR 2003 Feb 6th) has been georeferenced to a South Pole Polar Stereographic projection. The study area is in evidence in the inset. The yellow dashed line indicates the grounding line profile, as estimated by InSar observations (Rignot 2002). The picture shows the deployment of the 2005–06 seismic network (green stars) and GPS receivers (red circles). TNB1 (yellow star) is the permanent geodetic Italian station operating close to the Italian scientific base MZS

It has been shown that the use of GNSS instrumentations can be very useful for the measurement of glaciers movements (Hulbe and Whillans 1994; Frezzotti et al. 1998; Doake et al. 2002; Anandakrishnan et al. 2003; Bindcshadler et al. 2003; King and Aoki 2003; Legresy et al. 2004) and it is one of the most reliable methods to monitor the position of selected points located on a glacier with high precision. Moving glaciers can also be monitored in a seismological perspective. Seismic signals related to glacier motion have been observed since early '70s(Neave

Table 1 Position and operativeness of GPS receivers and seismographic stations used in this study

		Latitude	Longitude	Height (m)	Station code	Operativeness (Julian days)
GPS sites	GPS HUGHES (master station)	75°23′52.68″ S	162°12′06.18″ E	165.667	VL19	316/05–038/06
	GPS Drygalski	75°21′45″ S	162°08′15″ E	37	DRY1	015/06–038/06
	GPS Cauldron	75°23′07″ S	160°54′06″ E	201	ICF1	313/05–038/06
SEISMOGRAPHIC sites	Cape Philippi	75°13′08.80 S	162°32′40.93 E	425	PHIL	318/05–365/05
	Hughes Bluff	75°23′53.10 S	162°12′08.155 E	215	HUGH	316/03–022/04 307/05–033/06
	Mc Daniel	75°47′57.83 S	161°47′53.74 E	780	MDAN	332/03–045/04
	Morris Basin	75°39′22.51 S	159°04′19.08 E	880	MORR	341/03–050/04
	Mt Joyce	75°37′05.55 S	160°53′27.80 E	1230	JYCE	320/03–022/04 305/05–033/06
	Mt Priestley	75°13′25.38 S	161°54′32.37 E	475	PRST	316/03–021/04 312/05–033/06
	Ohg	75°07′57.18 S	161°07′50.05 E	630	OHG	337/03–050/04 321/05–033/06
	Starr Nunatak	75°53′56.36 S	162°35′33.235 E	70	STAR	332/03–present
	Trio Nunatak	75°29′52.23 S	159°41′19.71 E	1145	TRIO	309/05–033/06

and Savage 1970; Van Wormer and Berg 1973) obtaining increasing interest over the last decades among the scientific communities. Actually, seismic activity originated beneath fast flowing ice streams has been recorded both in Greenland and in Antarctica with fairly different characteristics. From the study of surface seismic waves Ekström et al. (2003, 2006a, 2006b) have identified a number of glacier-related earthquakes, with magnitude higher than 4.5 that can be modelled with glacial-sliding slips mainly located beneath flowing glaciers. They have observed a remarkable increase in the number of such events between 2002 and 2005 as a consequence of significant changes in the glacial flow owing to late global warming. Differently then in Greenland, in Antarctica low-magnitude seismicity is most commonly recorded and located beneath ice streams either associated with repeating stress release across small sticky patches or basal drag regime (Bahr and Rundle 1996; Alley 2000; Smith 2006).

2 Seismological Campaign

2.1 Data Collection and Analysis

The best-studied system of Antarctic ice streams is located in the Siple Coast, the western coast of the Ross Ice Sea. Over the past 20 years (Blankenship et al. 1986; Alley et al. 1986; Anandakrishnan and Bentley 1993) a number of geophysical observations have been held in the area providing deeper understanding about the dynamics of fast flowing ice streams and climatic implications in a global sense (Alley et al. 2005; Horgan and Anandakrishnan 2006).

In the David Glacier area, on the eastern side of the Ross Embayment, a low-energy but significant seismic activity was revealed during the 1999–2000 seismological campaign that was carried out by a joint Australian and New Zealand project in the central Transantarctic Mountains area (Bannister and Kennett 2002). The glacier was more than 200 km far from the network target area, therefore the events were initially located with some uncertainties (owing to the unbalanced azimuthal coverage of the array) and possibly ascribed either to the presence of a large regional right-lateral lineament beneath the David Glacier (Salvini and Storti 1999) or to stick-slip motion at the base of the ice stream basically driven by basal shear stress (Bannister and Kennett 2002).

Afterward, the David Glacier was the target area for 2 seismological campaigns: the first was jointly held by Italy and New Zealand during the 2003–04 Austral Summer (19th Italian Expedition) (Danesi et al. 2007 – DBM07 hereinafter), the second was carried out by the Italian group during the 2005–06 Austral Summer (21st Italian Expedition) with simultaneous seismic and GPS observations.

Figure 1 shows the sites where autonomous remote seismographic stations were installed in 2005–06 (green stars). The stations were sited on rock outcrops surrounding the glacier (some were re-occupied after the first expedition) and operated continuously over 3 months recharged by photovoltaic panels. Each site

Fig. 2 Typical station installation. (**a**) All the seismographic stations were sited on rocky outcrops around the glacier and equipped with photovoltaic panels for autonomous recharging. In the foreground the seismic sensor Trillium40 sheltered with a cover and anchored to the rock. Digitizer/data-logger RefTek 130-01, batteries and electronic regulator were protected inside the insulated grey box behind the panel. (**b**) Temporary GPS site on the Drygalski. A dual frequency GPS Trimble5700 receiver was equipped with 1 Gb removable memory cards for data logging and batteries recharged by photovoltaic panels; a Zephyr antenna was tightened on the head of an aluminium pole well driven into the ice

was equipped with a 3-component broadband seismometer Trilium40 and a Reftek 130-01 digital recorder/data logger with 125sps fixed (Fig. 2a). The array covered an area of about 100×150 km^2 around the David Glacier, from the coastline toward the first peaks of the Transantarctic Mountains (Fig. 1).

After data retrieval and detection of seismic episodes with trigger algorithms, we identified more than 10000 low energy events distributed on three main classes.

The array successfully recorded teleseismic signals, mainly originated along the peri-Antarctic spreading oceanic ridges.

Then, we principally recorded episodes originating within the ice layer that we call *icequakes* hereinafter. Their seismic signal is strongly impulsive, short in duration, rich in high frequencies and rapidly attenuating (Fig. 3), reason why they are usually recorded by only one or two close stations and can be hardly or not at all located. Hundreds of icequakes per day testify that these are the most common and frequent occurrence in glacial environments: fractures can originate and propagate through the ice owing to crevasses, steep changes in bedrock topography, walls and asperities dragging effects, inhomogeneous strain in the flow, iceberg calving or combinations of several causes. Typically, icequakes are local effects of the glacier flow.

However, the most interesting events are of a third class. They occurred at the bedrock-ice interface with more than 150 occurrences that will be referred to as basal events hereinafter. The characteristic signal of basal events shows weak P and S arrival, duration of a few tens of seconds and spectra rich in low frequencies (Fig. 4).

Fig. 3 Icequake seismic signal. (**a**) Vertical component of seismic signal (in velocity) vs. time. Icequake are frequent events, have short duration and impulsive onset. (**b**) Amplitude of frequency spectrum of icequake signal vs. frequency in hertz: maximum amplitudes are reached for frequencies higher than 10 Hz

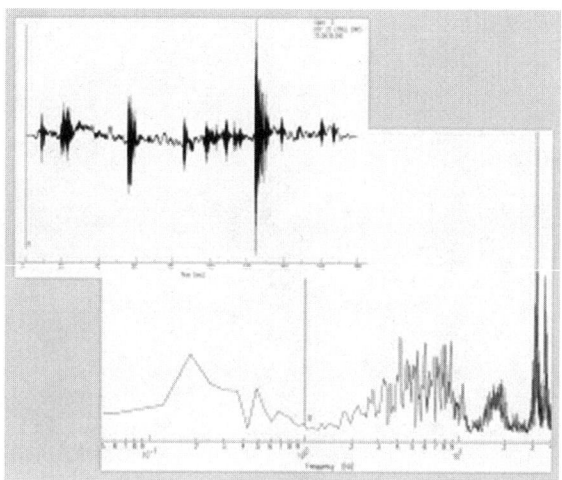

After a preliminary determination of hypocentres using the HYPOINVERSE algorithm (Klein 2002), we determined the accurate location of basal events with the Double-Difference (DD) method of Waldhauser (2001) which provides high precision relative locations (see method description in DBM07). The spatial distribution of epicentres is plotted in Fig. 5: red and green stars represent events recorded respectively during the first (Nov 2003–Jan 2004) and the second experiment (Nov 2005–Jan 2006). Note that all the episodes occurred along the main branch of the David Glacier, that is the southern, faster, stream draining DomeC and no events were detected in the slower tributary flowing from Talos Dome.

Seismic occurrences are clustered in space and generally concentrated around the main icefall (Fig. 5), allowing us to exclude a tectonic origin eventually related to the presence of an active fault and pointing to a glacial-related source. Following DBM07, we distinguish 3 main groups of basal events depending on their location referred to the icefall: Upstream North (UP-N), Upstream-South (UP-S) and Downstream (DW).

Fig. 4 Basal earthquake seismic signal. (**a**) Vertical component of seismic signal (in velocity) vs. time for typical basal event. The weak onset of body waves is followed by high amplitude surface waves that are amplified as reverberations within the ice layer. (**b**) Amplitude of frequency spectrum is rich in low frequencies, typically lower than 1 Hz

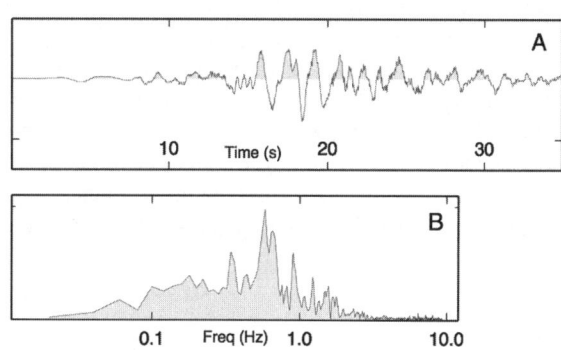

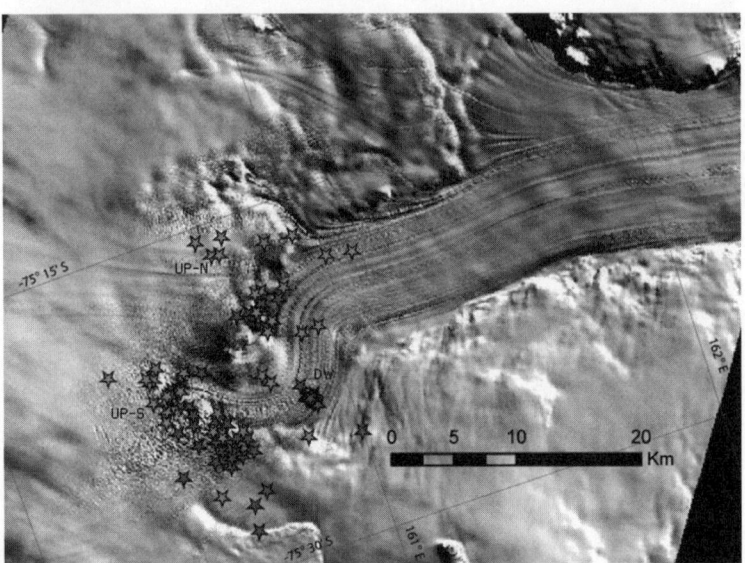

Fig. 5 Epicentral distribution. Spatial distribution of basal event epicentres related to local topography. Satellite image of the David Glacier (ASTER_VNIR 2003 Feb 6th) georeferenced to a South Pole Polar Stereographic projection. Red and green stars represent the location of events recorded respectively during the first (Nov 2003–Jan 2004) and the second experiment (Nov 2005–Jan 2006). Clusters of events are labelled following DBM07: UP-N and UP-S mark respectively Upstream North and South clusters, DW is the Downstream cluster referring to the main icefall. Note that all the events occurred during the 2005–06 campaign were located at the top of the icefall, and none of them occurred in the DW area

2.2 Upstream Events

The UP-N and UP-S clusters count together 74 events occurring sparsely at the top of the icefall, in correspondence of the area where the ice flows fast (more than 500 m/y from Frezzotti et al. 1998), the bedrock topography drops 300–400 m and the mean slope is around 60–80% (Rignot 2002).

Following the procedure explained in DBM07, we found the Mw magnitude for all the events as shown in Fig. 6 where orange and green bars represent the magnitude distribution vs. event count for UP clouds (2003–04 and 2005–06 respectively).

The events observed in this study have magnitude ranging between 1.1 and 2.3, generally do not indicate evident foreshock/aftershock pattern and do not follow the typical Gutenberg-Richter distribution of magnitude–frequency scaling (Gutenberg and Richter 1944).

Changes in topography have a major influence on glacier dynamics, as the driving shear stress at the glacier bed σ_d depends on both ice thickness h and surface slope α as

$$\sigma_d = \rho g h \sin \alpha \qquad (1)$$

where ρ is the density of ice and g is gravity.

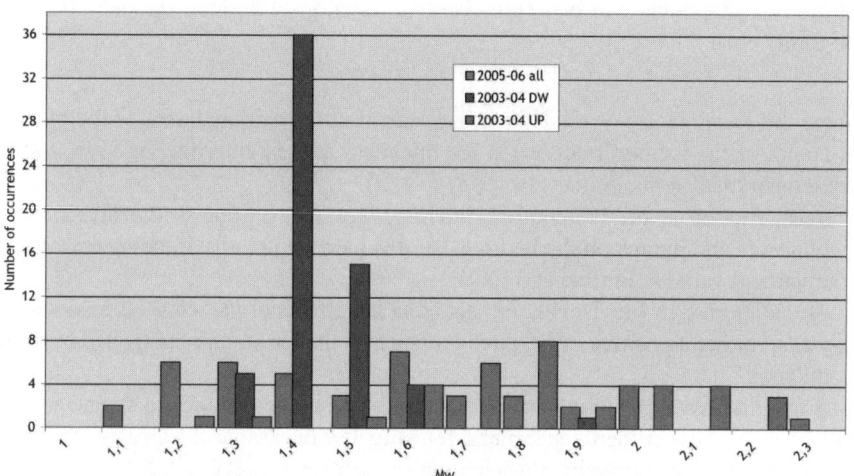

Fig. 6 Magnitude distribution. Distribution of occurrences with respect to moment magnitude Mw of all the basal events
Colours are different for events of different clusters: red bars refer to DW events, orange bars refer to UP events (first Campaign), and green bars refer to UP events of the second Campaign

The coupling between a moving glacier and its bedrock is generally a highly non linear problem with net basal shear stress depending upon ice plastic properties, till deformation, characteristics of sediments beneath the till, basal melt rate, basal topography and roughness at the ice-till interface.

Nevertheless, if we assume a remarkable effect of abrasion over the bedrock at the top of the icefall, owing to the steep slope and the fast sliding flow, we can neglect the sediments and their deformation and expect the movement of a lubricated glacier on the bedrock will be primarily controlled by ice deformation.

Provided realistic values for parameters involved in the equation, the driving shear stress at the ice-rock interface is:

$$\sigma_d = \rho g h \sin \alpha \approx 2.25 MPa$$

where:
$\rho = 917 \text{ kg} \cdot \text{m}^{-3}$, $g = 9.81 \text{ m} \cdot \text{s}^{-2}$, $h = 500 \text{ m}$, $\sin \alpha \approx 0.5$ assuming $\alpha = 30°$.

In fact, besides weight of ice and gravity, several forces brake heavily the ice stream flow.

The dominant resistive stress, which may support more than 50% of the driving stress (Raymond et al. 2001), is the side drag exerted by slow flowing ice at borders. Hooke (2001) demonstrates that the effect of drag on the valley side may yield a reduction of the flow velocity at depth up to a factor 8.

Streaming in a deep and rugged valley, David Glacier is actually heavily crevassed along its edges, and lateral drag might be reduced but not negligible.

Basal drag is another decelerating stress in a valley glacier. It largely depends on the bed shape and can be expressed as a function of depth z (Hooke 2005) in the form:

$$\sigma_b(z) = -S_f \rho g z \sin \alpha$$

where the *shape factor* S_f is defined in terms of cross sectional area of the glacier A, length of the ice-bed interface P, ice thickness at the centerline H: $S_f = A/PH$ (for a theoretical semicircular glacier $S_f = 1/2$).

Basal friction at the ice-bedrock interface can also be due to the presence of roughnesses and bumps on the bedrock, so it is hard to quantify it, being extremely irregular and variable in time and space.

An outlet glacier, like David, will also feel the effects of the ice shelf load which may exert either a positive or negative pressure on the ice stream controlled by tidal modulation.

A quantitative approach to estimating the force balance on the ice stream would require *ad hoc* experiments and measurements, but the result of all resistive stress might reduce the effective basal shear stress by orders of magnitude.

It is worth noting that an increase in basal shear stress may result in further formation of melt-water at the interface and may favour sliding speed, eventually causing a partial decoupling of the glacier from the bed. On the other hand, obstacles distributed on the bedrock will experience higher drag forces and may originate stick-slip seismic episodes.

2.3 Downstream Events

75 epicentres in the DW cluster occurred in an extremely tight cluster of less than $2\,\text{km}^2$, where glacier likely deviates its flow forced by the presence of topographic boundary (Fig. 5).

Moment magnitudes Mw are plotted in Fig. 6 with red bars revealing a Gaussian-like frequency distribution centred at Mw ≈ 1.4.

Cross-correlations between waveforms for event pairs above 0.95 (DBM07) indicate extremely small separations between the hypocentres and impressing similarity in waveforms (Fig. 7).

The extremely high similarity of hypocentres, waveforms and magnitudes for DW events, implies that the source process cannot be purely ascribed to the sliding friction on the bedrock, but rather suggests the presence of an asperity on the ice–bedrock interface, that repeatedly breaks with a characteristic length.

An isolated asperity, resistant to the ice flow and acting as a sticky-spot, would account for local higher concentration of shear stress and occurrences of recurring, almost identical, episodes.

Our estimate of the cumulated slip provided by the DW events over a full year would be in the order of 16 m (DBM07) representing no more than 2.5% of the annual glacier motion inferred from GPS measurements. Therefore, the basic process by which the flow moves past the asperity is not the seismic slip but its plastic

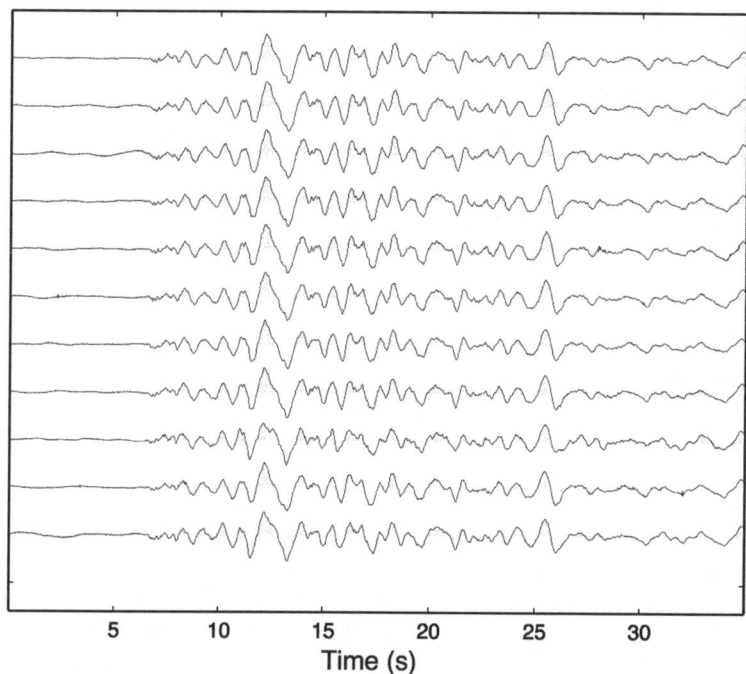

Fig. 7 DW seismic signals. Vertical components of seismograms for 11 different events, as recorded by a seismometer station 34 km WSW of the cluster. The similarity between waveforms for the different events is noteworthy

ice deformation around the obstacle. Episodically, the DW patch experiences brittle failure and, when it slips seismically, the seismic episode is always associated with the same energy and characteristic length.

Interestingly, none of the events recorded during the 2005–06 campaign (green stars in Fig. 5) was located in the area of the DW cluster, suggesting that the sticky-spot would be no longer active either due to the possible abrasion and smoothing of the patch or to higher basal lubrication.

In the latter case, we might expect an increasing flow speed with respect to previous horizontal velocity observations. GPS data collected in the David Cauldron (site ICF1 in Fig. 1) during the 2005–06 austral summer reveal that the flow speed in this point was 1.6 m/day (see Sect. 2) but, unfortunately, we don't have any direct GPS measurement available for the 2003–04 ice flow velocity. Frezzotti et al. (1998) measured the mean ice speed between 1991 and 1994 some km downstream of ICF1 founding a slightly lower value of 1.39 m/day (their station Da4).

The two measures cannot be directly compared because the glacier traces a sharp S-shaped turn between ICF1 and Da4 but we can reasonably think that the flow will reduce its speed before bending and will slowly accelerate again after the corner. Bearing in mind that it is not possible to correlate the two measures exactly, the essential point at this level is their compatibility.

There is the possibility that some DW events have not been detected. This might not be due to the seismic network (during the second experiment on 2005–06 stations were closer to the Cauldron, they worked for a longer period, we used identical instrumentation and software for the data analysis) but we cannot exclude that the DW cluster could have been active before or after the austral summer.

3 GPS Campaign

In order to follow the kinematics of the David-Drygalski, two temporary GPS geodetic stations were installed on the glacier at points ICF1 and DRY1 (Fig. 1), in November 2005 and in January 2006 respectively.

We used dual frequency Trimble5700 GPS receivers equipped with Zephyr and Zephyr Geodetic antennas and 1 Gb removable memory cards; each antenna was tightened to the head of an aluminium pole (well driven into the ice) and acquisition parameters were set with measurement rate 15 s, elevation mask 13° and daily sessions (Fig. 2b).

The equipment was prepared to operate continuously, using solar energy for battery charging. The first point (ICF1) was installed at the bottom of the icefall, as close as possible to the epicentral area of the DW cluster of earthquakes previously described, compatibly with severe crevassing; the second point (DRY1) was installed on the floating ice tongues at the bottom of Hughes Bluff outcrop (red circles in Fig. 1).

As primary master station for the kinematic processing, a GPS receiver was installed on a 3-D benchmark fixed into the rocky outcrop Hughes Bluff, but in order to insert the measurements within the International Terrestrial Reference Frame (ITRF2000), we also used simultaneous GPS acquisitions from the GPS permanent station located at the Mario Zucchelli Station (TNB1).

Kinematic GPS is a satellite-based relative survey method that allows for the evaluation of a moving receiver trajectory with respect to a reference station. Using this method, accuracies in a range of few centimetres in relative positioning are achievable. In a classical carrier phase differenced approach, positioning reliability is mainly related to the correct setting of unknown phase ambiguities as integer numbers. Therefore the problem of the area coverage is related to ambiguity resolution, which is typically possible in the range of limited distances (generally speaking 15–30 km) from a reference station. On wider ranges the solution becomes more disturbed owing to the de-correlation of systematic effects in the GPS measurements and particularly at high latitudes, where the presence of large ionospheric activities causes significant scintillation effects on the GPS signal.

The Drygalski Ice Tongue fluctuates forced by the Ross Sea tide. One of our goals was the investigation on the effect of the ocean tide on the ice tongue vertical and horizontal movements. Unfortunately, we couldn't observe any direct correlation between tidal forcing and seismic occurrences because we didn't register any earthquake in the DW area during the GPS campaign.

An analysis of discontinuous measurements acquired in 1991, 1994 and 1995 by tide gauges installed at MZS, reveals that the stability values of the major components of ocean tides (O_1, K_1, Q_1, M_2, S_2 and N_2) ranges from 99.6 to 100%. The other components (M_4, M_3, MN_4, $\mu_2 M_1 MO_3 L_2$, M_6 and $2MS_6$) show stability values in the range 8097%. These 15 harmonic constants can nearly completely account for amplitude values (Capra et al. 1999). The tide character expressed by the ratio of the sums of amplitudes (O_1 and K_1 with respect to M_2 and S_2) is 2.72, therefore the dominant pattern of the tide is diurnal, with small semi-diurnal components. This means that during a lunar day (24 h and 50 min) there is only one high tide and only one low tide with a clear spring-to-neap tidal cycle.

Unfortunately a technical problem occurred at the reference station Hughes Bluff in the period 024/06-030/06 and so we processed the ionosphere-free (LC) observable double-differenced with respect to the Italian permanent station TNB1 which is continuously operative at the Italian scientific station MZS. Due to the long distance between the David-Drygalski and TNB1 (more than 100 km), the noise level observed in relative positioning is about one decimetre.

Data processing of Antarctic long-range kinematic acquisition requires the development of strategies especially tuned for each experiment. In order to manage the effects of such stringent constraints, we adopted the open scientific GPS data processing software Gamit/Globk package (King and Bock 2005) and in particular in this work we analyzed twelve days of contemporaneous GPS observations at DRY1 and ICF1 using the kinematic module TRACK version 1.15 (Chen 1998; Herring 2002; Herring et al. 2006).

In the data analysis process we computed site positions by means of RINEX format (Receiver INdependent Exchange) data files and precise ephemeris postcalculated by the IGS (International GNSS Service) using global data.

In the last version (1.15) of TRACK, a tool for antenna phase-centre mapping has been implemented so that the software is also able to read the file of absolute calibration (e.g. igs05_1402.atx). TRACK allows for the use of different strategies of analysis depending on the site distance from the reference master station. In our case, the length of the baselines is about 100 km and the differential ionospheric and tropospheric delays are very high. For this reason, the software uses a floating-point estimate technique and ionospheric delay constraint to LC observable (the Melbourne-Webbena linear combination is often used to resolve L1–L2 ambiguities and several different approaches to determine L1 and L2 cycles separately).

In Fig. 8 we report the comparison between vertical components derived by GPS kinematic solutions (computed every 15 s) at points ICF1 and DRY1 and the predicted ocean tide values, computed in the same period adopting the 26 harmonic constants estimated at MZS.

The vertical motion at point DRY1 with periodic undulations up to 70 cm (Fig. 8) demonstrates the clear response of floating ice-tongue to the sea tide. On the other hand, the amplitude of vertical motion at point ICF1 is strongly reduced, suggesting that ICF1 is constrained and located close to the grounding line, possibly 1–2 km upstream of the grounding line if compared with InSAR interpretations (Rignot 2002).

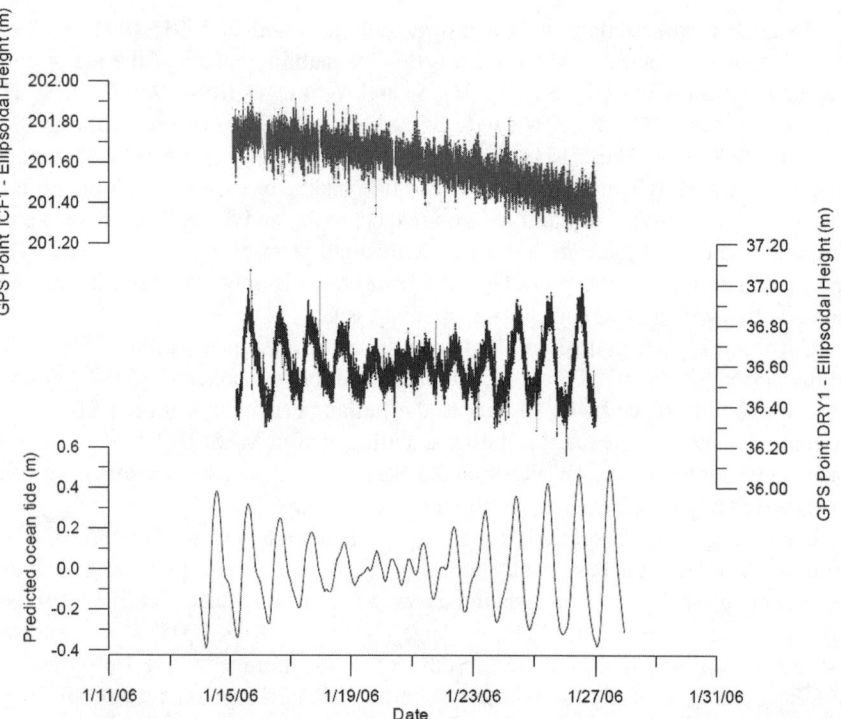

Fig. 8 GPS Vertical Displacements. ICF1 and DRY1 ellipsoidal heights derived by kinematic GPS solutions as a function of time, with respect to predicted sea tide

The horizontal velocities of the Drygalski Ice-tongue vary as well as a function of time with respect to tidal amplitudes: in the two-week acquisition time, the extent of speed amplitude with respect to spring-to-neap cycles decreased till a neap tide and increased during the period that preceded the spring tide (Fig. 9).

GPS data collected during the observed period of the 2005–06 austral summer reveal that the average horizontal flow speed was 1.6 m/day (ca 580 m/y) at point ICF1 with azimuth 42°57′, and 1.48 m/day (ca 540 m/y) with azimuth 101°15′ at point DRY1.

4 Discussion and Conclusions

The deployment of seismographic and geodetic temporary stations around the David Glacier allowed us to collect simultaneous time series of data and to analyze them jointly.

Low-magnitude seismic occurrences at the ice-bedrock interface are common, but high concentration of shear stress is purely due to basal friction or to the presence of bedrock asperities acting as sticky-spots. In fact, epicentres are either spread at

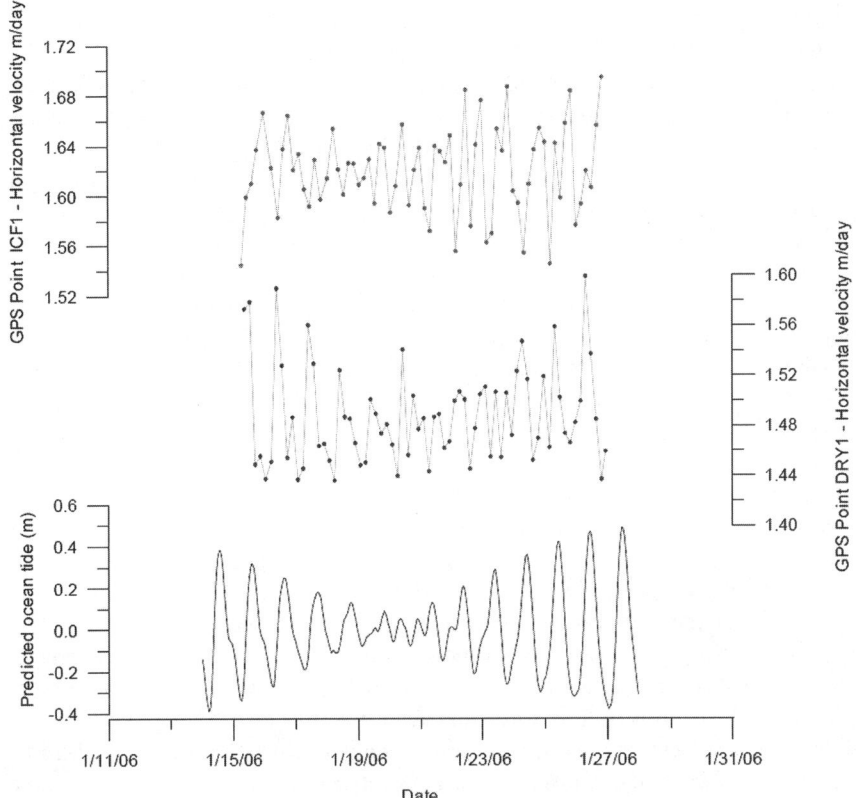

Fig. 9 GPS Horizontal Displacements. ICF1 and DRY1 horizontal velocity measured by kinematic GPS approach. The picture compares predicted ocean tide and horizontal speed of GPS points, showing that speed variation with time depends on tidal amplitude

the top of the icefall (where the mean slope is around 60–80%) or focused in small clusters, which would reveal the presence of basal patches. Nevertheless, isolated asperities would not affect significantly the stability of the ice stream, but rather sliding, creeping and basal deformations would account for most of the glacier motion.

GPS observations at the two sites placed after the icefall reveal that both vertical and horizontal glacier displacements are largely forced by the Ross Sea tides. Unfortunately, none of the events occurring during the GPS campaign were located in the same area, so we couldn't verify possible correlations between the tide and the seismic triggering.

As expected, the vertical motion of the floating tongue (site DRY1) is wholly correlated, in phase and amplitude, with the predicted tide (Fig. 8). On the contrary, the major movement at ICF1 is descendent with only moderate oscillations, so we can infer that the glacier is still anchored at the bedrock in that point, upstream of the grounding line.

Interestingly, the dominant diurnal ocean tide also regulates horizontal velocities of the ice stream with a clear temporal variation correlated with the spring-to-neap tidal cycle (Fig. 9). Longitudinal perturbations propagate rapidly over large distances, a conclusion that is supported by several previous studies (Payne and others 2004; Gudmundsson 2006). In fact, the tidal oscillation acts as a vertical stress modulator: neap and spring tide correspond respectively to reduction and increment of vertical stresses (and glacier weight) on the bedrock, making the ice flow faster or slower. It is worth noting that amplitude variations at ICF1 are not in phase with the tide: the delay might be introduced as combined effect of plastic necking upstream of the grounding line and possible variable transfer of water filling cavities where the glacier decouples from its bed and begins to float.

Our estimate of ICF1 horizontal velocity is in good agreement with InSAR evidences: the site would be located inland, between the InSAR equal-velocity 500 and 600 m/y contour lines reported in Rignot (2002) which is confirmed by our estimate of 580 m/y for horizontal motion.

GPS data collected during the 2005–06 austral summer reveal that the flow speed generally agrees with data collected between 1991 and 1994 by Frezzotti et al. (1998): in particular, the horizontal velocity measured in 1991–1994 at point Da2 (1.51 m/day) located at S 75°21′45″, E 162°08′58″, is quite confirmed by the new GPS measurements carried out at point DRY1 eleven years later (Fig. 9).

In line with recent observations (Shepherd and Wingham 2007), these results may suggest that the average flow of the glacier has not varied significantly over the last decade in terms of flow horizontal velocities: in a time when polar areas are generally experiencing increasing instability owing to late global warming, the East Antarctic Ice Sheet still exhibits positive mass balance, contributing to moderate the sea-level rise.

Acknowledgments The manuscript was significantly improved by comments from anonymous reviewers. We thank all the people involved in preparation, deployment and retrieval of stations and data. This research was conducted with field logistic and financial support provided by the Italian national organization for research in Antarctica (PNRA S.C.r.l.).

References

Alley RB (1992) Sticky spots under ice streams. Antarctic Journal of the U.S. 28 Vol 5 pp 50–51.
Alley RB, Blankenship DD, Bentley CR, Rooney ST (1986) Deformation of till beneath ice stream B West Antarctica. Nature vol 322 6074 pp 57–59. doi:10.1038/322057a0.
Alley RB, Clark PU, Huybrechts P, Joughin I (2005) Ice-sheet and sea level changes. Science 310 pp 456–460.
Anandakrishnan S, Bentley CR (1993) Micro-earthquakes beneath Ice Streams B and C, West Antarctica: Observations and implications. J. Glaciol. vol 39 133 pp 455–462.
Anandakrishnan S, Voigt DE, Alley RB and King MA (2003) Ice stream D flow speed is strongly modulated by the tide beneath the Ross Ice Shelf. Geophys. Res. Lett. vol 30 7 pp 1361–1364 doi:10.1029/2002GL016329.
Bahr DB, Rundle JB (1996) Stick-slip statistical mechanics at the bed of a glacier. Geophys. Res. Lett. vol 23 16 pp 2073–2076.

Bannister S, Kennett BLN. (2002) Seismic Activity in the Transantarctic Mountains – Results from a Broadband Array Deployment. Terra Antarctica 9 vol 1 pp 41–46.

Bindschadler RA, King MA, Alley RB, Anandakrishnan S, Padman L (2003) Tidally Controlled Stick-Slip Discharge of a West Antarctic Ice Stream. Science 301 pp 1087–1089 doi: 10.1126/science.1087231

Blankenship DD, Bentley CR, Rooney ST Alley RB (1986) Seismic measurements reveal a saturated porous layer beneath an active Antarctic ice stream. Nature vol 322 6074 pp 54–57. doi:10.1038/322054a0.

Capra A, Gandolfi S, Lusetti C, Stocchino C, Vittuari L (1999) Kinematic GPS for the study of tidal undulation of floating ice tongue. Bollettino di Geodesia e Scienze Affini IGM LVIII no. 2 pp 151–173.

Chen G (1998) GPS Kinematic Positioning for the Airborne Laser Altimetry at Long Valley California. Ph. D. Thesis Massachusetts Institute of Technology pp. 173 Cambridge MA.

Danesi S, Bannister S, Morelli A (2007) Repeating earthquakes from rupture of an asperity under an Antarctic outlet glacier. Earth Planet. Sci. Lett. 253 vol 1–2 pp 151–158.

Doake CSM, Corr HFJ, Nicholls KW, Gaffikin A, Jenkins A, Bertiger WI and King MA (2002) Tide-induced lateral movement of Brunt Ice Shelf, Antarctica. Geophys. Res. Lett. vol 29 8 pp 1226–1229 doi:10.1029/2001GL014606

Ekström G (2006b) Global detection and Location of Seismic Sources by Using Surface Waves Bull. Seismol. Soc. Am. 96 4A pp 1201–1212.

Ekström G, Nettles M, Abers GA (2003) Glacial earthquakes. Science 302 pp 622–624.

Ekström G, Nettles M, Tsai VC (2006a) Seasonality in Increasing Frequency of Greenland Glacial Earthquakes. Science 311 pp 1756–1758.

Frezzotti M (1993) Glaciological study in Terra Nova Bay Antarctica inferred from remote sensing analysis. Ann. Glaciol. 17 pp 63–71.

Frezzotti M, Capra A, Vittuari L (1998) Comparison between glacier ice velocity inferred from GPS and sequential satellite images. Ann. Glaciol. 27 pp 54–60.

Frezzotti M, Tabacco IE, Zirizzotti A (2000) Ice discharge of eastern Dome C drainage area Antarctica determined from airborne radar survey and satellite image analysis. Ann. Glaciol. 46 vol 153 pp 253–264.

Gudmundsson GH (2006) Fortnightly variations in the flow velocity of Rutford Ice Stream West Antarctica. Nature 444 vol 7122 pp 1063–1064.

Gutenberg B, Richter CF (1944) Frequency of earthquakes in California. Bull. Seismol. Soc. Am. 34 pp 185–188.

Herring TA (2002) Track GPS kinematic positioning program. Version 1.07. Cambridge MA: Massachusetts Institute of Technology.

Herring TA, King R, McClusky S (2006) Gamit Reference Manual release 10.3 Report Massachusetts Institute of Technology Cambridge MA.

Hooke RLeB (2005) Principles of Glacier Mechanics, Cambridge University Press, Cambridge UK

Horgan HJ, Anandakrishnan S (2006) Static grounding lines and dynamic ice streams: Evidence from the Siple Coast Antarctica. Geophys. Res. Lett. 33 L18502 doi:10.1029/2006GL027091.

Hulbe CL and Whillans IM (1994) Evaluation of strain rates on Ice Stream B Antarctica obtained using GPS phase measurements. Ann. Glaciol. 20 pp 254–262.

King M and Aoki S (2003) Tidal observations on floating ice using a single GPS receiver. Geophys. Res. Lett. vol 30 3 pp 1138–1141, doi:10.1029/2002GL016182.

King RW and Bock Y (2005) Documentation for the GAMIT GPS processing software release 10.2 Mass. Inst. of Technol. Cambridge.

Klein FW (2002) User's guide to HYPOINVERSE-2000 a fortran program to solve for earthquake locations and magnitudes. U.S. Geol. Survey Open File Report 2–171.

Legresy B, Wendt A, Tabacco I, Remy F and Dietrich R (2004) Influence of tides and tidal current on Mertz Glacier, Antarctica. J. Glac. 50 170 pp 427–435.

Neave KG, Savage JC (1970) Icequakes on the Athabasca Glacier. J. Geophys. Res. 75 vol 8 pp 1351–1362.

Payne AJ, Vieli A, Shepherd AP, Wingham DJ, Rignot E (2004) Recent dramatic thinning of largest West Antarctic ice stream triggered by oceans. Geophys. Res. Letts. 31 L23401.

Raymond CF, Echelmeyer KA, Whillans IM, Doake CSM (2001) Ice stream shear margins. In Alley RB, Bindschadler RA (Eds.), The West Antarctic Ice Sheet: Behaviour and Environment. Antarctic Research Serier, vol 77. Am. Geophys. Union, Washington DC, pp 137–155.

Rignot E (2002) Mass balance of East Antarctic glaciers and ice shelves from satellite data. Ann. Glaciol. 34 pp 217–227.

Rignot E, Thomas RH (2002) Mass Balance of Polar Ice Sheets. Science 297 pp 1502–1506.

Rignot E, Jacobs S (2002) Rapid bottom melting widespread near Antarctic Ice Sheet grounding lines. Science 296 pp 2020–2023.

Salvini F, Storti F (1999) Cenozoic right-lateral strike-slip tectonics in the Ross Sea region Antarctica. In: Ricci C.A. (ed.) The Antarctic region: Geological Evolution and Processes. Terra Antarctica Publ, Siena, pp 585–590.

Shepherd A, Wingham D (2007) Recent Sea-Level Contribution of the Antarctic and Greenland Ice Sheets. Science 315 pp 1529–1532 doi:10.1126/science.1136776.

Smith AM (2006) Microearthquakes and subglacial conditions. Geophys. Res. Lett. 33 L24501 doi:10.1029/2006GL028207.

Van Wormer D, Berg E (1973) Seismic evidence for glacier motion. J. Glaciol. 12 vol 65 pp 259–265.

Waldhauser F (2001) hypoDD — a program to compute Double-Difference hypocentre locations. U.S. Geol. Survey, Open File Report, 1–113.

Geomagnetic Observatories in Antarctica; State of the Art and a Perspective View in the Global and Regional Frameworks

L. Cafarella, D. Di, S. Lepidi and A. Meloni

1 Introduction

The Earth is immersed in a planetary magnetic field. The field is generated in the Earth's core and can be measured at its surface. It shows mainly a typical dipolar profile with the dipole axis roughly parallel to the Earth's rotation axis (tilting about 12°). At low latitudes the field reaches its minimum, while its maximum intensity is observable in polar regions, reaching there almost three times its equatorial value. The region around the Earth where the geomagnetic field extends is known as the Earth's *magnetosphere*. This region contains a very low density gas of electrically charged particles and is the space around the Earth where many electric and magnetic phenomena happen.

A fast outflow of hot plasma is emitted from the Sun in all directions in the interplanetary space. This 'solar wind' at the Earth's orbit carries a variable low-intensity interplanetary magnetic field (IMF). When the IMF reaches our planet it encounters the magnetosphere and, owing to this interaction, most of the solar wind particles are deflected around the Earth, twisting the magnetosphere in a comet type shape. The Earth's magnetic field extends for about 10 Earth's radii in the solar direction and has a long tail on the other side, in the anti-sunward direction. Some solar wind particles leak through the magnetic barrier and are trapped inside. They can also rush through funnel-like openings (cusps) at the North and South polar regions, releasing tremendous energy when they hit the upper atmosphere. The northern and southern lights, known as auroras, are a visible evidence of this energy transfer from the Sun to the Earth. The polar regions are key areas for this

---

L. Cafarella
Istituto Nazionale di Geofisica e Vulcanologia, Rome, Italy, e-mail: cafarella@ingv.it

D. Di
Istituto Nazionale di Geofisica e Vulcanologia, Rome, Italy, e-mail: dimauro@ingv.it

S. Lepidi
Istituto Nazionale di Geofisica e Vulcanologia, Rome, Italy, e-mail: lepidi@ingv.it

A. Meloni
Istituto Nazionale di Geofisica e Vulcanologia, Rome, Italy, e-mail: meloni@ingv.it

energy transfer. The structure and behavior of the magnetosphere is determined not only by its internal source, the Earth's main magnetic field, but also by the solar wind and their interaction.

On the Earth the magnetic field varies both in space and time. Spatial variations are related to the dipolar geometry (which is only the first order of the mathematical model) but also to higher order terms of core origin that generate very large scale (thousands km) magnetic field structures; spatial variations also include, at smaller scale, wavelengths due to a magnetic crustal contribution. Time variations are present in a wide frequency range. They have both internal and external origins with respect to the Earth's surface. Variations of internal origin cover time scales longer than a few years (2–5 years or more, generally); they are known as 'secular variation' and show up in unpredictable time patterns. The secular variation also causes a slow drifting of the magnetic poles, which can then be assumed constant in position only over a few years time interval.

The diurnal variation of the geomagnetic field is due to the photoionization of the upper atmosphere and to atmospheric tides, while the more rapid time variations are caused by the interaction between the solar wind and the magnetosphere. The solar wind is not constant being influenced by solar disturbances (flares, coronal mass ejections, etc ...) and their fallouts in the interplanetary space. The magnetosphere dynamics can be heavily influenced by the solar wind and IMF variable conditions. In particular the electrical current systems within the magnetosphere can be enhanced causing a general perturbed state of the Earth's magnetic field. Although generally irregularly distributed, the magnetic fluctuations on Earth can exhibit a 27-day recurrence because some magnetic perturbations are related to persistent active regions on the Sun and this is the period of the rotation of the Sun as seen from Earth. Magnetic pole positions show also a more rapid dynamic behavior: a daily motion of the pole is due to the diurnal variation of the Earth's magnetic field and also to solar activity. The distance and speed of these displacements depend on the state of perturbation of the magnetosphere.

Several books on geomagnetism are available for the interested reader: see for example Chapman and Bartels (1940); Parkinson (1983); Backus et al. (1996); Merril et al. (1996); Lanza and Meloni (2006), and for space physics aspects Hargreaves (1992); Kivelson and Russell (1996). Finally Campbell (2001 and 2003) have given a very nice summary of all geomagnetism aspects.

The continuous monitoring of the Earth's magnetic field is the only mean to study the different features of the field and especially of its time variations. This activity is regularly undertaken by geomagnetic observatories all over the world. Observatory recordings reveal geomagnetic variation features which give important information about the geomagnetic field nature, its internal source dynamics and its interaction with the external sources. In this paper the important contribution of geomagnetic observatories in Antarctica is shown and a perspective view of their future is also discussed.

2 Geomagnetic Observatories in Antarctica

All over the world, when and where this was possible, geophysicists have installed permanent structures devoted to recording magnetic field time variations. The datasets obtained from these structures are the basis of many geomagnetic field investigations. For example, following Gauss spherical harmonic analysis methods, the regular measurements of the magnetic field from all over the world are systematically used for a mathematical representation of the geomagnetic field at each epoch (usually every 5 years), known as the International Geomagnetic Reference Field (IGRF).

The geomagnetic field is monitored and consequently recorded in a geomagnetic observatory by means of so called magnetic elements: the horizontal magnetic field intensity, H, the angular difference between geographic north and magnetic north, called Declination, D, and Inclination or dip, I, the angle that the field vector makes with the horizontal plane. Other recorded elements are the intensive elements indicating the total field intensity F and the three perpendicular components X (geographic South-North), Y (West-East) and Z (vertical, positive downward). The most commonly used magnetic field unit is nT.

Two main categories of instruments generally operate in an observatory. The first includes variometers used for continuous measurements of elements of the geomagnetic field vector in arbitrary units (for example electrical voltage in the case of fluxgate instruments). The second category comprises absolute instruments, which can make measurements of the magnetic field in terms of absolute physical basic units and universal physical constants. The most common kind of absolute instrument are the fluxgate theodolite for measuring D and I and the proton precession magnetometer for measuring the field intensity F. In the fluxgate the basic unit is a dimensionless angle, while the proton precession magnetometer is based on the use of the universal physical constant, the gyromagnetic proton ratio, and the basic unit is frequency. Measurements with a fluxgate theodolite can only be made manually whilst a proton magnetometer can operate automatically. A detailed report on magnetic observatory operations and instruments can be found in Jankowsky and Sucsdorff (1996).

In order to facilitate data exchanges and to make geomagnetic products available nearly in real time, an international coordination programme called INTERMAGNET is in operation among many world magnetic observatories (Kerridge 2001). The programme has allowed the establishment of a global network of cooperating digital magnetic observatories, adopting modern common specifications for measuring and recording equipments. An INTERMAGNET Magnetic Observatory (IMO) is a modern magnetic observatory, full filling the following requirements: to provide one minute magnetic field values measured by a vector magnetometer, and an optional scalar magnetometer, all with a resolution of 0.1 nT and an original sampling rate of 5 s. Vector measurements performed by a magnetometer include the best available baseline reference measurement (see http://www.intermagnet.org/).

The global distribution of geomagnetic observatories is strongly unbalanced in favor of the northern hemisphere and leaves the southern hemisphere poorly covered. In Fig. 1a the geomagnetic observatory world distribution is reported for INTERMAGNET observatories. In Fig. 1b all the Antarctic geomagnetic observatories are reported in a polar view. In Table 1 the Antarctic observatories with their geographic and corrected geomagnetic coordinates (IGRF2005) are reported. As it is clear from figures and table, most of the Antarctic observatories are located, for practical and historical reasons, along the coast. For this reason they are consequently subject to coast effects (a strong influence from horizontal electrical conductivity contrast between continent and ocean) and crustal field contamination. To decrease the influence of coast effects on data a more uniform observatory distribution is necessary; it would be very important also to improve the definition of the mathematical global models of the field in polar areas. The only continental stations are South Pole and Vostok that through the years have produced very significant data. Recently (from 2004) a geomagnetic observatory is working at Concordia station (code DMC) built as the result of an agreement between the French and Italian Antarctic Programs (IPEV and PNRA respectively) at Dome C (Lepidi et al. 2003). This site is particularly interesting for geomagnetic studies for many reasons. In particular since it is very close to the present position of the geomagnetic south pole (its geomagnetic latitude is nearly 89°S)

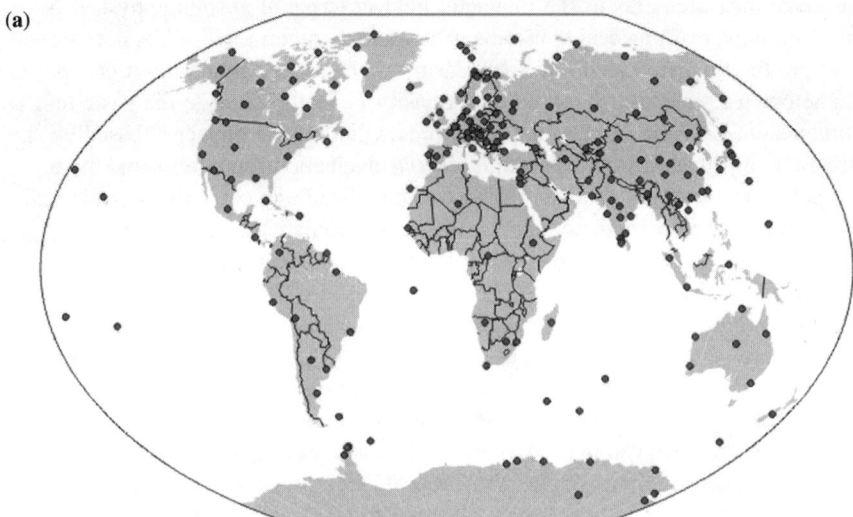

Fig. 1 (**a**) INTERMAGNET network of magnetic observatories (picture from BGS website (http://www.geomag.bgs.ac.uk/); (**b**) Magnetic observatories over Antarctica (picture adapted from http://www.geoscience.scar.org)

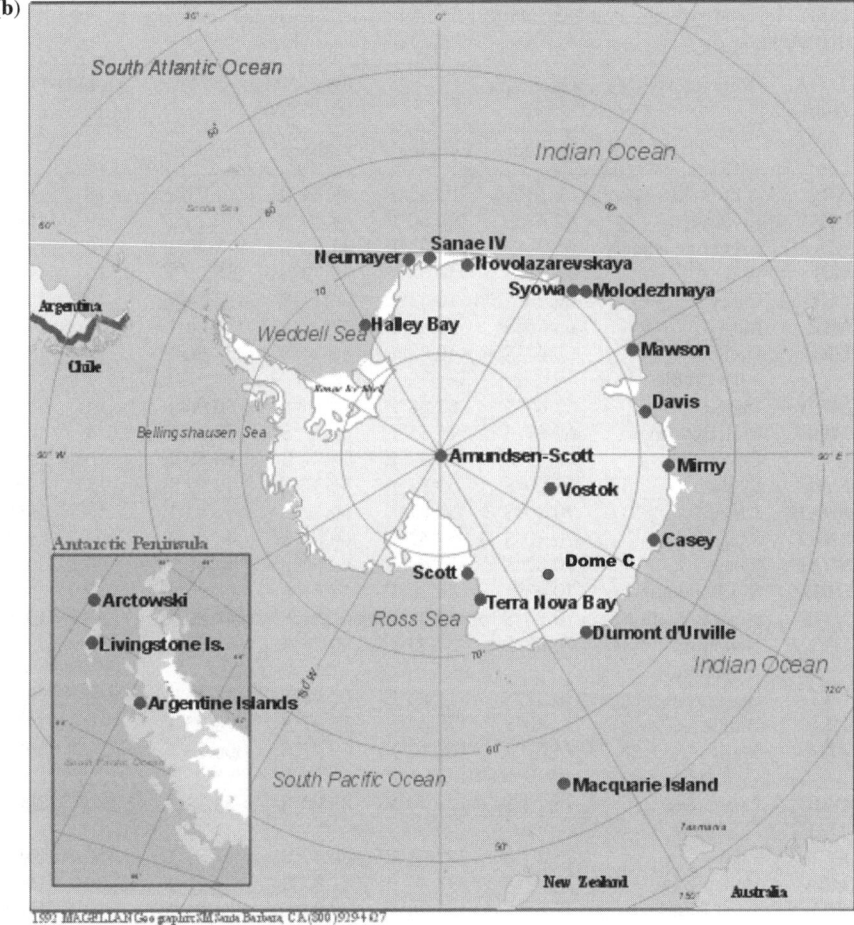

Fig. 1 (continued)

and because of its location on a very thick (>3,000 m) ice cap. This observatory, being so far from the bedrock, is less sensitive to the crustal magnetic anomalies that always bias measurements, for example the vertical component of the field.

The importance of magnetic observatories in Antarctica, is therefore the capability of continuously monitoring the geomagnetic field. In fact, time variations of both internal and external origin in polar regions show very peculiar features that can be used to address general problems related to the Earth's magnetic field. In the following, we will describe some examples of the important results obtained from the high latitude geomagnetic data set in the last years.

Table 1 Antarctic observatories with geographic and corrected geomagnetic coordinates (IGRF2005)

IAGA code	Observatory	Geographic		Corr. Geomagn IGRF 2005		Altitude (m)	Data from
		Latitude	Longitude	Latitude	Longitude		
ARC	Arctowski	62.17°S	301.52°E	47.84°S	11.71°E	16	1978.8
LIV	Livingston	62.67°S	299.60°E	48.19°S	10.70°E	19	1997.0
AIA	Argentine Islands	65.25°S	295.73°E	47.63°S	8.40°E	10	1960.5
WIL	Wilkes	66.25°S	110.58°E	80.79°S	158.05°E	10	1960.5
CSY	Casey	66.28°S	110.53°E	80.73°S	157.84°E	40	1978.5
MIR	Mirny	66.55°S	93.02°E	77.31°S	124.04°E	20	1960.5
DRV	Dumont D'Urville	66.67°S	140.02°E	80.34°S	235.90°E	30	1960.5
MAW	Mawson	67.60°S	62.88°E	70.44°S	90.96°E	3	1960.5
MOL	Molodyozhnaya	67.67°S	45.85°E	66.77°S	78.38°E	854	1965.5
DVS	Davis	68.58°S	77.97°E	74.71°S	100.80°E	0	1979.4
SYO	Syowa	69.00°S	39.58°E	66.37°S	72.23°E	15	1960.5
SNA1	Sanae 1	70.30°S	357.63°E	60.71°S	45.03°E	50	1962.7
SNA2	Sanae 2	70.30°S	357.42°E	60.68°S	44.90°E	50	1971.7
SNA3	Sanae 3	70.32°S	357.42°E	60.69°S	44.90°E	50	1980.5
RBD2	Roi Baudouin 2	70.43°S	24.30°E	64.61°S	60.92°E	39	1964.7
NWS	Norway Station	70.50°S	357.47°E	60.82°S	44.80°E	80	1960.5
GVN2	Neumayer Station	70.67°S	351.73°E	60.19°S	41.29°E	0	1993.6
NVL	Novolazarevskaya	70.77°S	11.82°E	62.94°S	53.15°E	460	1961.5
HLL	Hallett	72.30°S	170.32°E	77.27°S	298.93°E	0	1960.5
TNB	Mario Zucchelli Station	74.68°S	164.12°E	79.97°S	306.70°E	28	1987.1
DMC	Concordia Station	75.10°S	123.40°E	88.90°S	54.38°E	3,280	2005.0
EGS	Eights	75.23°S	282.83°E	60.32°S	5.89°E	450	1963.7
HBA	Halley Bay	75.52°S	333.40°E	61.85°S	29.31°E	30	1960.5
SBA	Scott Base	77.85°S	166.78°E	79.94°S	326.54°E	10	1964.5
VOS	Vostok	78.45°S	106.87°E	83.65°S	54.93°E	3,500	1960.5
PTU	Plateau	79.25°S	40.50°E	71.95°S	52.37°E	3,620	1966.5
BYR1	Byrd Station 1	79.98°S	240.00°E	68.29°S	353.43°E	1,515	1960.5
BYR2	Byrd Station 2	80.00°S	240.51°E	68.24°S	353.64°E	1,515	1962.5
SPA	Amundsen-Scott South Pole	90.00°S	0.00°E	74.00°S	18.90°E	2,800	1960.5

3 Analysis and Modeling of the Earth's Magnetic Field and Secular Variation

Global mathematical models are used to make world maps of the Earth's magnetic field. Models are made for different epochs, also to make maps of secular variation (isoporic charts). Antarctica is characterized by one of the isoporic foci (areas of maximum rate of change of the main field elements) so the monitoring of the absolute level of all magnetic field elements here is fundamental.

The International Geomagnetic Reference Field (IGRF) is a global model of the Earth's magnetic field based on the international co-operation among geomagnetic field data contributors and modelers. IGRF represents the main magnetic field (of core origin) without external sources. The model is composed of a series of spherical harmonics with their associated coefficients. The first three terms describe the geomagnetic field of a dipole. Every five years the IGRF is revised and during the 5-year intervals between consecutive versions of the model, linear interpolation of the coefficients is recommended. The most recent IGRF set of coefficients is reported in McMillan and Maus (2005). It is generally agreed that the IGRF achieves an overall accuracy of better than $1°$ in declination. To attain these accuracies must be taken into account possible crustal field contaminations, daily variations and magnetic storms. For this reason the contribution of magnetic observatories is very important.

Torta et al. (2002) and De Santis et al. (2002) used a slightly different technique to represent, in a more accurate way, the field in a cap area delimited by latitude $60°$ South encircling the Antarctic continent and large part of the Southern Ocean. The model is based on the use of the Spherical Cap Harmonic Analysis (SCHA). Introduced by Haines (Haines 1985 and 1990), SCHA is a regionalization of the global model reduced to a spherical cap by means of non integer Legendre polynomials and Fourier series. The Antarctic reference model based on the use of this technique was called Antarctic Reference Model (ARM). In the most recent ARM version (Gaya-Piqué et al. 2004 and 2006), annual means of X, Y and Z components registered at Antarctica observatories as well as a selected subset of satellite total field value data, have been used to develop a model, formed by 123 coefficients. In Fig. 2 the maps of the magnetic field components for 2005 epoch from this model are reported for Antarctica.

The Antarctic region is the area where the South geomagnetic and magnetic poles are located. The location of the first one is determined by the dipole part of the IGRF global model, whereas the second one is the point on the Earth's surface where the IGRF magnetic field is purely vertical (i.e., inclination is $-90°$). In Fig. 3 the location of the two poles from 1900 to 2010 (prediction) based on IGRF2005 model is reported; in the same figure also the position of the measured south magnetic pole is indicated by stars starting from 1840. Its measured coordinates and their observers are reported in Table 2.

Observations of secular variation in Antarctica, by magnetic observatories, show a rapid decrease in the total magnetic field, as recently reported for example by Rajaram et al. (2002). The authors accurately report on this effect noting that a large region of the southern hemisphere has suffered a strong decrease in F over the past five decades. The maximum decrease occurs in a belt covering Argentine island, Sanae, Maitri, Novolazarevskaya, Syowa and Hermanus. Some authors have speculated that this rapid decrease would be of global relevance implying a progress towards a dipole reversal as happened several times in the Earth's history (see for example De Santis et al. 2004) although others have denied this possibility (Gubbins et al. 2006).

One of the most unusual features of the temporal change of the magnetic field for a given element, is the geomagnetic jerk. This is a rapid change in the secular

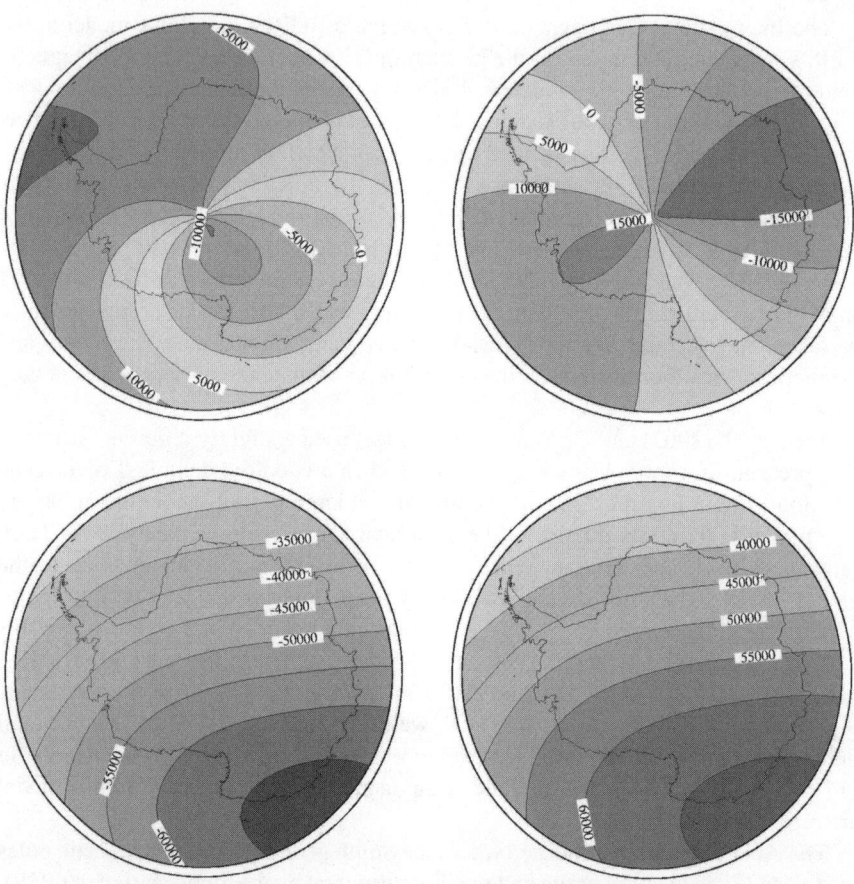

Fig. 2 Magnetic maps, in nT, produced by using the Antarctic Reference Model at epoch 2005.0 (sea level) for X (top left), Y (top right), Z (bottom left), and F (bottom right) magnetic elements

variation slope that takes place in periods of one or two years. More precisely, it is an abrupt change (a step-function) in the second time derivative (the secular acceleration) of the geomagnetic field. Jerks are of internal origin (Malin and Hodder 1982) and represent a reorganization of the secular variation while their short time scale implies that they could be due to a change in the fluid flow at the surface of the Earth's core (Waddington et al. 1995). Recent papers try to explain the physical origin of jerks: Bloxham et al. (2002) suggest that jerks can be explained by oscillatory flows (torsional oscillations) in the Earth's core. Moreover, Bellanger et al. (2001) show a correlation between geomagnetic jerks and the Chandler wobble (the motion due to the flattening of the Earth, that appears when the Earth rotation axis does not coincide with the polar main axes of inertia). From the historical magnetic records, there is evidence of some jerks (mainly at European observatories) which have occurred during the last forty-five years in approximately 1969, 1978, 1991, and 1999.

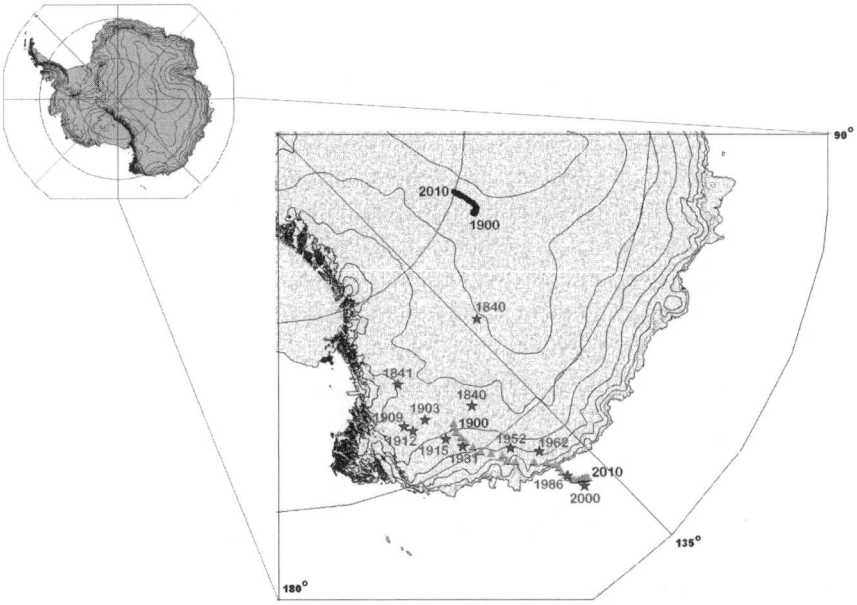

Fig. 3 Locations of geomagnetic and magnetic poles based on IGRF model from 1900 to 2010. South geomagnetic pole position is reported as black dots while south magnetic pole is reported as blue triangles. Red stars indicate south magnetic pole positions from measurements

Recently, Meloni et al. (2006), investigated the presence of jerks in Antarctica by means of both single and multi-station analyses applied to the longest available magnetic observatory time series. The existence of jerks was verified in some of the studied observatories, but not in all. The difficulties in the managing Antarctic observatories (magnetic measurements restricted in many cases to Austral summer for instance) limit the amount of high quality data, necessary for these kind of studies. Figure 4 shows the intensity maps relative to the 1969, 1978, 1991 jerks in the Antarctic region, obtained from merging the analysis of the trend of the secular

Table 2 Measured coordinates of south magnetic pole

Year	Observer	Latitude	Longitude
1840	*Doumlin, Coupvert*	75°20′ °S	132°20′ °E
1840	*Wilkies*	71°55′ °S	144°00′ °E
1841	*Ross*	75°05′ °S	154°08′ °E
1899	*Bernacchi, Colbeck*	72°40′ °S	152°30′ °E
1903	*Chetwynd*	72°51′ °S	156°25′ °E
1909	*Mawson*	72°25′ °S	155°16′ °E
1912	*Webb*	71°10′ °S	150°45′ °E
1931	*Kennedy*	70°20′ °S	149°00′ °E
1952	*Mayaud*	68°42′ °S	143°00′ °E
1962	*Burrows, Hanley*	67°30′ °S	140°00′ °E
1986	*Quilt, Barton*	65°20′ °S	139°10′ °E
2000	*Barton*	64°40′ °S	138°20′ °E

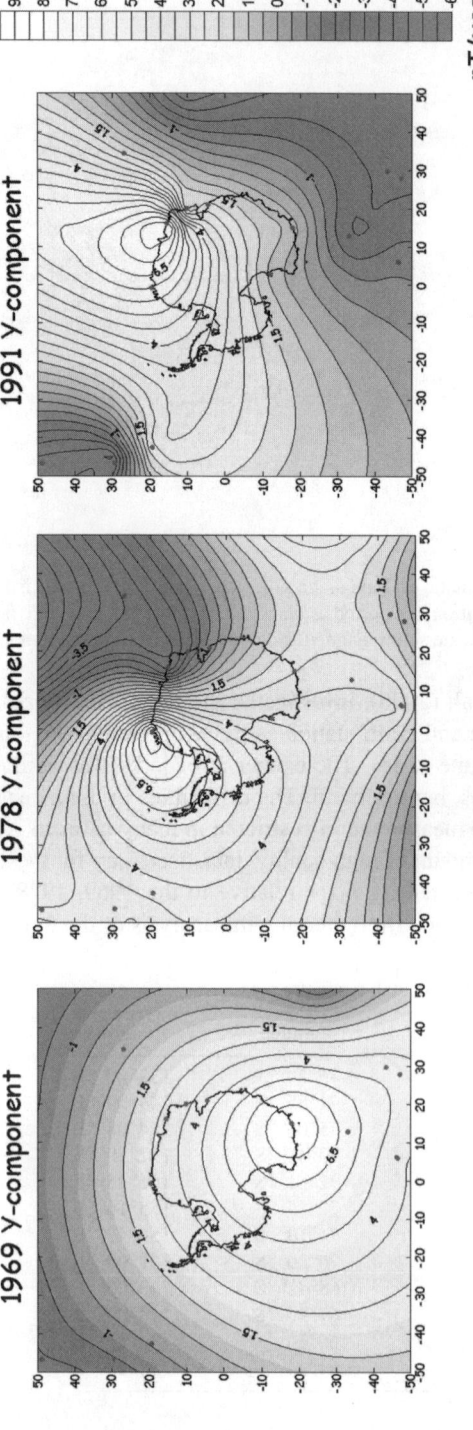

Fig. 4 Azimuthal equidistant projection of the spatial distribution of 1969, 1978 and 1991 jerks intensity. The maps, expressed in nT/year², are relative to the Y component

variation recorded in Antarctic and Southern hemisphere observatories (indicated by points in the figure). As is evident from the figure, the three events present similar structure with a focus of maximum intensity that shows a longitudinal rotation in time, from 120°E to 20°W and finally to 20°E. However, the authors do not exclude that this result could be influenced by low data quality and by bad data distribution coverage.

4 Data Analysis for External Earth Studies

Geomagnetic field measurements in polar regions are very important for investigating magnetospheric dynamics and dynamic processes controlling the energy transfer from the solar wind to the Earth's magnetosphere. Indeed, a direct access of solar wind particles and fields is possible, under particular conditions, only in polar areas and in particular through the polar cusps, which are the regions typically near the 77° geomagnetic latitude (see Fig. 5) separating sunward field lines, closed in the dayside magnetosphere, from open field lines, which are swept back on the tailward side and can be connected to the IMF.

The direct connection of polar regions to the magnetopause (the boundary between magnetosphere and the solar wind) and to the geomagnetic tail, is one of the causes of some phenomena like dayside auroral emissions and the driving of

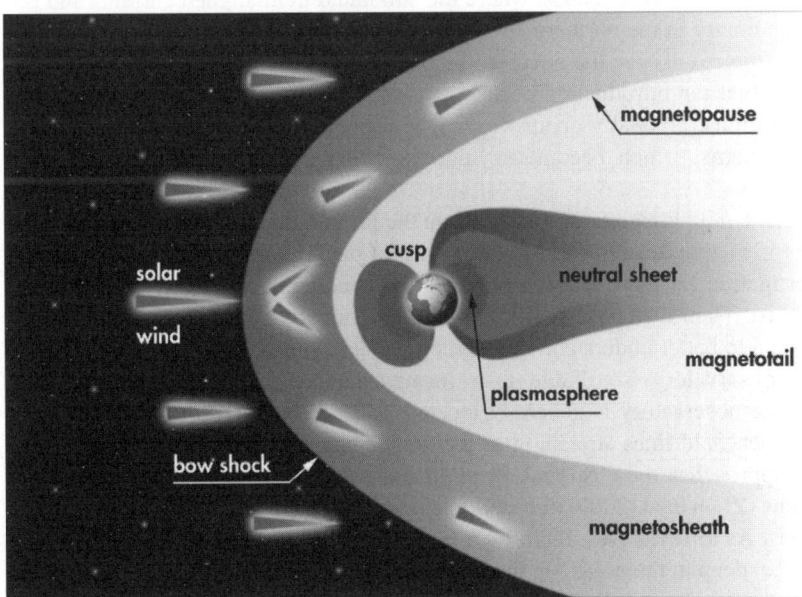

Fig. 5 The Earth magnetosphere, the space in which the Earth's magnetic field is confined (picture taken from the website "*Window to the Universe*", (http://www.windows.ucar.edu/), University Corporation for Atmospheric Research)

magnetic substorms, which are geomagnetic disturbances due to the reconnection between geomagnetic and IMF lines. The auroral zone is also the region where field aligned currents (FACs) flow into and away from the ionosphere, giving rise also to particular geomagnetic phenomena. Auroras and substorms are probably among the most important magnetic phenomena of the polar region. Understanding these phenomena is in fact leading to a more complete view of the interaction and underlying processes that exist among the various components of the ionosphere-magnetosphere-solar wind system.

The coordinate system used to study the coupling between the IMF and the magnetosphere is the Geocentric Solar Magnetospheric System (GSM), in which the three orthogonal components are defined as follows: the X-axis is directed from the Earth to the Sun; the Y-axis is perpendicular to the Earth's magnetic dipole so that the $X - Z$ plane contains the dipole axis; the positive Z-axis is in the same sense as the northern magnetic pole. The basic interplanetary parameter used to determine the interaction with the solar wind is the IMF B_z vertical component: for southward directed IMF (negative B_z) there is an interconnection between interplanetary and magnetospheric field lines, which provides entry to solar wind particles and triggers the biggest global perturbations of the magnetosphere, i.e. geomagnetic storms (Gonzalez et al. 1999); for northward directed IMF (positive B_z) the magnetosphere is essentially closed and the global geomagnetic activity is reduced.

At high latitude also the IMF east-west component B_y is very important, particularly for the geometry and the currents of the polar cap, i.e. the region at the footprint of open field lines, connected to the IMF. This feature clearly emerges from Fig. 6 (from Vennerstrom et al. 2005), where the simulated field aligned currents and polar cap geometry in the northern hemisphere are shown for different orientations of the IMF (the results for the southern hemisphere should be the mirror image). It is evident that the introduction of an eastward IMF component, during northward IMF conditions, gradually opens the magnetosphere and changes the ionospheric current patterns, which become asymmetric with respect to the noon-midnight meridian.

Magnetic field fluctuations measured on the ground in polar regions by magnetic observatories are a result of ionospheric current patterns variability and thus a tool of investigation for many plasma processes. In Antarctica, only a few observation points exist. Their data have been used both individually and with conjugate hemisphere data to better understand the topological behavior of the magnetosphere.

TNB observatory (see Table 1) is located at corrected geomagnetic latitude 80.0°S; the observatory is generally located in the polar cap, i.e. under magnetospheric open field lines stretching in the geomagnetic tail. However, around local noon it approaches the cusp and, in particular magnetospheric and interplanetary conditions (Zhou et al. 2000), the station could be situated at the footprint of closed field lines. As to DMC (see Table 1), it is located at corrected geomagnetic latitude 88.9°S, i.e. deep in the polar cap through the whole day.

Among the variations linked with external sources, the daily variation, with a periodicity of 24 h, is one of the earliest recognized in magnetic records. Many studies have shown that the shape of this field variation has a spatial dependence related to

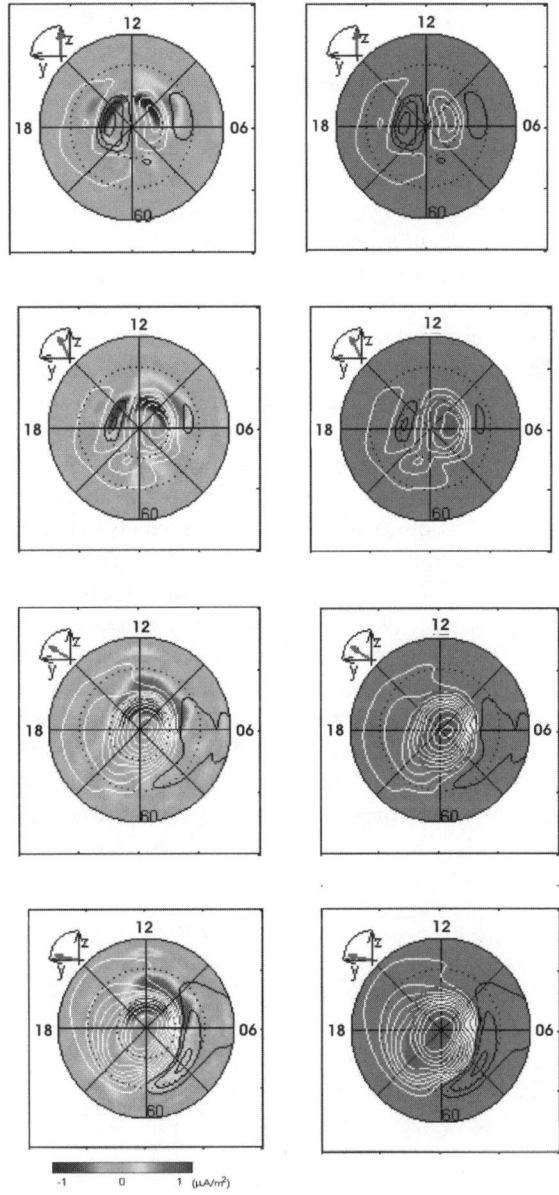

Fig. 6 Polar view of the effect of the IMF (shown as a red arrow in the upper left corner of each plot) turning from northward to eastward on simulated field aligned currents (color-coded; blue = upward; red = downward; left panels), on the geometry of the polar cap (indicated as the red region; right panels) and on the electric potential (contour lines in all plots) (adapted from Vennestrom et al., 2005)

geographic and geomagnetic latitude. At low to mid latitude, it is related to electric currents in the upper atmosphere that flow at altitudes between 100 and 130 km. There the atmosphere is ionized by the Sun's ultraviolet and X-radiation and atmospheric motions in the Earth magnetic field create a natural dynamo with two current cells: one in the sun-lit northern hemisphere flowing counter-clockwise, the other in the sun-lit southern hemisphere flowing clockwise. In the polar regions the daily variation, besides being due to the extension of the mid latitude system, is mainly due to FACs flowing along the geomagnetic field lines and connecting the magnetosphere to the ionosphere. The FACs are strongly dependent on the IMF and are believed to be the primary source for the high latitude daily variation, especially during local winter when the ionospheric ionization is strongly reduced. For this reason the study of the diurnal variation in Antarctica could provide important information on the complex current system flowing both in the polar ionosphere and along field lines. A recent paper of the diurnal variation at TNB through several years of observations has shown a seasonal as well as a solar cycle effect (Santarelli et al. 2007). In Fig. 7 (from Cafarella et al. 2007) the average daily variation at DMC in 2005–2006 during northward IMF conditions, is shown separately for positive and negative B_y. The asymmetry related to the east-west IMF component is evident, as the pattern of the diurnal variation for negative B_y values is shifted at about 3 hours earlier with respect to positive B_y. This result clearly indicates the contribution of FACs on the diurnal variation deep inside the polar cap.

A different type of geomagnetic field variations of external origin are geomagnetic pulsations, with periods from seconds to minutes. The generation mechanisms for low frequency pulsations (known as Pc3-Pc4 for $f = 7$–10 mHz and Pc5 for $f = 2$–7 mHz) is basically related both to flow instabilities along the flanks of the magnetopause, waves generated upstream of the Earth's bow shock (see Fig. 5) by solar wind ions reflected back, and to local phenomena in the auroral oval region. In addition, low frequency pulsations can be generated by interplanetary shocks impacting the magnetopause; such impact can trigger cavity/waveguide modes (e.g. Kivelson and Southwood 1985) localized between an outer boundary, such as the magnetopause, and an inner turning point (Fig. 8). The peculiar feature of these modes is that they are characterized by discrete frequencies and have a global

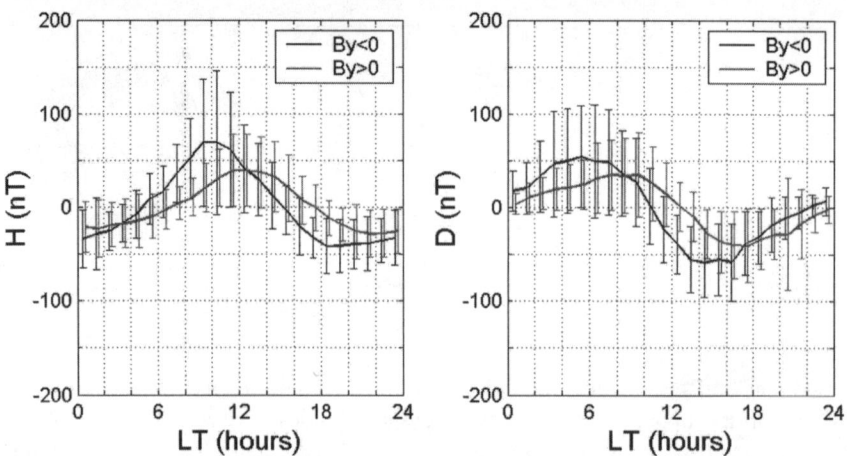

Fig. 7 Daily average variation at DMC as function of local time for the two horizontal components H and D during northward IMF conditions, separately for positive and negative By (from Cafarella et al. 2007)

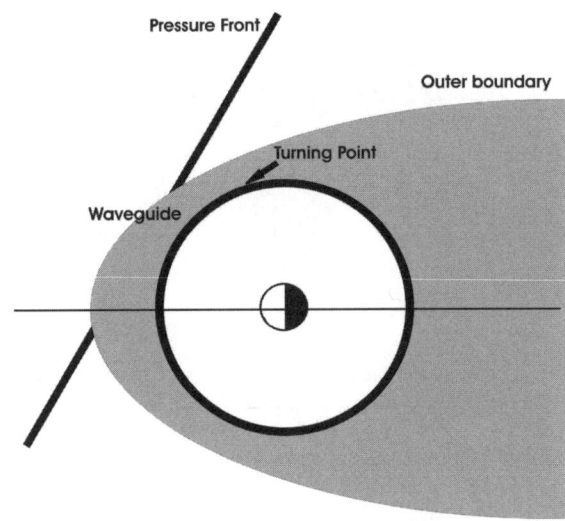

Fig. 8 Schematic view in the equatorial plane of the impact of a shock front on the magnetopause

character within the magnetosphere; besides being observed at auroral and low latitude, they have been observed also deep in the polar cap, i.e. at the footprint of open field lines stretching in the geomagnetic tail (Villante et al. 1997).

In Fig. 9 (from Lepidi et al. 2007) we show a pulsation event simultaneously detected at the three Antarctic stations TNB, SBA and DRV (see also table 1), at the Canadian auroral station Cambridge Bay (CBB) and at a latitudinal chain of European low latitude stations (corrected geomagnetic latitude from 38°N to 48°N). This event occurs on October 30, 2003, during a strong geomagnetic storm triggered by complex interplanetary structures passing near the Earth. Evident in the figure is the presence of simultaneous wave packets at discrete frequencies, the same at all the stations, which can be interpreted in terms of global magnetospheric oscillations; this event occurs during open magnetospheric conditions, i.e. when the Antarctic stations are deep in the polar cap. So it is interesting to note that the same oscillations are present both on field lines closed in the inner magnetosphere and on field lines connected to the IMF.

From the study of global oscillations of the magnetosphere, which at selected latitudes can couple with local field line resonances, information on the magnetospheric structure, in particular Alfven gradients and density profiles, can be obtained.

5 Perspective View in the Regional and Global Framework

Geomagnetic observatories record continuously and over long term the time variation of the Earth's magnetic field maintaining an accurate absolute standard. The history of magnetism is probably thousand years long but the insight of Earth's

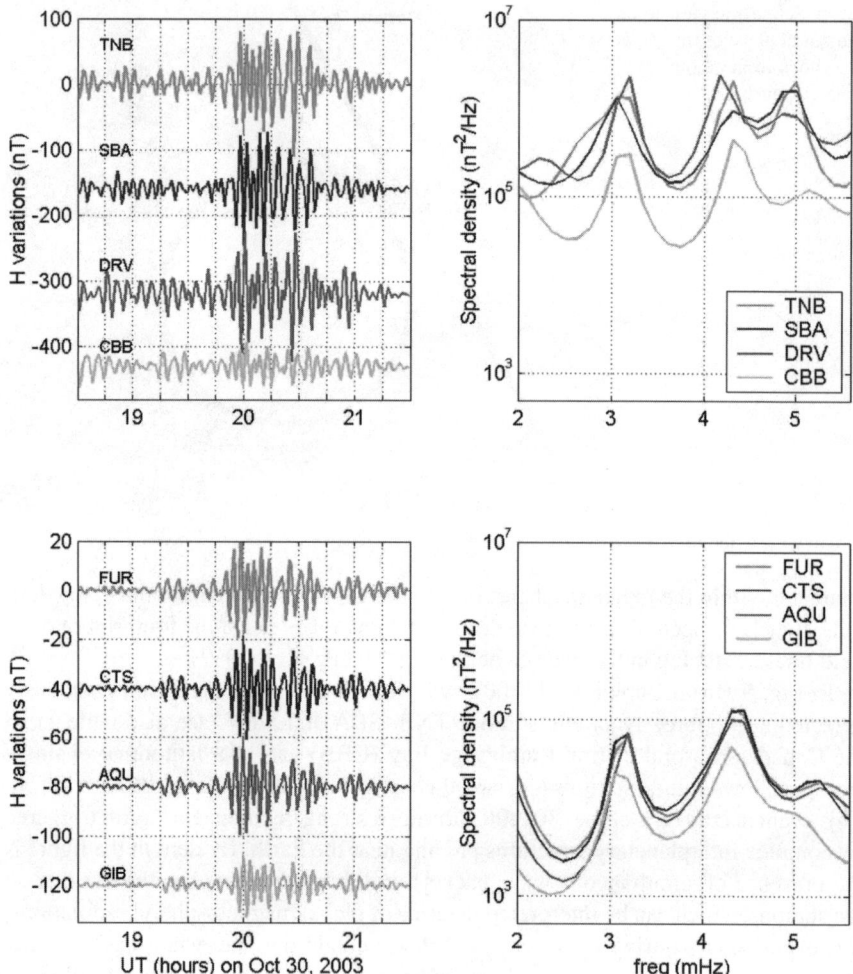

Fig. 9 Pulsation event simultaneously observed at high latitude stations (upper panels) and low latitude stations (lower panels). Left: filtered (2.5–5 mHz) data; right: power spectra (from Lepidi et al. 2007)

magnetism came only after a quantitative use of magnetic measurements. On the global scale this was firstly done by Gauss at the beginning of the XIX century with the introduction of the Spherical Harmonic Analysis. Moreover Gauss himself gave his strong push to the achievement of magnetic measurements and observatories all over the world. Only magnetic observatory data allow a 'comprehensive modeling' of the geomagnetic field, not only from a mathematical standpoint but also on a physical basis, where all sources of the measured field can be correctly distinguished and characterized.

Long-term magnetic observations, both old and new, can lead to unexpected results and more generally to new science. In Antarctica this story is only a little less than 50 years long, since the first magnetic observatory installations date back to the IGY (1957–58) when a real effort was put in the Antarctic exploration. In spite of this limited time window, there are results of global relevance as for example finding of the strong decrease shown by all intensive magnetic elements in Antarctica. Moreover, only the high-accuracy of secular variation data, as obtained from magnetic observatories, can be used to increase our knowledge of long term geomagnetic variation and jerks and, as reported previously, we cannot exclude a relevant role of the polar areas in this phenomenon.

A regional magnetic observatory is needed usually as base station for magnetic surveys. Observatory data are then used for the correction of magnetic time variations (diurnal, storm and other) during the operation of magnetic surveys. This role is probably even more important in Antarctica than elsewhere in the world, given the strong and rapid fluctuations of magnetic elements in polar regions and for the strong interest in determining the magnetic crustal anomalies in the continental area of Antarctica.

Observatories can provide data which are useful for the analysis of aeromagnetic surveys and satellite magnetic observations, now easily available. All these observations are complementary showing different and balancing roles. Observatories, through long-term and consistent recordings at a stable site (with careful absolute control) report time variations at one position in space. Satellites can survey spatial variation rapidly but report intrinsically an ambiguity between space and time variations.

The INTERMAGNET standard is now the excellence for geomagnetic observatories. All Antarctic observatories should progressively reach this standard for best quality data access, notwithstanding their importance of yielding absolute measurements. Magnetic observatories should try to achieve 'broadband' geomagnetism measurements with higher sampling rates. This improvement would be of high importance for satellite data users and for the space physics community.

For Antarctica a special magnetic activity index is missing. In particular, for the polar cap a Polar Cap Magnetic Activity Index (PC) was introduced about 15 years ago and since then it was extensively used in scientific studies. Two near-pole magnetic observatories are selected to derive this dimensionless index: Thule (now Qaanaaq) in Greenland at 85.4°N corrected geomagnetic CGM latitude, and Vostok in Antarctica at 83.4°S. More exactly, a northern PC index (PCN) and a southern PC index (PCS) are now both derived as 1-min data using a "unified" procedure. The concept is published in Troshichev et al. (2006). Both indices are expected to be available on-line via the World Wide Web for the space science community. The definition of specific indices for the southern polar areas that are still to be defined, will significantly help the space science community to characterize the observed phenomena in this area of the planet.

In conclusion, the best observatory distribution would be a uniform coverage of the continent. This is of course very difficult in Antarctica as a thick ice layer covers the inner continent. In any case, a special effort will be necessary to fill the inland gap with more observatories.

Acknowledgments We would like to thanks the anonymous referee and Dr. Stephen Monna for their fruitful comments and suggestions. The research activity at TNB and DMC are supported by the Italian Antarctic funding agency PNRA. DMC is also supported by the French Antarctic funding program IPEV.

References

Backus G., Parker R., Constable C. (1996) Foundations of Geomagnetism, Cambridge University Press, Cambridge.
Bellanger E., Le Mouël J.-L., Mandea M., Labrosse S. (2001) Chandler wobble and geomagnetic jerks. Phys. Earth Planet. Int., 124, 95–103.
Bloxham J., Zatman S., Dumberry M. (2002) The origin of geomagnetic jerks. Nature, 420, 65–68.
Cafarella L., Di Mauro D., Lepidi S., Meloni A., Pietrolungo M., Santarelli L., Schott J.J. (2007) Daily variation at Concordia station (Antarctica) and its dependence on IMF conditions, Ann. Geophysicae, 25, 2045–2051.
Campbell W.H. (2001) Earth Magnetism, A Guided Tour Through Magnetic Fields. Harcourt Academic Press, San Diego, CA, USA.
Campbell W.H. (2003) Introduction to Geomagnetic Fields, Cambridge University Press, Cambridge.
Chapman S., Bartels J. (1940). Geomagnetism, Oxford University Press, Oxford.
De Santis A., Torta J.M., Gaya-Piqué L.R. (2002) The first Antarctic geomagnetic Reference Model (ARM). Geophys. Res. Lett., 29, N. 8, 33.1–33.4.
De Santis A., Tozzi R., Gaya-Piqué L., (2004) Information content and K-entropy of the present Earth magnetic field, Earth Planet. Sci. Lett., 218, 269–275.
Gaya-Piqué L.R., De Santis A., Torta J.M. (2004) Use of champ magnetic data to improve the Antarctic geomagnetic reference model. Proceedings of the 2nd Champ Scientific Meeting.
Gaya-Piqué L.R., Ravat D., De Santis A., Torta J.M. (2006) New model alternatives for improving the representation of the core magnetic field of Antartica. Antarct Sci., 18, 101–109.
Gonzalez W.D., Tsurutani B.T., Clua de Gonzalez A.M. (1999) Interplanetary origin of geomagnetic storms, Space Sci. Rev., 88, 529–533.
Gubbins D., Jones A.L., Finlay C.C. (2006). Fall in Earth's magnetic field is erratic, Science, 312, 900–902.
Haines G.V. (1985). Spherical cap harmonic analysis. J. Geophys. Res. 90 (B3), 2563–2574.
Haines G.V. (1990) Regional magnetic field modelling: a review. J. GEOMAG. Geoelect. 42, 1001–1007.
Hargreaves K.J. (1992) The Solar-Terrestrial Environment, Cambridge University Press, Cambridge.
Jankowsky J., Sucsdorff C. (1996) IAGA Guide for Magnetic Measurements and Observatory Practice, Warsaw.
Kerridge D. (2001) Intermagnet: worldwide near-real-time geomagnetic observatory data. Proceedings of the Workshop on Space Weather, ESTEC.
Kivelson M.G., Russell C.T. (1996). Introduction to Space Physics, Cambridge University Press, Cambridge.
Kivelson M., Southwood D. (1985) Resonant ULF waves: a new interpretation, Geophys. Res. Lett., 12, 49–52.
Lanza R., Meloni A. (2006) The Earth's magnetism, an introduction for geologist, Springer, New york.

Lepidi S., Cafarella L., Francia P., Meloni A., Palangio P., Schott J.J. (2003). Low frequency geomagnetic field variations at Dome C (Antartica), Annales Geophysicae, 21, 923–932.

Lepidi S., Cafarella L., Santarelli L. (2007) Low Frequency Geomagnetic Field Fluctuations at cap and low latitude During October 29–31, 2003, Ann. of Geophys., 50, 249–257.

Malin S.R.C., Hodder B.M. (1982) Was the 1970 geomagnetic jerk of internal or external origin? Nature, 296, 726–728.

McMillan S., Maus S. (2005) Modelling the earth's magnetic field: the 10th generation IGRF – Preface Earth Planets and Space, 57(12), 1133–1133.

Meloni A., Gaya-Piqué L.R., De Michelis P., De Santis A. (2006) Some recent characteristics of geomagnetic secular variation in Antarctica, in: Fütterer D.K., Damaske D., Kleinschmidt G., Miller H., Tessensohn F. (eds.). Antarctica: Contributions to Global Earth Sciences. Springer-Verlag, Berlin Heidelberg New York, 377–382.

Merril R.T., McElhinny M.W., McFadden P.L., (1996). The magnetic field of the Earth: Paleomagnetism, the Core and the Deep Mantle, Academic Press, San Diego, CA, USA.

Parkinson W.D. (1983) Introduction to Geomagnetism. Scottish Academic Press, Edinburgh.

Rajaram G., Arun T., Dhar A., Patil G. (2002). Rapid decrease in total magnetic field F at Antarctic stations – its relationship to core-mantle features. Antarct. Sci., 14, 61–68.

Santarelli L., Lepidi. S., Di Mauro D., Meloni A. (2007). Geomagnetic daily variation studies at Mario Zucchelli Station (Antartica) through fourteen years, Ann. of Geophys., 50, 225–238.

Torta J.M., De Santis A., Chiappini M., von Frese R.R.B. (2002). A model of the Secular Change of the Geomagnetic Field for Antarctica. Tectonophysics, 347, 179–187.

Troshichev O., Janzhura A., Staunming P. (2006) Unified PCN and PCS indices: method of calculation, physical sense, and dependence on the IMF azimuthal and northward components, J. Geophys. Res., 111, A05208.

Vennerstrom S., Moretto T., Rastatter L., Raeder J. (2005) Field-aligned currents during northward interplanetary magnetic field: Morphology and causes, J. Geophys. Res., 110, A06205.

Villante U., Lepidi S., Francia P., Meloni A., Palangio P. (1997) Long period geomagnetic field fluctuations at Terra Nova Bay, Geophys. Res. Lett., 24, 1443–1446.

Waddington R., Gubbins D., Barber N. (1995) Geomagnetic field analysis-V. determining steady core-surface flows directly from geomagnetic observations. Geophy. J. Int. 122, 326–350.

Zhou X.W., Russell C.T., Le Fuselier S.A., Scudder J.D. (2000) Solar wind control of the polar cusp at high altitude, J. Geophys. Res., 105, 245–251.

Structure of the Wilkes Basin Lithosphere along the ITASE01 Geotraverse

F. Coren, N. Creati and P. Sterzai

Abstract During the 2000/2001 scientific expedition a new gravity, magnetic and airborne radar dataset has been acquired along the regional geotraverse ITASE01A/B to improve the knowledge about the tectonic setting and geodynamic origin of East Antarctica. This new traverse crosses the Northern Wilkes Basin, the Webb Subglacial Trench, the Southern Cross Subglacial Highlands and the Astrolabe Trench. The gravity data have been converted to free air anomaly. First, the Free Air gravity anomalies has been modelled assuming two isostatic models, Airy and flexural, and later using a raw 2D forward model along the ITASE01A/B traverse. The isostatic approach fails to predict the observed gravity anomaly due to the oversimplification of the assumed model, the lack of extensive information such as heat flow, age, mantle viscosity, rheology and Moho depth, etc. The raw 2D forward gravity model points to a thick crust, of about 40 km, and the presence of a sedimentary body along the traverse and yields a good fit between the observed and predicted gravity anomaly. The thickness of this sedimentary body is variable and ranges from 6 to 7 km in the Astrolabe Trench to 5 km in the Wilkes Basin and decreases further in the Southern Cross Subglacial Highlands. There is no evidence of recent rifting along the traverse considering the time decay of thermal anomalies following a rifting episode and the state of relative compression of the whole Antarctic continent. The shape and origin of the East Antarctic basins should be traced back before the Gondwana disruption in Early Cretaceous. At present, the studied area behaves as a stable area even if this conclusion can be affected by the limited spatial extension of known data, mainly recorded along traverse, the lack of any geological hints about

---

F. Coren
Istituto Nazionale di Oceanografia e Geofisica Sperimentale (OGS), Borgo Grotta Gigante, 34010 Sgonico (Trieste), Italy

N. Creati
Istituto Nazionale di Oceanografia e Geofisica Sperimentale (OGS), Borgo Grotta Gigante, 34010 Sgonico (Trieste), Italy, e-mail: ncreati@inogs.it (Nicola Creati)

P. Sterzai
Istituto Nazionale di Oceanografia e Geofisica Sperimentale (OGS), Borgo Grotta Gigante, 34010 Sgonico (Trieste), Italy

the basins and the ridges in the East Antarctic craton and the bidimensional nature of the proposed model.

1 Introduction

East Antarctica is considered an ancient craton (Bentley 1999) characterized by a long, 3500 km, N-S elongated chain, the Transantarctic Mountains (TAM), that separate East Antarctica from the Ross Sea and West Antarctica, and several large and small basins that develop just west of TAM. Bedrock outcrops only in the highest part of TAM and much of the surface geology is buried under an ice cap (up to 2 km) that uniformly covers the East Antarctica craton. The subglacial relief shows evidence of a large long-wavelength down warp called the Wilkes Basin (WB) that runs parallel to the TAM and is characterized by a mean depression of 500 m under sea level. The basin is about 400 km wide near the GiorgioV coast at north-, and narrows to 100 km at south. The northern sector is characterize by a prominent subglacial high known as the Southern Cross Subglacial Highlands (SCSH). West of this high it develops the narrow Webb Subglacial Trench which reaches depths of 1250 m bs.l and at east the deep Astrolabe Trench can be found (AT) (around 2000 m bs.l). In southern WB East Antarctic craton is characterized by two elevated zones, the Resolution and Belgica Subglacial Highlands, divided by the narrow Adventure Subglacial Trench that resembles the northern Astrolabe Subglacial Basin. The origin of TAM and western basins is subject to an ongoing debate. In particular, the TAM show no evidence of contractional structures and are considered a gently tilted to block-faulted mountain range (ten Brink et al. 1997). Since it is an anorogenic feature a number of possible uplift mechanisms have been proposed in the past: thermal uplift due to conduction/advection from East Antarctica to West Antarctica (Stern and ten Brink 1989; ten Brink and Stern 1992; ten Brink et al. 1993), isostatic rebound due to normal faulting (Bott and Stern 1992), simple shear extension (Fitzgerald et al. 1986), plastic necking (Chéry et al. 1992), elastic necking (van der Beek et al. 1994), isostatic erosional rebound (Stern and ten Brink 1989). It seems that none of the previous models is able to explain the uplift mechanism and more data would be needed to better address the genesis (Studinger et al. 2004).

Gravity, magnetic and seismic surveys are generally restricted to the TAM-WB transition. The surface geology in poorly known, there are no heat flow measures, and the geometry of the Moho, mainly from tomographic studies (Bannister et al. 2003; Ritzwoller et al. 2001), is not known with sufficient details to be coupled with the known subice main units. Anyway, the WB has been described as a rift basin (Drewry 1976; Drewry 1983; Steed 1983; Ferraccioli et al. 2001) or as a flexural outer low whose genesis is coupled with the near TAM uplift. The AST has been interpreted as a rift (Ferraccioli et al. 2001) or associated with some contractional structures in the Resolution Subglacial Highland (Studinger et al. 2004).

During the 2000/2001 scientific expedition an oversnow traverse (ITASE01A/B) was carried out, crossing the northern part of the WB close to 70°S. Gravity

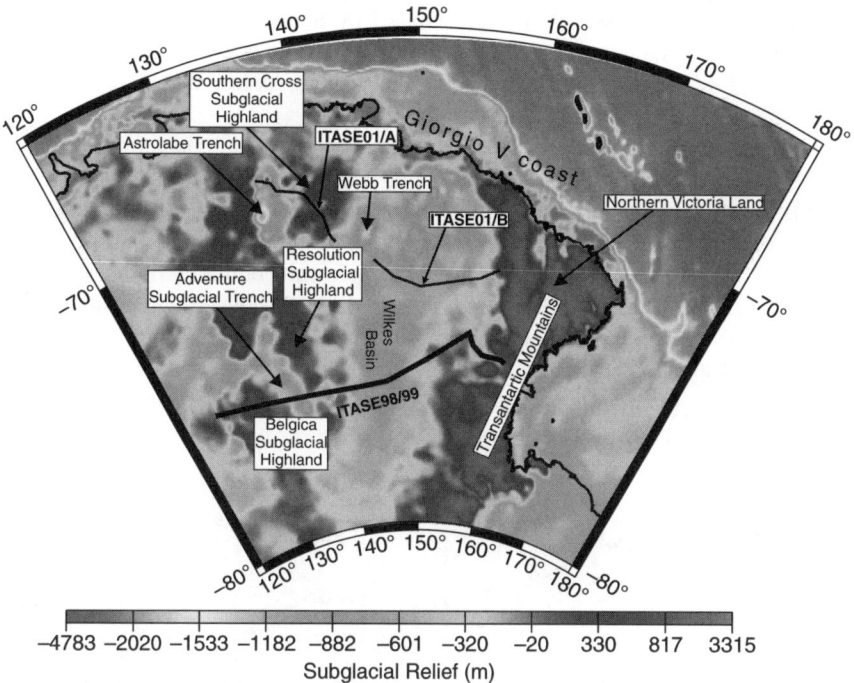

Fig. 1 Subice topography (Lytte and Vaughan, 2000) and bathymetry of East Antarctica with traces of the traverse ITASE98/99 and ITASE01 A/B and location of the main areas cited in the paper

and magnetic data could be collected and combined with airborne radar data (M. Frezzotti, pers. comm.) (Fig. 1). This new dataset was combined with the already known results of ITASE98/99 (Ferraccioli et al. 2001) and can help to fix more constraints and depict the structure and origin of West Antarctica basins.

2 Geophysical Dataset

The ITASE01 traverse is composed by 173 measure points spaced by 5 km distributed along two main profiles. Potential field stations were positioned using a differential GPS with an accuracy of ±5 m.

2.1 Gravity Dataset

The data has been acquired with a LaCoste Romberg gravimeter. Each measure has been corrected for instrumental drift and tide effects. The stations were connected to the absolute gravity station at Terra Nova Bay (Cerrutti et al. 1992). The free air anomaly has been computed at the geoid using the Geodetic references System Formula of 1980 while the GPS elevation data for the geoid undulation have been corrected using the EGM96 model.

The free-air gravity anomaly (Fig. 2) nicely reflect the subglacial relief topography except in the central part along the ITASE01/B. It's almost negative over the WB with a minimum of -100 mGal even if the presence of relative peaks and troughs, such as at 200 km, and the large increase of the free-air anomaly from 300 to 500 km, should reflect that there is no uniqueness and continuity of the structures that characterize the basin. The eastern border of the WB shows a high positive gravity gradient since the anomaly shifts from around -70 mGal to more than 0 mGal in less than 100 km. This step coincides with the positive anomaly of the SCSH. A sharp negative gradient mark the boundary between the SCSH and the AT where the free-air anomaly drops to less than -100 mGal.

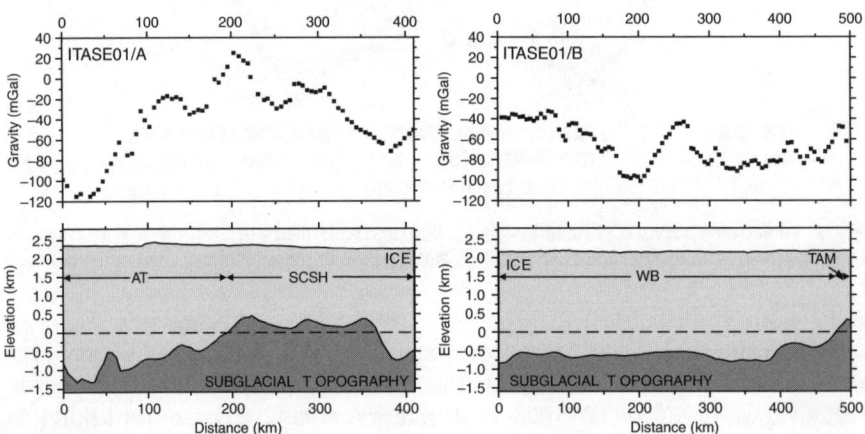

Fig. 2 Free-Air anomaly calculated along the ITASE01A/B traverse, ice thickness and subice relief topography have been measured by Airbone Radar. Key: AT = Astrolabe Trench; SCSH = Southern Cross Subglacial Highland; TAM = Transantarctic Mountains; WB = Wilkes Basin

2.2 Subglacial Topography

The subglacial topography has been derived from the processing of airborne radar data (unpublished data, courtesy of M. Frezzotti). It is almost below zero (Fig. 2). It is well below -500 m in the WB and reach around -1500 m in AT. WB is large and characterized by a sharp step near the TAM and a gentle rise in the SCSH. This is a more than 200 km wide chain characterized by downwarps and ranges whose amplitude are not larger than 500 m. The eastern side of the profile crosses the northern portion of AT characterized by a step western border and a deep-narrow profile.

3 Geological Settings

3.1 Transantarctic Mountains

TAM include a Precambrian and Cambrian basement deformed and intruded during the Ross Orogeny in the Cambro-Ordovician (510–470 Ma) (Fitzgerald 1994). The trend of the Ross Orogeny structures is subparallel to the TAM. During this deformation the basement was exhumated as much as 15–20 km and eroded to form the Kukri Peneplain (Capponi et al. 1990). The Beacon Supergroup overlays the Kukri Peneplain and comprises 2.7–3.5 km of undeformed sequence of Devonian to Jurassic age, fluvial, glacial and shallow marine deposits (Barret et al. 1986). It has been interpreted as a passive margin in the Devonian that transformed in a foreland basin bordering the East Antarctica Craton (Collinson 1991). Sills of the Ferrar Dolerite and flows of the Kirkpatrick Basalt flooded the length of the TAM (Elliot 1975) as part of the large Karoo flood basalt province that followed the large-spread Gondwana magmatism and break-up (Elliot 1992). The TAM is characterized by different uplift velocities and phases that took place at around 55 Ma (Fitzgerald 1994), 30 Ma after the large continental extension between New Zeland and West Antarctica occurred in the late Cretaceous (Lawver et al. 1992). The uplift is large near the coast where it is up to 10 km (Fitzgerald and Gleadow 1988) in Northern Victoria Land and decrease at west. Central and Southern Victoria Land are characterized by an uplift of 7 and 6 km respectively (Fitzgerald 1992; Fitzgerald 1994). The elevation is larger near the cost and decreases at west. The area affected by vertical movements is characterized by high angle normal faults that show an offset of 40–1000 m.

3.2 Wilkes Basin

The whole basin is covered by a thick ice cap and information (better speculations) come from geophysical studies such as radio echo sounding surveys. Its origin is a matter of debate. Estimates of the sediment thickness are puzzling since they range from less than 1 km (Studinger et al. 2004) to 3–4 km (Drewry 1976; Ferraccioli et al. 2001). Sediments could be of marine origin since in late Miocene the collapse of the East Antarctica ice cap favoured the establishment of a marine environment in some zones of the WB (Webb 1990). Classically the ice cap is considered uniform all over the Miocene and Pliocene (ten Brink et al. 1997).

4 Crustal Modelling of the Wilkes Basin

The most important density contrasts along the ITASE01A/B traverse are at the ice-crust boundary and at the crust-mantle boundaries. The depth of the ice-crust boundary has been measured by airborne radar method while the Moho depth is only indirectly known by means of seismic studies (35–40 km) and with a low resolution. To better address the Moho discontinuity along the traverse two possible isostatic models can be used for the mechanics of the lithosphere: the Airy and the flexural model. In the Airy approach the crustal thickness changes as a response to ice and surface topographic loading according to:

$$\Delta moho = -\frac{h_{ice}\rho_{ice} + z_{bedrock}\rho_{bedrock}}{\rho_{mantle} - \rho_{crust}} \qquad (1)$$

where h_{ice} is the ice thickness, $z_{bedrock}$ is the elevation of the subglacial topography, ρ_{ice} is the density of ice, ρ_{crust} is the density of crust and ρ_{mantle} is the density of the mantle below the Moho. Densities of 900, 2800 and 3400 kg/m³ have been assumed for the ice, the crust and the mantle, respectively. The flexural hypothesis assumes that the loads are supported by stresses in the lithosphere. The compensation spreads over a wider area and is described by the 4th order differential equation of a bending beam:

$$D\nabla^4 W(x) + \Delta\rho g W(x) = P \qquad (2)$$

where D is the flexural rigidity of the plate (i.e. the lithosphere), $\Delta\rho$ is the density contrast of material above and below the flexure, g is the gravitational acceleration, $W(x)$ is the vertical deflection and P the applied vertical load. A common way to express the strength of the plate is to convert the flexural rigidity D to the equivalent elastic thickness T_e through the relation:

$$D = \frac{ET_e^3}{12(1-\nu^2)} \qquad (3)$$

where E is the Young Modulus (typically assumed to be 10^{11} Pa) and ν is the Poisson ratio (typically assumed to be 0.25).

The differential equation has been solved assuming a broken plate condition (ten Brink et al. 1999; Studinger et al. 2004). The traverse has been extended up to Northern Victoria Land coast where the plate has been assumed to break. Plate flexure is due to the action of loads such as the ice cap, an isostatic end load and an erosional load, that takes in account the mass removed from the original TAM topography, at the eastern edge. The flexural profile has been tied to the extrapolated maximum height of the TAM that is around 10 km near Northern Victoria Land coast (Fitzgerald and Gleadow 1988). Equation (2) has been solved numerically using a finite difference method (TAO, Garcia-Castellanos et al. 1997) allowing for a lateral change of the flexural rigidity D. The best model (Fig. 3) is fitted by a broken plate characterized by a sharp increase of T_e from 5 km near the coast to 130 km inland. The model allows the variation of the end load magnitude and the preflexural elevation. The least-squares estimation points to an end load of $7 \cdot 10^{11}$ N/m and a pre-elevation of 400 m. The Airy and flexural derived Moho have been used in a simple 2D forward gravity model and compared to the observed anomaly along the ITASE01A/B (Fig. 4). The two supposed isostatic end-members clearly fail to completely model the Free Air anomaly. The flexural model shows an overall lower

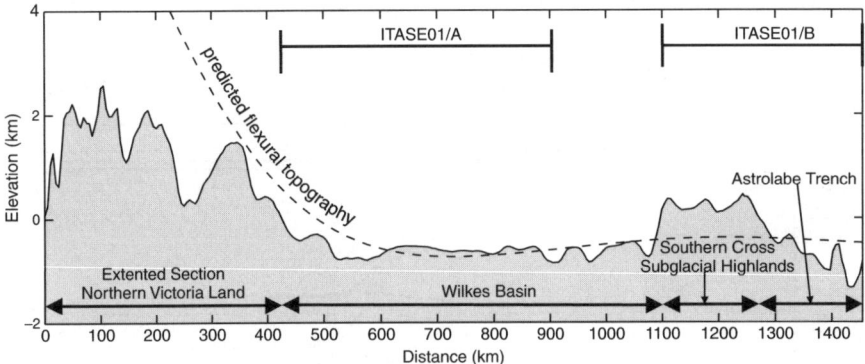

Fig. 3 Isostatic flexural model of the ITASE01A/B traverse. The best fit model has been found solving with finite difference method Equation (2) assuming a broken plate model, an end load of $7 \cdot 10^{11}$ N/m, a preflexural topographic elevation of 400 m and an erosional load that take in account the mass removed from the TAM considering an original elevation of 10 km near Northern Victoria Land coast (Fitzgerald and Gleadow, 1988). The Moho has been initially assumed equal to 37.5 km according to the mean known values in this area. The elastic thickness increase from 5 km near the coast to 130 inland inside the Antarctic Craton

rms than the Airy model along both profiles. It is remarkable that both models misfit at the western side of the traverse near the TAM eastern front.

The anomaly is clearly due to some unmodelled features inside the lithosphere that this raw force balance calculation cannot account for. Since there are no constraints and information about the geology and the geometry of the crust below the ice sheet a simple 2D forward gravity modelling along the two segments of the ITASE01 traverse has been performed. The ice, crust and mantle densities have been fixed to 900, 2750 and 3400 kg/m^3 respectively after the modelling. The crustal density has been lowered from 2800 to 2750 kg/m^3 since it decreased the overall misfit in the model. The modelled Moho has a mean depth of around 40 km. Both the geometry of the crust-mantle boundary and its depth control the long-wavelength components of gravity anomaly. The contribution of the Moho calculated by the isostatic (both Airy and flexural) and 2D forward gravity modelling is similar but cannot predict short wavelengths gravity components. The introduction of a near-surface element, possibly of sedimentary origin, with a density of 2400 kg/m^3 greatly improves the fit (Fig. 5). The body is not a continuous feature but shows a maximum depth of 6–7 km in the AT and a thinning in the SCSH. In the WB the sedimentary body has a maximum thickness of around 5 km.

5 Discussion

This analysis serves as a test of genetic models proposed in the past for the development of the WB and near basins. Two models have been proposed in the past to explain the development of the WB: rifting (Ferraccioli et al. 2001;

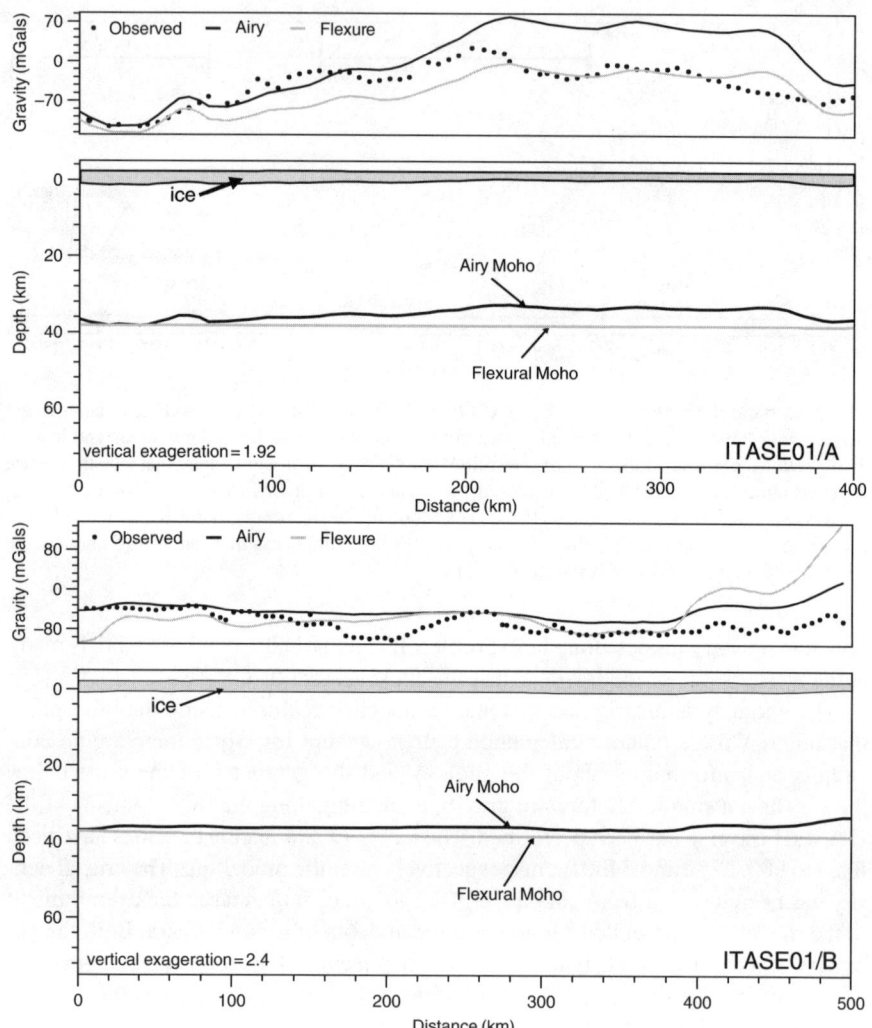

Fig. 4 Comparison between the observed Free Air gravity anomaly measured along the ITASE01A/B traverse and those predicted assuming an Airy or flexural isostatic deformation model

Drewry 1976, 1983; Steed 1983) and flexure (Stern and ten Brink 1989, ten Brink and Stern 1992, ten Brink et al. 1997). According to the rift hypothesis the WB and AT lows are rift structures characterized by thick sediments infill and thin crust. Sediments range from 3 to 4 km in the AT and from 6 to 14 km in WB. The crust is thinner than the surrounding Antarctic Craton, from 20 to 30 km and an average of 30 km in AT and WB respectively. Rifting is coupled by an anomalous low density mantle just below the thinned crust (Ferraccioli et al. 2001). In the flexural approach the uplift of the TAM led to the development of the WB.

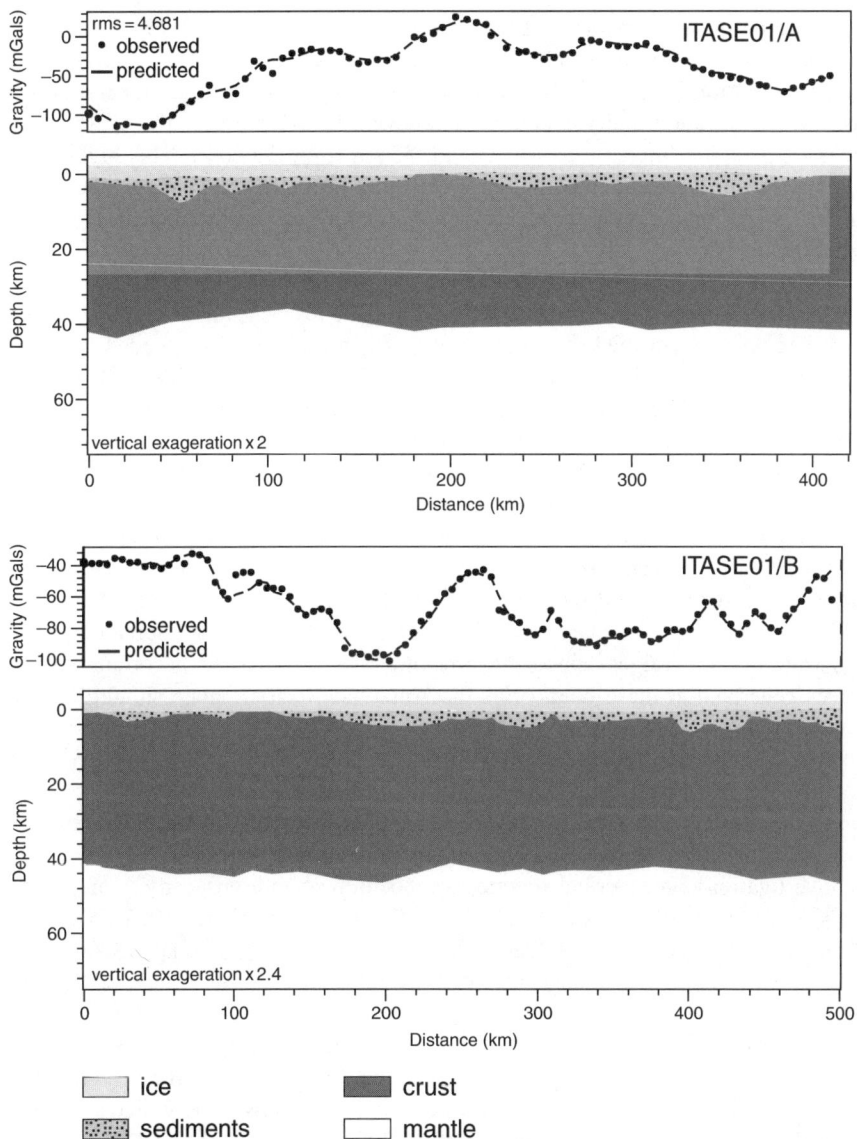

Fig. 5 2D forward gravity model along the ITASE01A/B traverse. The Moho has a mean depth of around 40 km its geometry controls the long-wavelength behaviour of the observed Free-Air gravity anomaly. Short wavelength gravity anomaly are mainly controlled by surface sedimentary bodies that spam along the traverse

The flexural model proposed in this work is more suited to explain the development of the northern WB. The predicted flexural low from 75 to 785 km matches well the extent of the basin and the position of the Southern Cross Subglacial Highlands. The best-fit model assumed a preflexural elevation of 400 m along the entire traverse that is smaller than the value of around 700 m found in sourthern WB (ten Brink et al. 1997). The mismatch between the observed and the predicted topography is obviously due to the presence of a preflexural structure whose elevation and geometry is unknown. T_e is 5 km near the coast (plate break) and rapidly increase to 130 inland, 35 km from the coast. This high value is comparable to measurements done over other cratonic areas (Watts 2001) but further inland model results are very close for any assumed value of T_e larger than 80 km.

Anyway both models fail to explain the genesis of the WB. Along the ITASE01A/B there is no evidence of rifting episodes beneath WB and AT. The crust has an uniform thickness and an anomalous mantle is not needed both for gravity modelling and tomographic analysis (Ritzwoller et al. 2004). The sediments that have been modelled along the traverse should have been deposited before the Miocene since at that time the present ice cover started to form. A rifting origin for the East Antarctica basins seems also improbable since there is no evidence of a mantle with low shear waves velocity under this region from tomographic studies (Ritzwoller et al. 2004). Generally the thermal anomaly associated to a rift lowers the velocity of shear waves in the mantle and the perturbation can last for several millions of years after the end of the rifting. Moreover the Antarctica tectonic plate boundaries have been marked by spreading centres (mid-oceanic ridges) since the breaking of Godwana, so according to the forces generated at these boundaries the Antarctica tectonic plate should be in a state of relative compression. All the previous considerations put a serious limit on the age and the occurrence of a rifting episode. The flexural hypothesis is not able to model the gravity anomaly nor the overall geometry of the crust along the traverse. The flexure of the lithosphere is controlled by many features that are generally unknown and reasonable values cannot be assumed in the area of investigation: age, thermal regime, composition, mantle viscosity, reology, inplane stress field (Burov and Diament 1995; Lowry and Smith 1995). The isostatic model cannot take the load effect of sediments or internal density contrasts into account that change the relation between gravity, surface and Moho topography.

The results of ITASE01A/B clearly contrast with those of ITASE98/99 where rifting was supposed for the genesis of WB and AT. The rifting conclusion along the ITASE98/99 has been recently rejected in view of new gravity, magnetic and radar data in the same area (Studinger et al. 2004). Since TAM and WB are long uniform features the occurrence of two different geodynamic processes for their formation can be invoked only if there is an important regional discontinuity whose evidence is absent in the Eastern Antarctic craton. The link between TAM, WB and the eastern Antarctica basins is an oversimplification of the complexity of geodynamic events. Each structure could have different age and then associated to different tectonic events whose expression is now buried below the ice cover. Such limit is due to the lack of an extensive regional dataset that covers Eastern Antarctica.

6 Conclusions

ITASE01A/B provides a new piece of information about the structural setting and history of the basins in Eastern Antarctica craton. Gravity, magnetic and radar data were analysed and combined to draw an improved picture.

1. Airy and flexural models fail to completely model the observed gravity anomaly along the traverse. The flexural hypothesis grant the best performance but it is affected by extreme simplification. The elastic thickness is large over the craton (130 km) and it's consistent with values from other cratonic areas in the world. Anyway, such a large value is restricted to few areas and the flexural solution must be carefully taken in account.
2. A more detailed 2D forward gravity model revealed the need of a thicker crust of around 40 km along the traverse. A thin layer of low density materials, may be sediments, can reconcile short wavelengths anomaly with model predictions. There are no constraints about the origin of this near-surface body. If it is a sedimentary body it should be not older than the establishment of the ice cover in Miocene.
3. There is no evidence of recent rifting in WB and AT. Mantle density is uniform and the crust has a uniform thickness along the traverse. This result certainly contrasts with the ITASE98/99 traverse (Ferraccioli et al. 2001) but agrees with more recent analysis (Studinger et al. 2004). Intracratonic rifting cannot be reconciled with the past and recent global geodynamic framework of Antarctica and should be happen before the Godwana disruption in Early Cretaceous ($\sim$115 Ma) (Stampfli and Borel 2002).
4. The lack of any evidence for recent tectonic activities points out to a stable area that inherits past deformations. Gravity and magnetic data analysis along the known geotraverses are not sufficient to trace the geologic structures and have a picture of the geodynamic evolution of the investigated area. The overall geology of the area is not known as well as the occurrence of faults, folds, etc. The logistic issues, the climatic conditions and the thick ice cap that covers the area make geophysical data a more suitable mean for the study. Unfortunately data are spatially limited and mainly measured along few long geotraverses, so all the models are mainly bidimensional and they depict very far areas. The availability of more data (mainly gravity and magnetic) along some other traverses or acquired over a large area would help to elaborate a more detailed tridimensional model and to verify the lateral and longitudinal continuity of the structures (basins and ridges) that characterize the East Antarctica craton. The general simplification of the model proposed in this paper and by other authors is probably a source for the conflict of the different results.

Acknowledgments We would thank Dott. M. Frezzotti for kindly providing the subglacial topography data. We are grateful to an anonymous reviewer that greatly helped to improve this paper.

References

Bannister S, Yu J, Leitner B, Kennet BLN (2003) Variations in crustal structure across the transition from West to East Antarctica, Southern Victoria land. Geophys. J. Int. 155:870–884.

Barret PJ, Elliot DH and Lindsey JF (1986) The Beacon Supergroup (Devonia-Triassic) and Ferrar Group (Jurassic) in the Beardmore Glacier area, Antarctica, in Geology of the Central Transantarctic Mountanis. In: Turner MD, Splettstosser JF (eds) Antarct. Res. Ser., vol 36. AGU, Washington D.C, pp 339–428.

Bentley CR (1991) Confguration and structure of the subglacial crust. Geol. Geophys., 14:335–364.

Bott MHP, Stern TA (1992) Finite element analysis of transantarctic mountain uplift and coeval subsidence in the Ross Embayment. Tectonophysics 201:341–356.

Burov E, Diament M (1995) The effective elastic thickness (Te) of continental lithosphere: What does it really mean? J. Geophys. Res. 100:3905–3927.

Capponi G, Messiga B, Piccardo B, Scambelluri M, Traverso G, Vannucci R (1990) Metamorphic assemblages in layered amphibolites and micaschists from the Dessert Formation (mountaineer Range, Antartica). Mem. Soc. Geol. Ita. 43:87–95.

Cerrutti G, Alasi F, Germak A, Bozzo E, Caneva G, Lanza R, Marson I (1992) The absolute gravity station and the Mt. Melbourne gravity network in Terra Nova Bay, North Victoria Land, East Antarctica. In: Yoshida Y, Kaminuma K, Shiraishi K (eds.) Recent Progress in Antarctic Earth Science, TERRAPUB, Tokyo, pp 589–593.

Chéry J, Lucazeau F, Daignieres M, Villotte JP (1992) Large uplift of rift flanks: A genetic link with lithospheric rigidity? Earth Planet. Sci. Lett. 112:195–211.

Collison JW (1991) The paleo pacific margin as seen from the East Antactica. In: Thomson JW, Crame JA, Thomson JW (eds) Geological Evolution of Antactica, Cambridge University Press, New York, pp 199–204.

Drewry DJ (1976) Sedimentary basins of the East Antarctic Craton from geophysical evidence. Tectonophysics 36:301–314.

Drewry DL (1983) Antarctica: Glaciological and Geophysical Folio. Scott Polar Res. Inst., University Cambridge, Cambridge, 9 sheets.

Elliot DH (1975) Tectonics of Antarctica: A review. Am. J. Sci. 275:45–106.

Elliot DH (1992) Jurassic magmatism and tectonism associated with Gondwana break-up: an Antarctic perspective. In: Storey B, Alabaster T, Pankhurst PJ (eds) Magmatism and the causes of Gondwana Break-up, Geol. Soc. Spec. Publ. London, 68:165–184.

Ferraccioli F, Coren F, Bozzo E, Zanolla C, Gandolfi S, Tabacco I, Frezzotti M (2001) Rifted(?) crust at the East Antarctic Craton margin: gravity and magnetic interpretation along a traverse across the Wilkes Subglacial Basin region. Earth Planet Sci. Lett., 192:407–421.

Fitzgerald PG (1994) Thermochronological constraints on post-Paleozoic tectonic evolution of the central Transantarctic Mountains, Antartica. Tectonics 13: 818–836.

Fitzgerald PG (1992) The Transantarctic Mountaine of Southern Vicotria Land: the application of apatite fission track analysis to a rift shoulder uplift. Tectonics 11:634–662.

Fitzgerald PG, Gleadow AJW (1988) Fission-track geochronology, tectonics and structure of the Transantertic Mountains in Northern Vicotria Land, Antartica. Chemical Geology 73:169–198.

Fitzgerald PG, Sandiform M, Barret PJ, Gleadow AJW (1986) Asymmetric extenzsion associated with uplift and subsidence in the Transantartic Mountains and Ross Embayment. Earth Planet. SCi. Lett. 81:67–78.

Garcia-Castellanos D, Fernandez M, Tornè M (1997) Numerical modeling of foreland basin formation: a program relating thrusting, flexure, sediment geometry and lithosphere rheology. Comput. Geosci. 23:993–1003.

Lawver LA, Gahagan LM, Coffin MF (1992) The development of paleoseaways around Antarctica. In: Kennet JP, Warnke DA (eds) The Antarctic Paleoenvironment: A Perspective on Global change, Antarct. Res. Ser., vol 56. AGU, Washington DC, pp 7–30.

Lowry AR, Smith RB (1995) Strength and rheology of the western U.S. Cordillera. J. Geophys. Res. 100:17947–17963.

Lytte MB, Vaughan DG (2000) BEDMAP-Consortium, BEDMAP - bed topography of the Antarctic, British Antarctic Survey, Cambridge.

Ritzwoller MH, Shapiro NM, Levshin AL, Leahy GM (2001) Crustal and upper mantle structures beneath Antartica and surroundings oceans. J. Geophys. Res. 106:30645–30670.

Steed RHN (1983) Structural interpretation of Wilkes Land, Antartica. In: Oliver RL, James PR, Jago JB (eds) Antarctic Earth Science Proceedings of the Fourth International Symposium on Antarctic Earth Sciences, Cambridge University Press, New York, pp 567–572.

Stampfli GM, Borel GD (2002) A plate tectonic model for the Paleozoic and Mesozoic constrained by dynamic plate boundaries and restored synthetic oceanic isochrons. Earth Planet. Sci. Lett. 196:17–33.

Stern TA, ten Brink US (1989) Flexural uplift of the Transantartc Mountains. J. Geophys. Res. 94:10315–10330.

Studinger M, Bell RE, Buck WR, Karner GD, Blankeship DD (2004) Sub-ice geology inland of the Transantartic Mountains in light of new aereogeophysical data. Earth Planet. Sci. Lett. 220:391–408.

ten Brink US, Stern TA (1992) Rift flank uplifts and hinterland basins: compartision of the Transantartic Mountains with the Great Escarpment of southern Africa. J. Geophys. Res. 97:569–585.

ten Brink US, Bannister S, Beaudoin BC, Stern TA (1993) Geophysical investigation of the tectonic boundary between East and West Antarctica. Science 216:45–50.

ten Brink US, Hackney RI, Bannister S, Stern TA, Makovsky Y (1997) Uplift of the Transantarctic Mountains and the bedrock beneath the East Antarctic ice sheet. J. Geophys. Res. 102:27603–27621.

Van der Beek P, Cloetingh S, Andriessen P (1994) Mechanism of extensional basin formation and vertical motion at rift flanks: constraints from tectonic modelling and fission track thermocronology. Earth Planet. Sci. Lett. 141:417–433.

Watts AB(2001) Isostasy and Flexure of the Lithosphere. Cambridge University Press.

Webb PN (1990) The Cenozoic history of Antarctica and its global impct. Antartic. Sci. 2:3–21.

Report on Photogrammetric Research Conducted at the Antarctic Station "Academician Vernadskyy"

V. Hlotov

Abstract The analysis of terrestrial digital stereofotogrammetric methods for determination of surface volumes of island glaciers of Argentine archipelago of the Antarctic coast is given. The data on several cycles of observations made in the period of season Antarctic expedition on Ukrainian Antarctic Station Academician Vernadsky are shown. The proper conclusions are made.

Paper presents technology of creation of large scale topographic plans of Argentine Islands as the base for geoinformation system of the station Academician Vernadsky region. Author describes theoretical substantiation and development of technologies of application of processes of navigational-digital photogrammetry for needs of development of large scale mapping of Antarctic territories and determination of quantitative parameters of objects.

1 Introduction

Observations on temperature and salt concentration in seawater in the region of Argentine Islands located near the Antarctic Peninsula testify about considerable increasing of these elements in summer period during last decade. Considerable increasing of water and air temperature had been observed in January 2001–2005. It is necessary to underline that average annual air temperature in the region of the archipelago had increased more than 2°C during last 40 years. This anomaly influenced on melting of island glaciers arisen on these islands more than 10 thousands years ago.

Intensive melting of island glaciers will considerably effect on biodiversity and evolution of adaptation of flora and fauna of Antarctic coast of this region. Therefore observations over dynamics of island glaciers will give possibility to research climate, glaciological and biological changes appeared in this part of the Antarctic.

Establishment of Antarctic station in 1996 allows Ukraine to enter into the group of Antarctic states and at the same time it put before scientific society tasks for

---

V. Hlotov
National University, Lvivska Politechnika, Ukraine

solving various problems when investigation of this region. During these years a large amount of varied results of scientific researches are being collected. And they undoubtedly have to be systemized. Therefore the idea about creation of GIS of the Argentine Islands has appeared.

The topographic map with the scale of 1:1000 were chosen a basis for making GIS because such map is very convenient to indicate in details all the information like biologic areas, colonies of birds, location of various sensors (radiometric, magnetometric, meteorological and the like).

Application of topographic plans of abovementioned scale can also be used for preliminary designing of proper buildings, communications etc.

2 Creation of Frontal Plans of Glaciers

Analysis of methods of glacier observations allows to make conclusions that geodetic researching methods have contact character and are not technological sufficiently. In respect to labour protection these methods are even dangerous for many kinds of works. At the same time methods of remote sensing (aerial and terrestrial phototheodolite surveys) completely satisfy to technical-technological requirements and exclude a danger during the work implementation. Unfortunately application of aerial survey in the Antarctic region is a technological problem and has very high cost price. (Hlotov et al. 2004).

Application of terrestrial phototheodolite survey excludes negative features of these methods and gives possibilities to determine the parameters of glacier, main of them are (Hlotov 2003):

- Space location of the glacier in a certain period of time;
- Data on morphometric characteristics of the glaciers;
- From the results of several cycles it is possible to derive space change of the glaciers, the change of their shape and size, the change of the volume of the ice.

For the survey there was proposed the digital phototheodolite where optical thodolite Theo-010B serves as orienting device and digital cameras Kodak DC-260 and Olympus E20p were used for photographing.

The problem of a priory precision of space coordinates of glaciers is also actual.

Precision of determination of space coordinates of object points using stereo pairs of terrestrial photographs depends on range of error sources. Errors of determination of coordinates and parallaxes on the photographs cause to errors m_{Xph}, m_{Yph}, m_{Zph} in photogrammetric coordinates of object points. For determination of coordinate precision there are used the formulas for normal and parallel cases of survey (Lobanov 1984).

In the Table 1 there are presented the values of errors m_{Xph}, m_{Yph}, m_{Zph}, for two cases of survey. For digital camera Kodak DC-260: $x_1 = 40$ mm, $z_1 = 25$ mm, $m_B = 5$ mm, $m_p = m_x = m_z = 0.005$ mm and for digital camera Olympus E20p: $x_1 = 40$ mm, $z_1 = 30$ mm.

Table 1 Estimation of a priory precision of photogrammetric coordinates when normal and parallel cases of survey

The name of glacier	Yph max, m	B, m	Camera	f, mm	Normal case of survey			Parallel case of survey, ($\varphi = 15$)°		
					m_{Xph}, m	m_{Yph}, m	m_{Zph}, m	m_{Xph}, m	m_{Yph}, m	$m_{Z}\varphi$, M
Galindez	200	40	Kodak DC-260	92	0.028	0.054	0.020	0.029	0.056	0.020
			Olympus E20P	85	0.032	0.061	0.025	0.034	0.061	0.025
			Kodak DC-260	280	0.006	0.018	0.004	0.006	0.018	0.004
			Olympus E20P	337	0.005	0.015	0.003	0.005	0.015	0.003

The length of focal distances explains sufficiently high precision. For such aperture plate they will be long focus.

Field works include the following processes:

- To reconnoitre the territory and to make the working project of survey;
- To fix the control and base points and to define their geodetic coordinates;
- To make a long-term centers of survey;
- Photographing (Hlotov 2003).

The long-term centers of survey were made of the thick-walled tubes with diameter 7 cm and length 160 cm. For instrument mounting there was used special stand with size 15×15 cm fixed to end part of the tube by welding. In this stand the aperture with diameter 16 mm was made for forced centering. With purpose to increase rigidity the stand was fixed additionally by slanting prop. There were made six such constructions: two for observations of glacier on island Galindez, two for observations of glacier on the island Winter and two for observations of glacier on the island Barchan. The centers were installed using concreting.

The complex monitoring consisted in the following. The devise and the mark are fixed on the left and right points of the base. The digital phototheodolite fixed on the left point and an sighting mark on the right one. The centering was done in a usual for geodetic devices way. After the digital phototheodolite and the mark have been fixed, the altitude of the tool i is being measured and object photographing was implementing.

The survey was done with different focal lengths of the camera, which were determined by the angles of object coverage (Hlotov 2004). The next step according to the technology was to find out the coordinates of the points of survey by means of GPS in static condition to raise the precision of finding the location of centers. In this way zero cycles of the glacier survey were completed. The surveys done in different time resulted in 12 stereopairs, both, normal and with equal deviation of the camera optical axis to the right and to the left ($15°$) with the focal length of the camera 92 and 280 mm (Fig. 1)

The cameral processing the materials was done on the digital photogrammetric station (DPS) "Delta".

In result the relief of glaciers outcrop was sketched in verticals, and longitudinal profiles of glaciers have been created. Figure 2a shows the panoramic photographs of glaciers on Galindez Island; Fig. 2b shows the glacier represented in verticals with relief section 1 m and longitudinal profile of western part of the glacier on Galindez island. Here the profile on year 2002 is shown by pink color; on year 2003 by green, on year 2004 by blue, and on year 2005 by red. State of profiles visually demonstrates melting of the glacier not only near the grotto but also near the vertical wall, the base of which is located on the island surface.

For visualization of glacier melting the dynamic 3-D models of western outcrop of the glacier on Galindez island have been created. These models are represented in video files (format .avi). For creation of these models and for producing video files it was used program packet 3-D MAX STUDIO.

Report on Photogrammetric Research Conducted 337

Fig. 1 General view of the Galindez Glacier

(a)

(b)

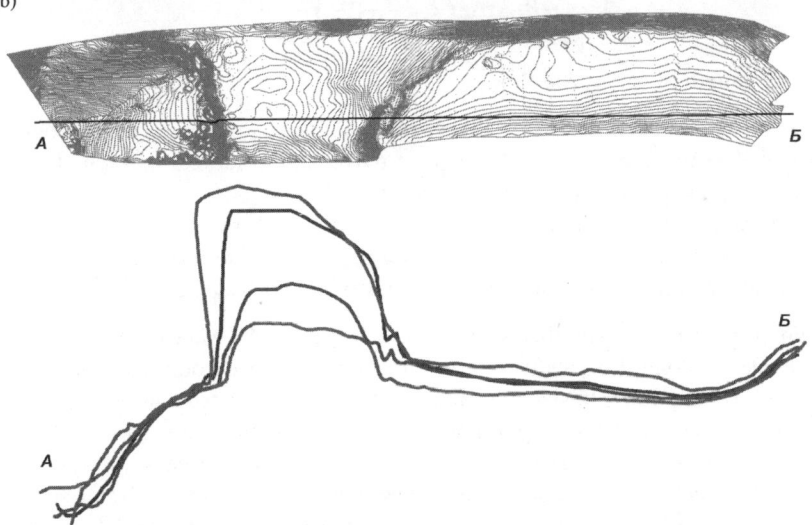

Fig. 2 (a) The panoramic photographs of glaciers on western outcrop of Galindez Island. (b) frontal plan and longitudinal profiles of western part of the glacier

3 Determination and Analysis of Precision of Surface Volumes of Island Glaciers using the Result of Digital Stereophotogrammetric Survey

For obtaining quantitative parameters of glaciers it is necessary to implement preliminary calculation of precision of surface volumes determination and also to analyze the materials produced using the digital survey. Application of digital phototheodolite survey for researching glaciers will give possibility:

- To determine the surface volumes of glaciers;
- To determine longitudinal profiles of glacier's cupola;
- To implement comparative analyses of the results of the measurements.

For determination of surface volumes of glaciers it was proposed method of vertical network, when position of glacier outcrop is projected on the plane which is a normal to optical axis of camera. (Hlotov et al. 2005).

In that way during 2002–2005 there were implemented four cycles of surveying western part of the glacier on Galindez island and three cycles of surveying glacier outcrops of southern part of Galindez island and islands Winter and Barchan. In result the materials for determination of surface volumes of glaciers have been collected (Hlotov et al. 2003). As it was mentioned above for calculation of volumes the method of vertical network has been proposed. Working formula of the method for calculation of total excavation is equal to sum of elementary volumes (Puzanov and Ivanov 1959):

$$V = \frac{1}{3}B^3 \Delta x \Delta z f \left\{ \left[\frac{1}{p_{av}^{'3}} \right] - \left[\frac{1}{p_{av}^{3}} \right] \right\}, \tag{1}$$

where B - length of survey baseline;

$\Delta x, \Delta z$ – adopted sizes of sides of elementary square (*rectangle*) on photograph which determine themselves the steps of movement of left photograph of stereo pair along axes of abscises and applicates;

f – focal distance of camera;

p_{av}, p'_{av} – parallaxes of zero and following cycles, they are determined as arithmetical mean from four parallaxes of vertex of network elements.

Wit purpose to estimate this method the geodetic works were implemented and coordinates of control points in control directions have been obtained. On surface of the glacier according to size of stereopair there were selected contour points and their coordinates have been determined by direct geodetic intersection. Measurements were implemented using theodolite Theo-010B. But because of some reasons (points had not been marked as it was impossible to get to glacier surface) error of sighting had occurred and mean square error of coordinate determination was 0.09 m.

Determination of volume was done in program packet "Digitals" by the following way. On the surface of glacier outcrop on characteristic surfaces of relief there were collected picket points (apr. 1500–2000 pickets). Then it was determined conditional volume of current cycle relatively to the surface formed by pickets and vertical plane specified from the edge of glacier. Later the next cycle was processed in a similar manner and difference of these volumes was determined. In result of calculations it was obtained the difference of surface volumes of adjacent cycles, i.e. change of this value inter seasons.

In the result of measurements of western part of glacier of Galindez Island it was investigated that surface volume is decreasing during 2002–2003 on 23000 м^3, during 2003–2004 – 28000 м^3, and during 2004–2005 – 17000 м^3.

Correspondently for south part of glacier the change of surface volume is following: during 2003–2004 – 1481 м^3, 2004–2005 – 360 м^3.

The similar researches were implemented for the glacier on islands Winter and Barchan. For the glacier Winter the surface volume is decreased during 2003–2004 – 1256 м^3, 2004–2005 – 4817 м^3.

For the glacier on Barchan Island the surface volume is decreased during 2003–2004 – 274 м^3, 2004–2005 – 1745 м^3.

Diagrams of changes of surface volumes of the glaciers relatively to implemented cycles are presented on Figs. 3–6. Here vertical lines represent volumes (m^3) and horizontal lines show years.

Using presented on diagrams results it necessary to make conclusions that even we can observe certain heterogeneity of surface volume melting but general tendency of decreasing of the volumes of island glaciers is a proved fact.

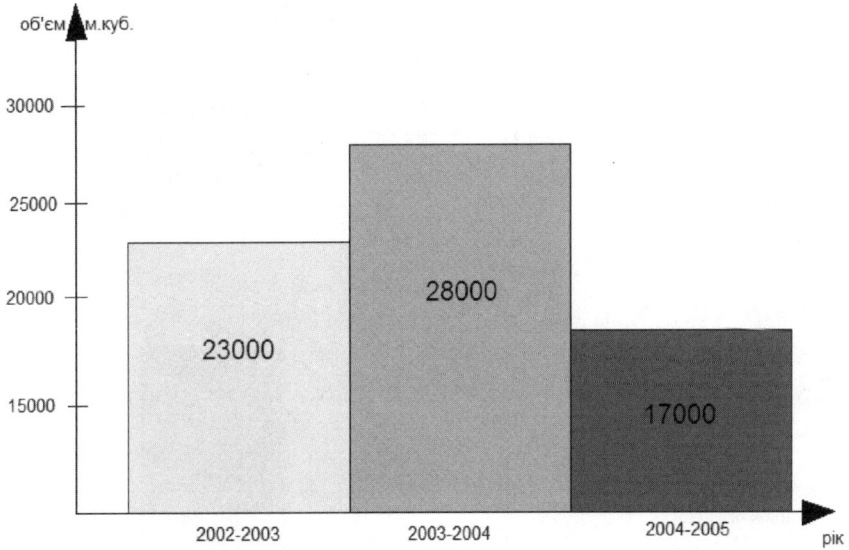

Fig. 3 Values of surface volumes of western outcrop of the glacier on Galindez Island

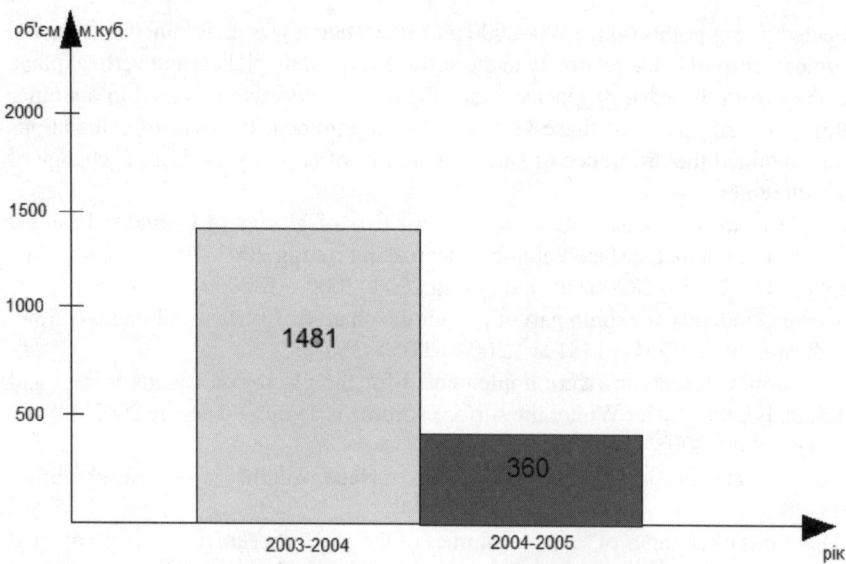

Fig. 4 Values of surface volumes of southern outcrop of the glacier on Galindez Island

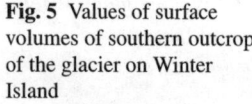

Fig. 5 Values of surface volumes of southern outcrop of the glacier on Winter Island

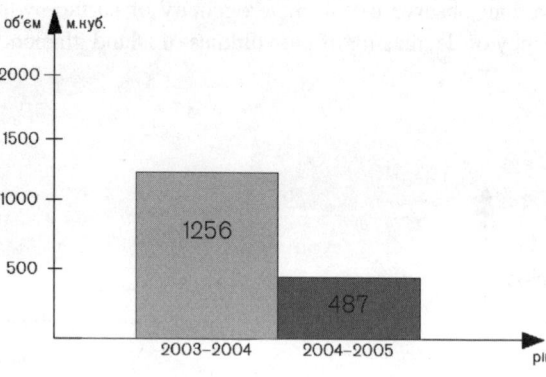

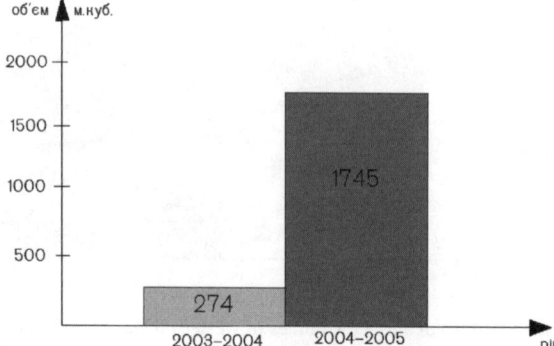

Fig. 6 Values of surface volumes of southern outcrop of the glacier on Barchan Island

4 Technology of Digital Phototheodolite Survey for Creation of Topographic Plans with the Scale 1:1000 of Antarctic Station "Academician Vernadskyy"

Analyzing topocinematic works in the Antarctic it should be noted that Russia shows great activity in this direction. A large amount of cartographic works had been done for creation of GIS on the territory of eastern Antarctic (Juskevich 2000). There were produced 24 trapeziums of topographic maps with the scale 1:100000 with total area – 34 thousands of sq.km. and 57 sheets with the scale 1:200 000 with total area – 330 thousands of sq.km.

By the Institute of Military geography and Antarctic Institute of Chile there were created topographic maps with scale 1:5000 in digital form (Barriga Rodrigo et al. 2001). The topographic plans were created by geodetic methods.

Therefore, analyzing the last cartographical works implemented in the Antarctic it should be noted firstly, tendency to large scale mapping and secondly, application of these plans and their forms. Considering such tendencies the actuality of this problem has to be underlined (Greku 2000).

Digital terrestrial phototheodolite survey was proposed to make the topographic maps considering that it has considerable advantages over the classic method.

The survey included field and cameral work. The field work aims to obtain photographic (digital) information (a stereogram) about the locality and data defining the elements of the absolute orientation (Dorozhynskyy et al. 2004, Hlotov and Chizhevsky 2004).

The cameral work consisted in processing of the terrestrial photographs on photogrammetric station "Delta" (Digital photogrammetric complex "Delta" 2000).

During the year 2004–2005. experimental-researching works on creating large scale plans of islands near Antarctic Station Academician Vernadsky were continuing. Totally there were produced more than 350 stereopairs from permanent and intermediate baselines and a boat. For field referencing and surveying control points there were used geodetic GPS receivers and so coordinates of more than Д240 points have been obtained.

In result of the implemented works the topographic plans have been created with the scale 1:1000 for Galindez, Winter, Skua Islands, western part of Peterman Island and Hat Point of the Antarctic coast. Figure 7 shows fragment of the plan of Galindez Island with the scale 1:1000.

Analysis of the proposed technology will enable to define needs for complication, improvement and modification of the researching method. The problem is to reveal peculiarities of the technological scheme of topographic plan creation (as the base for GIS) when striving for maximum low cost price not only for coastline but for the total area of the islands in this region. Totally, as it was mentioned above, there were produced more than 350 stereopairs using digital cameras Kodaк DC 260 , Olympus E 20p and Sony DCS. After processing of these photography's plans with the scale 1:1000 were created and analysis of the technology was implemented.

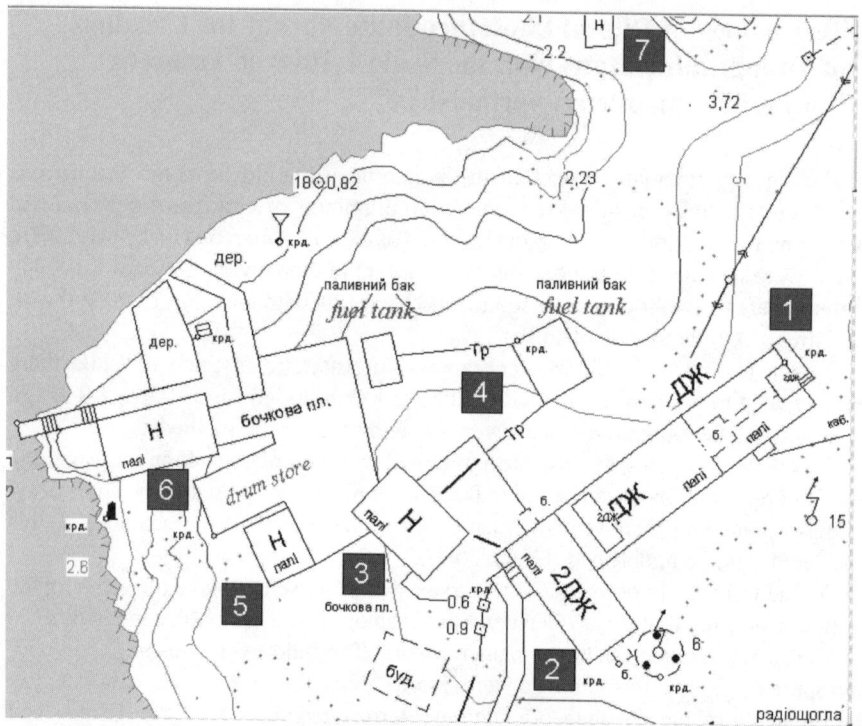

Fig. 7 Fragment of the plan of Galindez Island with the scale 1:1000

Tilting of average plane of surveyed object was considered when Baselines measurements. With this purpose the baseline was located above the object as far as possible and inclined case of survey was used. As distances to average plane do not exceed 100–300 m this case does not considerably effect on the precision of coordinate determination (Hlotov 2002).

Coordinates of baselines were determined with geodetic GPS (like Leica (SR 530), Topcon (ODYSSEY), Trimble (4600LS) etc.) and it gave possibility to calculate coordinates with sufficient precision (Hlotov 2002).

For determination of coordinates of control points the referencing was fulfilled during our researching process. Mainly it was sparse referencing as later it was planned the phototriangulation.

There were some peculiarities of these works:

- sighting on the not marked points is complicated as it is impossible to direct on the micro object both orienting device telescope and adjusting mark of DPS identically;
- marked points allows to increase precision of image orientation on DPS.

There were defined advantages of the survey of coast line and some remote territories implemented from the boat (Fig. 8) comparing with survey from the baseline obvious for this region:

Fig. 8 Survey of Skua Island coastline from the boat

- there is situations when it is impossible to locate baseline on the surveying object;
- when application of the boat the sequential survey close to normal case is implemented;
- there is also possibility to determine coordinates of camera projection center using GPS.

The main aspects of proposed method for determination of linear elements of absolute orientation are the following (Hlotov 2002, 2005). On land surface near surveying object the GPS receiver is installed and on the boat there are mounted the digital camera and on-board GPS-receiver with antenna connected with the camera. On the on-board GPS it is preliminary set the epochs corresponding to photographing intervals. Signal of epoch change from on-board GPS switches on the camera and at the same time the coordinates of camera projection center are determined.

For the method approbation the following experiment was done. On the coast line there were marked 47 control points and their coordinates have been determined by GPS in cinematic mode with precision: $m_X = m_Y = 0,01$ m, $m_Z = 0,04$ m. After that the survey from the boat and data processing on DPS using classic scheme (phototriangulation, relief and situation representation, etc.) were implemented. The distance to the object was 120–150 m. Velocity of the boat – 1 m/sec.

Then comparative analysis of the results of phototriangulation and results of coordinate determination comparing with data measured using GPS (they were considered as theoretical values) was done. Table 2 shows the results of mean square errors (MSE) of this analysis

Table 2 Values of MSE of control point coordinates obtained after calculation when the triangulation and the proposed method

MSE (phototriangulation)			MSE (proposed method)		
m_X (m)	m_Y (m)	m_Z (m)	m_X (m)	m_Y (m)	m_Z (m)
0,17	0,28	0,13	0.12	0.18	0.10

As it is shown the value of MSE when application of the proposed algorithm is approximately 1.5 times less comparing with classic method.

During the data processing on DPS the problems with orientation appeared. To avoid this it is necessary to implement survey in such way that to provide stereopairs overlap. It is possible when application of parallel case of survey and selection of proper angles of object coverage changing the focal length of camera.

Conclusions

1. The experimental-research work proved it is possible to use terrestrial phototheodolite survey for determination of quantitative parameters of Antarctic coast glaciers.
2. It is worthwhile to present the relief of the glacier outcrop as verticals.
3. Developed technological schemes of determination of glacier surface volumes give possibility to solve this problem with sufficient precision by both method of pickets and method of vertical grid.
4. In the results of research the surface volumes of glaciers located on Argentine islands Galindez, Winter and Barchan have been determined by two and three cycles. It is necessary to note that considerable dynamics of the glacier surface is observed.
5. The experimental-research work proved it possible to survey, process and make big-scale maps by means of digital phototheodolite.
6. There was developed the technology and implemented the approbation application digital camera in connection with GPS for determination of linear elements of absolute orientation, when survey of coastlines of Argentine islands from the boat.
7. The technology of processing digital terrestrial photographs on DPS "Delta" was developed.
8. It was created the topographic maps with the scale 1:1000 of Galindez, Winter, Skua Islands, western part of Peterman Island and Hat Point of the Antarctic coast.

References

Barriga R, Montero J, Villanuela V, Klotz J, Bevis M (2001) Geodesy and digital cartographic survey in Fildes Peninsula, Rey Jorge Island, Antarctica // Geo-spt. Inf. Sci. – 2, pp. 57–62.

Digital photogrammetric complex "Delta". 2000. Software for creation of digital plans and maps. Version 5.0. Manuals, Part 2.11, Scientific-production corporation "Geosystems".

Dorozhynskyy O, Minilevskyy H, Glotov V (2004) Photogrammetric research conducted at the Antarctic station " Academican Vernadskyy. XXth ISPRS Congress, Istambul, Turkey, pp 642–644.

Greku R (2000) Researching on international project SKAR Geodetic infrastructure in the Antarctic. Bulletin of Ukrainian Antarctic center, vol. 3, pp 141–149.

Hlotov V (2002) Coupling device of GPS and camera. Proc. Modern achievements of geodetic science and production Lviv, pp 346–348.

Hlotov V (2003) Creation of frontal plans of glaciers of the Antarctic coast. Proc. Modern achievements of geodetic science and production Lviv, pp 264–268.

Hlotov V (2004) results of monitoring of island glaciers of the Antarctic coast. Proc. Geoinformation environment monitoring – GPS and GIS technologies. Lviv, pp 20–23.

Hlotov V (2005) Investigation of the method of determination of spatial coordinates of the projecting center of camera. Proc. Modern achievements of geodetic science and production Lviv, pp 125–129.

Hlotov V, Chizhevsky V (2004) Analysis of technological scheme of creation of topographic plans by digital stereophotogrammetric method as the base of GIS of Vernadsky Station (GS17) Second Ukrainian Antarctic Meeting, Kyiv, pp 33–34.

Hlotov V, Kovalenok S, Milinevskyy G, Nakalov E, Fulitka J (2003) Monitoring of small glaciers as indicators of changes of climate in region of the Antarctic Peninsula. Ukrainian Antarctic Journal, #1, pp 93–99.

Hlotov V, Chizgevsky V, Milinevsky G, Kovalonok S (2004) Using of the digital stereo photogrammetric survey for the ice cap dynamic research (GS21) Second Ukrainian Antarctic Meeting, Kyiv, pp 37–38.

Hlotov V, Kovalenok S, Milinevskyy G, Chyzhevskyy V (2005) Determination of dynamics of surface volumes of island glaciers as a component of GIS "Antarctic". Proc. Geoinformation environment monitoring – GPS and GIS technologies. Lviv, pp 172–176.

Instructions on topographic surveys with scales 1:5000, 1:2000, 1:1000, 1:500. Phototheodolite survey. (1977). Nedra, Moscow.

Instruction on exploitation of digital camera DC-260.

Instruction on exploitation of digital camera OLYMPUS E-20p. OLYMPUS Optical CO., LTD. Tokyo, Japan.

Juskevich A (2000) Topographic surveys in the Antarctic. Journal Geodesy and cartography 6, pp 12–16.

Lobanov A (1984) Photogrammetry. Nedra, Moscow.

Puzanov B, Ivanov N (1959) Methodology of determination of volumes and areas using terrestrial stereo photographs. Orgenergostroy.

The Contribution of Russian Geodesists and Topographers to Antarctic Mapping

Alexander V. Yuskevich

Abstract The history of Russian mapping activities in Antarctica dates back to the First Russian Antarctic Expedition 1820. Since 1955, intensive investigations have been carried out by a number of Soviet, then Russian institutions in order to set up geodetic and topographic foundations for a detailed mapping of Antarctica. This paper describes the different Russian activities, among others astronomical observations, geodetic networks, height and gravity surveys, aerial photography, and GPS observations. The concept of topographic mapping will be discussed, examples for maps of different scale will be given, and it will be shown that, eventually, internationally coordinated activities aim on the provision of digital data and maps. The Geographic Information System (GIS) of East Antarctica can be given as a good example for these developments.

Keywords Geodesy · topography · GIS · Antarctica

1 Introduction

Russian mapping activities in Antarctica started immediately after the discoveries of the First Russian Antarctic Expedition, lead by F. F. Bellingshausen and M. P. Lazarev 1820. Since 1955 an intensive, detailed exploration of the sixth continent is being carried out. Scientists and technicians of the Soviet Union (until 1991) and of the Russian Federation made great contributions to the investigation of Antarctica in all fields of science, especially in geography, geology, glaciology and climatology. These investigations have been carried out by a number of institutions, among others institutions of the Ministry of Geology (SEVMORGEO company, SOYUZMORNIIPROJEKT research institute, and others), institutions of the Hydrometeorological Service (Arctic-Antarctic Research Institute AARI, and others), and institutions of the Soviet resp. Russian Academy of Science (Institute

---

Alexander V. Yuskevich
Department of foreign relations, Neftegazgeodeziya, 195112 St. Petersburg, Prospekt Utkina 15, Russia, e-mail: yuskevich@ngg.ru

of Geography, Institute of Physics of the Earth, Institute of Glaciology, Institute of Ocean Science, and others), as well as institutions of the civil aviation and of the Ministries of Defence and of Shipping. The vast amount of work was done during 34 Soviet Antarctic Expeditions (SAE) and 18 Russian Antarctic Expeditions (RAE). Among the expeditions leaders well-known polar explorers have to be mentioned: M. M. Somov, E. I. Tolstikov, or A. F. Treshnikov. Topographic-geodetic work and hydrographical surveys started 1955.

In 1970 the responsibility for all major topographic and geodetic investigations in Antarctica was given to the Central Administrative Board of Geodesy and Cartography, which further entrusted the following institutions with particular tasks: Enterprise No10 (from 2001 AEROGEODEZIYA Federal State Unitary Enterprise), PRIRODA State Centre, CARTOGRAPHY production and commercial union, and Central Research Institute of Geodesy and Cartography. A great amount of activities was carried out by specialists of Enterprise No. 10 / AEROGEODEZIYA, who participated in eighteen SAEs and eighteen RAEs. The first group of topographers and geodesists was lead by G. M. Muradov, later General Director of the enterprise. A detailed overview on all aspects of the geodetic-topographic works is given by Yuskevich (2004).

In the following, different aspects of the geodetic and topographic work, carried out by Soviet and Russian specialists in Antarctica, should be concisely discussed.

2 Geodetic Foundations

2.1 Astronomical Surveys

1955, during SAE-1, a first-order astronomical station was established in the Antarctic station Mirny, later followed by further first-order stations, mostly set-up by Enterprise No. 10. The first-order astronomical stations became a basis for diverse geodetic survey like terrestrial triangulation and trilateration networks or polygonal traverses. At these stations, the latitude was generally determined applying the Horrebow-Talkott method, while the longitude was determined by the Zinger method. The instrumentation to carried out these observations included the AU-2/10 multipurpose astronomical instrument, sidereal chronometer, R-311 radio, and further equipment. The accuracy (rms-value) of the obtained astronomical coordinates for all first-order stations was in the order of 0.1" for the longitude, and 0.2" for the latitude. The first-order stations were complemented by additional stations of lower order, where zenith distances and the Somner method were applied to determine latitude and longitude, resp.

2.2 Geodetic Networks

Geodetic networks were established beginning at Mirny station 1955. For their realization classical geodetic methods like triangulation, trilateration and polygonometry were used. The previously established astronomical station served as reference points for these networks. The triangulation networks were mostly created to realize the anticipated topographic mapping at scales 1:100 000 and 1:200 000. Transit traverses were realized in order to provide a geodetic foundation for the airfields on ice and other landing strips as well as for the large-scale topographic mapping of the surroundings of the Antarctic stations.

2.3 Radiogeodetic Methods

In order to carry out the mapping of vast areas in Antarctica the radiogeodetic method was applied. This method was first utilized in Antarctica during SAE 1 (1955/56) for the horizontal connection of airborne photogrammetric surveys performed in the area of Mirny station. The Russian radionavigation system RYM-S was used in this survey. During SAE 14 (1968/69) radiogeodetic measurements were carried out to support the Soviet-French glaciological expedition at the traverse from Mirny to Vostok. During SAE 17 (1971/72) experimental works were carried out to test the POISK radiogeodetic system. However, most of the works were carried out using the RDS airborne radio distance measuring equipment. In order to realize the scale of the fundamental radiogeodetic network for the subsequent mapping of Antarctica a space basis was measured between the locations Vechernyaya and Zenit using the RDS airborne system, installed on board of an IL-14 FKM aircraft. Simultaneously, data from the RVTD radar altimeter, S-51 pressure altimeter and TSAO meteograph were recorded. The aircraft altitudes were determined with an accuracy of a few meters. A principle sketch of these measurements is given in Fig. 1. The entire basis length could be determined with an accuracy of about 2 m.

2.4 Height Measurements

In order to realize elevation control points in Antarctica, the methods of geodetic levelling, radio-altimeter levelling and barometric levelling were applied. A number

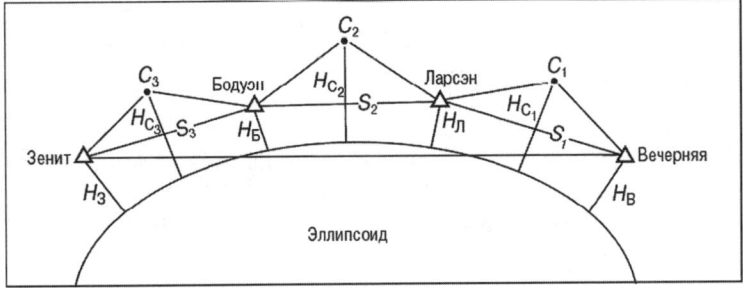

Fig. 1 Vechernyaya – Zenit radiogeodetic basis profile

of levelling traverses of 3rd and 4th order were carried out. One of the first levelling traverses was realized between the stations Mirny, Pionerskaya, Vostok and Komsomolzkaya during the years 1957–1961. Tank-type vehicles ATT-14 and ATT-15 were used for the travel as well as a basis for the levelling observations. The accuracy of the height determination with respect to the astronomical station in Mirny is in the range of 1–1.5 m.

3 Topographic Surveys

3.1 Aerial Photo-Topographic Surveys

The first Russian aerial photo-topographic survey was carried out in the area of Mirny station, using an AFA-37/70 camera on board an AN-2 aircraft. Subsequent surveys were conducted until 1962. Based on these surveys, topographic maps in the scales 1:500 000, 1:200 000 (42 sheets) and 1:100 000 (25 sheets) were published. These works on topographic mapping were intensified by Enterprise No. 10 / AEROGEODEZIYA during the years 1970–1978. For an area of more than 500,000 km^2 photo-topographic surveys provided data for publishing maps in the scales 1:200 000 and 1:100 000. Astronomical stations and stations of the geodetic networks were used as reference points for the surveys (Yuskevich, 2000b). Vertical control was realized by the radio-altimeter levelling method. The photogrammetric analysis employed numerical stereo triangulation and network reduction. For publication, the Gauss-Krüger projection was used, and the heights were determined with respect to the level of the Southern ocean. The height accuracy could be determined to be in the range of 4–6 m (for center of photographs) and 6–8 m at change points. Furthermore, for publication attention was given to different aspects like respective coloring and relief, geographic objects and naming conventions. Since 1970, about 80 sheets at scale 1:100 000 and about 100 sheets at scale 1:200 000 have been published. Figure 2 shows an example of the topographic map at scale 1:100 000. These maps provide an excellent basis for further studies in glaciology, geophysics and all environmental studies. Their information content and accuracy have been widely acknowledged not only in Russia, but also by the international community (Atlas of Antarctica, Part I and II, 1969).

3.2 Detailed Topographic Surveys

A first detailed topographic survey was carried out in the area of Mirny station 1955. 1975 specialists of Enterprise No10 / AEROGEODEZIYA started participating in such surveys. Topographic maps in the scales 1:1 000, 1:2 000 and 1:10 000, respectively, were made for the areas of the stations Bellingshausen,

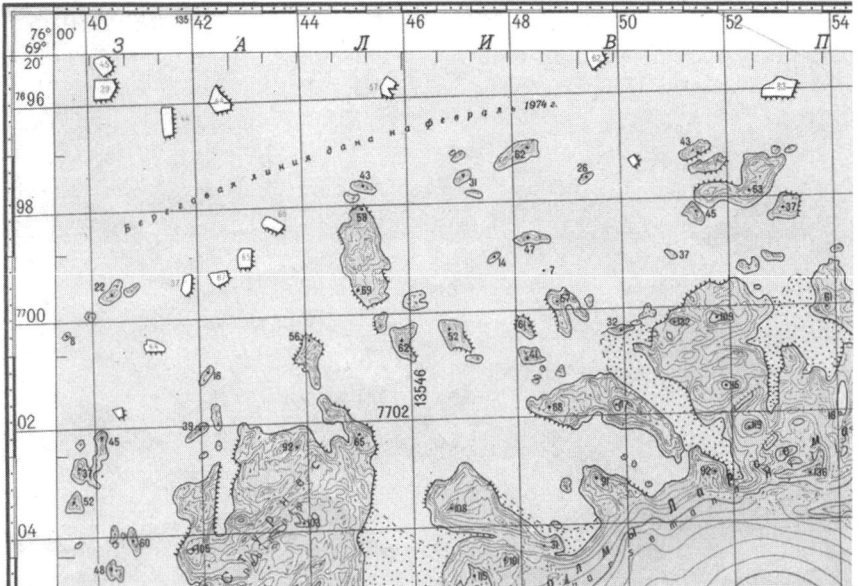

Fig. 2 Detail of the topographic map at scale 1:100 000 of Lassermann Hill area

Mirny, Molodezhnaya, Novolazarevskaya, Progress, Russkaya, and others (Muradov and Safronov, 1988). Figure 3 shows an example of a topographic map of the area of Novolazarevskaya station at scale 1:2 000.

3.3 Special Topographic Mapping

1985 specialists of Enterprise No10 / AEROGEODEZIYA started to develop special topographic maps utilizing radar methods. This type of survey was carried out with the help of IL-14, IL-18 and AN-2 aircrafts and MI-8 helicopter. The subglacial relief and ice thickness were determined applying radar ranging, which allows to measure the travel time of the radar signals reflected at the ice surface, at the internal ice layers and at the subglacial bedrock. A maximum ice thickness of 2,300 m was recorded, with an rms error of about 19 m. Positioning of the radar points was carried out using a Magnavox 1400 satellite receiver and DISS-013 Doppler aircraft speed indicator. Simultaneously, an AFA TE-10 aerial camera was used to obtain aerophotographs, and a RV-1 radio altimeter to carry out elevation measurements, which had an rms error of about 12 m. The data of these radar surveys were used to edit maps at scales 1:500 000 (contour interval 100 m) and 1:1 000 000 (200 m) applying Gauss-Krüger projection in the WGS-72 system.

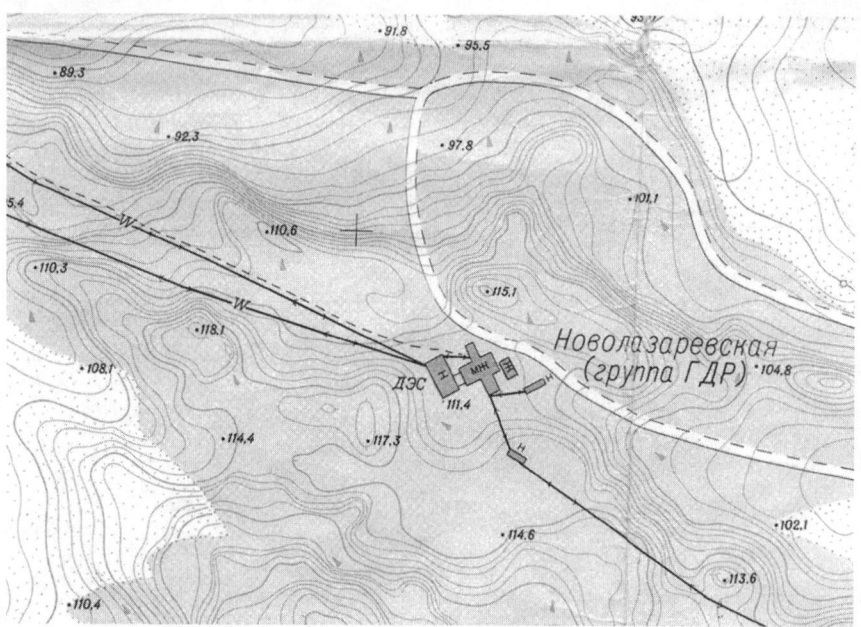

Fig. 3 Detail of the topographic map at scale 1:2 000 of Novolazarevskaya station

4 Gravity Measurements

Until 1970 only seven fundamental gravimetry stations existed in Antarctica, set up by scientists from the United States, Australia and Japan. The observations were carried out with the help of pendulum instruments and had an accuracy of about 0.5 mGal. Extensive Soviet / Russian gravimetric surveys at the Antarctic continent were realized in the years 1977–1990s (Yuskevich, 2000c). During this time 10 fundamental gravimetry stations were set up, located mostly in the area of Russian Antarctic stations. Gravity observations were made using the pendulum instruments OVM and AGAT. The additional equipment comprised, among others, Quartz chronometers. Above the concrete pillar a pavilion was erected in the event of gravity measurements in order to ensure the required environmental conditions. The gravity values were related to the central gravity reference station MOSCOW A, while realizing the gravity tie when travelling aboard an aircraft from Moscow to Antarctica and return. The accuracy of the gravity stations was found to be 0.25 mGal (average rms). The fundamental gravity stations were positioned with respect to the geodetic network with an accuracy of about 1 m.

Field gravity surveys started in the late 1950s, when gravity observations were carried out along the 60 km traverse Mirny – Sovetskaya using the gravity meters CH-3 and GAK-3M. Later, extensive field gravity measurements were realized with a density of about 1 point per 100 km^2 over an area of about 34,500 km^2. Additional field gravity observations were realized with the help of MI-8 helicopter and AN-2 aircraft, landing on site. The data processing was done at the Russian Antarctic station Druzhnaya.

5 Recent and Future Prospects of Geodetic and Mapping Activities

5.1 GPS Observations

During the mid-1990s international efforts lead to the set-up of geodetic GPS network in Antarctica and its regular observation, known as the SCAR Epoch GPS Campaigns (now: SCAR Epoch Crustal Movement Campaigns) (Dietrich et al. 1998, 2001). This activity was and is coordinated within the SCAR program Geodetic Infrastructure in Antarctica (GIANT) (Manning, 2001). In recent years, the Russian activities have been aimed towards participation in this program and linkage of the first-order astronomical stations and geodetic networks to the SCAR GPS network. Within this context, GPS observations have been carried out at the Russian stations Bellingshausen, Vostok, Mirny, Novolazarevskaya and Progress. These measurements served also as a basis for an update of local station networks and their transformation to the ITRF datum.

Furthermore, GPS observations could be utilized for scientific applications like the investigation of deformations and flow velocities at the subglacial lake Vostok. In the framework of a multi-year joint project, AEROGEODEZIYA and Dresden University of Technology carried out precise GPS measurements in the vicinity of Vostok station as well as along traverse routes heading from Vostok to Mirny station. The results were reported e.g. by Wendt et al. (2005, 2006).

5.2 Topographic Mapping and GIS Developments in East Antarctica

The realization of a geoinformation system (GIS) for East Antarctica has been investigated since the 1980s. 1988, a first version of the Antarctic Digital Database was implemented by SCAR based on the existing topographic map at scale 1:1 000 000 (ADD Consortium, 2000). Recently, version 4.0 has been released.

2000, at the SCAR business meeting in Tokyo it was decided to appoint Australia, China and Russia into a working group for a GIS-project of East Antarctica, which was then lead by AEROGEODEZIYA. Figure 4 shows the layout of available topographical maps for East Antarctica (Yuskevich, 2000a). 2003 AEROGEODEZIYA has started to develop a digital topographic base of the Amery ice shelf and Lamber glacier region at the scales 1:200 000 and 1:100 000. An overview on the available sheets at these scales is also given in Fig. 4. For this, Aerophotographs and

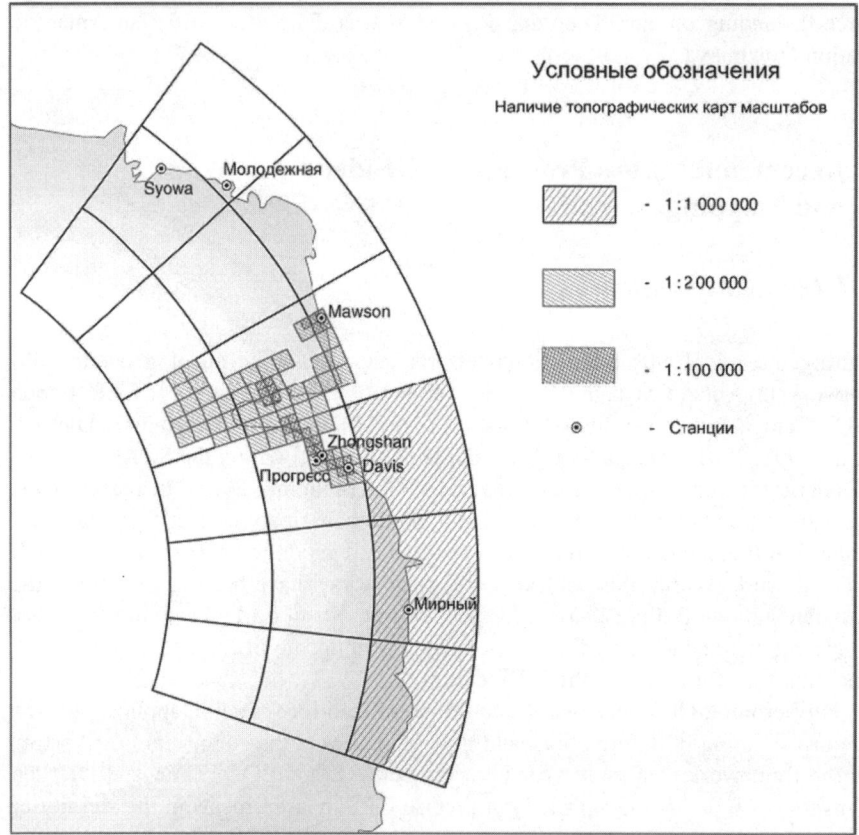

Fig. 4 Overview of available Russian topographic maps in East Antarctica

maps produced by the former SAEs were used as an initial basis. An extensive checking, including place names, has been carried out, the WGS84 datum and UTM projection have been used. Hence, the GIS of East Antarctica can be used for further investigations and scientific applications, but needs also further development (Yuskevich, 2002).

6 Conclusions

Soviet and Russian intensive activities in the fields of geodesy and topography started 1955. A great number of scientists and technicians took part in these activities during 34 Soviet Antarctic Expeditions (SAE) and 18 Russian Antarctic Expeditions (RAE), coming from a number of different institutions. The investigations included set-up, observation and realization, respectively, of astronomical stations, of geodetic networks based on triangulation, trilateration and, recently, on GPS, of gravimetric surveys, of aerial photo-topographic surveys and subsequent topographic mapping, and of a GIS for East Antarctica. The international cooperation in all fields of these activities will be a key for further successful scientific projects and developments in Antarctica.

References

ADD Consortium (2000): Antarctic Digital Database, Version 3.0. Database, manual and bibliography. Scientific Commitee on Antarctic Research, Cambridge, manual and bibliography, 93 pp.
Atlas of Antarctica (1969a): Part I. Hydrometeoizdat, 225 maps. (In Russian language: Atlas Antarktidy, t. 1, Gidrometeoizdat, 225 kart.)
Atlas of Antarctica (1969a): Part II. Hydrometeoizdat, 598 p. (In Russian language: Atlas Antarktidy, t. 2, Gidrometeoizdat, 598 s.)
Dietrich, R.; Dach, R.; Perlt, J.; Schenke, H.-W.; Schöne, T.; Pohl, M.; Soltau, G.; Engelhardt, G.; Mikolaiski, H.-W.; Seeber, G.; Menge, F.; Niemeier, W.; Salbach, H.; Lindner, K.; Kutterer, H.-J.; Mayer, M. (1998): The SCAR GPS Campaigns: Accurate Geodetic Reference in Antarctica.In: Forsberg, R.; Feissel, M; Dietrich, R. (eds.): Geodesy on the Move – Gravity, Geoid, Geodynamics, and Antarctica, Proc. IAG Scientific Assembly, Rio de Janeiro 1997, *Springer Series: IAG Symposia*, Vol. 119, pp. 474–479, Springer Verlag, Heidelberg.
Dietrich, R., R. Dach, G. Engelhardt, J. Ihde, W. Korth, H.-J. Kutterer, K. Lindner, M. Mayer, F. Menge, H. Miller, C. Müller, W. Niemeier, J. Perlt, M. Pohl, H. Salbach, H.-W. Schenke, T. Schöne, G. Seeber, A. Veit, C. Völksen (2001): ITRF coordinates and plate velocities from repeated GPS campaigns in Antarctica – an analysis based on different individual solutions. *Journal of Geodesy*, 74 11/12, pp. 756–766.
Manning, J. (2001): The SCAR Geodetic Infrastructure in Antarctica (pp. 22–30), Geodesy in Antarctica (pp. 30–33), SCAR report 20.
Muradov, G. M., Safronov, I.G. (1988): Topographic-geodetic investigations in Antarctica, Geodesy and Cartography 2, pp. 17–20 (in Russian language: Topografo-geodesicheskiye raboty v Antarktide, Geodesiya i Kartografiya 2, 1988, c. 17–20).
Wendt, A., R. Dietrich, J. Wendt, M. Fritsche, V. Lukin, A. Yuskevich, A. Kokhanov, A. Senatorov, K. Shibuya, and K. Doi (2005): The response of the subglacial Lake Vostok, Antarctica, to tidal and atmospheric pressure forcing. *Geophysical Journal International*, 161: 41–49, doi:10.1111/j.1365-246X.2005.02575.x
Wendt, J., R. Dietrich, M. Fritsche, A. Wendt, A. Yuskevich, A. Kokhanov, A. Senatorov, V. Lukin, K. Shibuya, K. Doi (2006): Geodetic observations of ice flow velocities over the southern part of subglacial Lake Vostok, Antarctica, and their glaciological implications. *Geophysical Journal International*, 166, p. 991–998, doi:10.1111/j.1365-246X.2006.03061.x.
Yuskevich, A. V. (2000a): Topographic maps in Antarctica, Geodesy and Cartography 6, 2000, pp. 12–16 (in Russian language: Topograficheskiye cjemki v Antarktide, Geodesiya i Kartografiya, 6, s. 12–16)
Yuskevich, A. V. (2000b): Set-up of horizontal and height references for photogrammetric investigations in Antarctica, Geodesy and Cartography, 7, 2000, pp. 41–45 (in Russian language: Sozdaniye planovo-vysotnogo obosnovaniya dlya sjemochnykh rabot v Antarktide, Geodesiya i Kartografiya, 7, s. 41–45)
Yuskevich, A. V. (2000c): Gravimetric surveys in Antarctica, Geodesy and Cartography, 11, 2000, pp. 9–11 (in Russian language: Gravimetricheskije raboty na territorii Antarktidy, Geodesiya i Kartografiya, 11, s. 9–11)

Yuskevich, A. V. (2002): Technological project for the development of an East-Antarctic GIS for topographic-geodetic and cartographic investigations, AEROGEODEZIYA 24 p. (in Russian language: Tekhnicheskiy project cozdaniya GIS topografo-geodezicheskoy i kartograficheskoy izuchennosti na territoriyu Vostochnoy Antarktidy, AEROGEODZIYA, 24 s.)

Yuskevich, A. V. (2004): The contribution of Russian geodesists and topographers to the mapping of Antarctica, Kartgeocentre, Moscow, 104 p. (in Russian language: Vklad geodezistov i topografov Rossii v kartografirovanije Antarktidy, Kartgeozentr, Moskva, 104 s.).